BRITISH CERAMIC PROCEEDINGS
NO. 52

ELECTROCERAMICS

PRODUCTION, PROPERTIES AND MICROSTRUCTURES

≠31089093

ELECTROCERAMICS

PRODUCTION, PROPERTIES AND MICROSTRUCTURES

Edited by

W. E. LEE
Department of Engineering Materials
The University of Sheffield
Sheffield, UK

and

A. BELL
Laboratoire de Céramique
Ecole Polytechnique Fédérale de Lausanne,
Switzerland

British Ceramic Proceedings
No. 52

THE INSTITUTE OF MATERIALS

Book 564
Published 1994 by
The Institute of Materials
1 Carlton House Terrace
London SW1Y 5DB

© The Institute of Materials 1994

British Library Cataloguing-in-Publication Data

Electroceramics: Proceedings of the Symposium
Held as Part of the Condensed Matter and Materials
Physics Conference, 20–22 December 1993,
University of Leeds.
– (British Ceramic Proceedings, ISSN
0268-4373; No. 52)
I Lee, W. E. II Bell, A. III Series
620.1404297

ISBN 0-901716-42-1

Typeset by Keyset Composition
Colchester, UK

Printed and bound at
The University Press
Cambridge, UK

D
621.381
ELE

Contents

ANDERSONIAN LIBRARY
WITHDRAWN
FROM
LIBRARY
STOCK
UNIVERSITY OF STRATHCLYDE

Introductory Statement

The papers contained in this volume of proceedings were presented at the electroceramics symposium of the Condensed Matter and Materials Physics '93 meeting of the Institute of Physics held at Leeds University, 20–22 December, 1993. The help of Lucy Bell of the IOP in London and Dr D. P. Thompson of Newcastle University with the symposium organisation and Cookson Group plc for financial support is gratefully acknowledged.

WEL/AB

The Influence of Crystal Chemistry on the Ferroelectric and Piezoelectric Properties of Perovskite Ceramics

NOEL W. THOMAS
School of Materials, University of Leeds, Leeds, LS2 9JT, UK

ABSTRACT
A strategy is defined for investigating and establishing structure–property relationships in electroceramics, with particular attention to the influence of crystal structure on the ferroelectric properties of perovskites. The geometrical constraints of the perovskite structure, of composition ABO_3, dictate that there are only three structural degrees of freedom: ionic displacement, anionic polyhedral distortion and octahedral tilting. The relative importance of each of these for the occurrence of ferrolectric properties is brought out, with the polyhedral volume ratio, $V_A:V_B$ introduced as a unifying structural determinant. The quantitative relationships between $V_A:V_B$ and tilting in rhombohedral and orthorhombic perovskites are described, and a method defined for the prediction of new ferroelectric–piezoelectric ceramic compositions.
A novel interpretation of the PZT phase diagram is given, with an extension of the methodology to related structural systems, Aurivillius phases and layered (slab-like) perovskites outlined. Both systems are potential piezoelectric materials for use at elevated temperatures. The effect of doping on the stability of alternative perovskite phases is also relevant to the cathodes and interconnects currently employed in solid oxide fuel cells.

1. INTRODUCTION

An essential part of materials science rests on the analysis of structure and its implications for the functional and mechanical properties of materials. This generalisation is also appropriate for a consideration of the functional properties of electroceramics. In this connection, the investigation of structure–property relationships often forms the basis on which the properties of such materials may be designed (Fig. 1).

Whereas the flowchart attaches equal weighting to the analysis of both crystal- and microstructure, the bias amongst ceramists tends to be in the area of microstructural analysis. The purpose of this paper, however, is to indicate how an understanding of crystal structure is also a crucial element of a comprehensive design strategy. Attention will be directed to computer-aided modelling techniques, as applied to the crystal structures of perovskites. Further, a parametrisation will be described which is relevant to ferroelectric and piezoelectric properties which can be shown by perovskite and related materials, i.e. tungsten bronzes, Aurivillius compounds and layered perovskites.

Note that the purpose of the modelling is not necessarily to predict the detailed ferroelectric properties of a given ceramic composition, just as one would not attempt to predict properties from an analysis of its microstructure. It is rather to

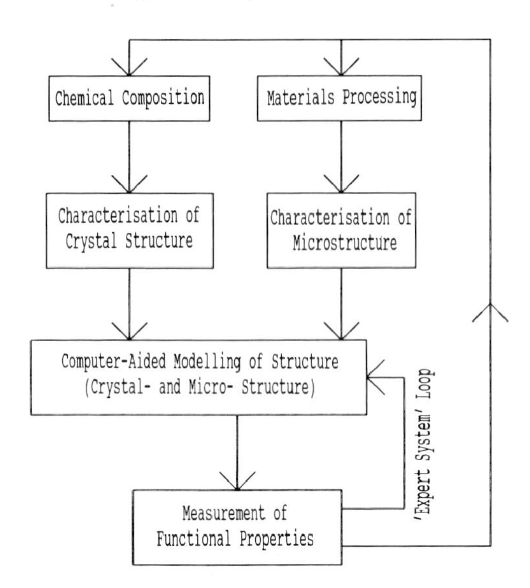

Figure 1 Flow-chart of stages involved in structural design of electroceramic materials.

provide a *framework* in which the interdependence of chemical composition, crystal structure and physical properties can be brought out, with a consequent deepening in understanding. Indeed, since the process of materials development is inevitably led by experimental work, an 'expert system' feedback loop is incorporated in Fig. 1, to reflect the desirability of including experimentally acquired information in any structurally based model of the physical properties of a ceramic.

3. METHOD

3.1 Identification of the Key Structural Variables in Perovskites

In essence, the perovskite structure, corresponding to a generalised composition ABO_3, consists of a three-dimensionally linked network of BO_6 octahedra, with A ions forming AO_{12} cuboctahedra, to fill the spaces between BO_6 octahedra. In view of these inviolable topological and geometrical constraints, there are only three structural degrees of freedom:

(a) displacement of cations A and B from the centres of their cation coordination polyhedra, AO_{12} and BO_6 respectively;
(b) distortions of the anionic polyhedra coordinating A and B ions;
(c) tilting of the BO_6 octahedra about one, two or three axes.

The first of these is most important for the occurrence of ferroelectricity, since a separation of the centres of positive and negative charges corresponds to an electric dipole moment.

In the cases of cubic (paraelectric), tetragonal and rhombohedral (ferroelectric) phases, the displacements (a) and the distortions (b) are straightforward to derive from the relevant crystallographic data.

In *cubic* barium titanate, $BaTiO_3$,[1] both Ba and Ti cations have zero displacements, with perfectly regular polyhedra of coordinating oxygen ions.

In *tetragonal* barium titanate,[2] both AO_{12} and BO_6 octahedra are elongated along the c-axis, as $c:a = 1.0098$. Displacements (a) are along the tetragonal (polar) axis, corresponding to 9.68 pm for the Ba^{2+} ion and 11.50 pm for the Ti^{4+} ion. Smaller displacements of the oxygen ions contribute to distortions (b), with the four oxygens in the BO_6 octahedron perpendicular to the tetragonal axis displaced by 3.63 pm, in the opposite direction to the Ti^{4+} displacement.

In *tetragonal* lead titanate,[3] by comparison, the $c:a$ ratio is larger (1.0651), as are displacements (a) and distortions (b). The c-axis displacements amount to 46.46 pm for the Pb^{2+} ion and 33.12 pm for the Ti^{4+} ion. Once again the displacements of the oxygen ions are smaller than those of the cations, with the four octahedral oxygens perpendicular to the tetragonal axis displaced by 2.33 pm.

In rhombohedral phases, displacements (a) and distortions (b) are correlated. In this case, displacements (a) are parallel to the threefold axis which passes through two opposite triangular faces of the octahedron, ABC and DEF (Fig. 2). Since face ABC has a larger area than DEF, this type of distortion favours displacements of the B cation in the direction of face ABC. Through this distortion, a ferroelectric rhombohedral perovskite is able to accommodate relatively larger ions in its BO_6 octahedra than in the tetragonal ferroelectric phase. Furthermore, electrostatic energy calculations in rhombohedral perovskites reveal that parallel A and B

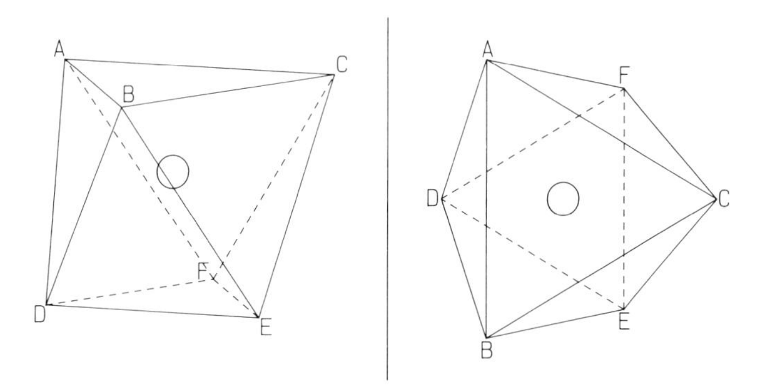

Figure 2 Octahedral distortions in rhombohedral perovskites: (a) viewed parallel to threefold axis, with area ABC > area DEF; (b) view perpendicular to threefold axis, indicating displacement of the B cation towards larger face ABC.

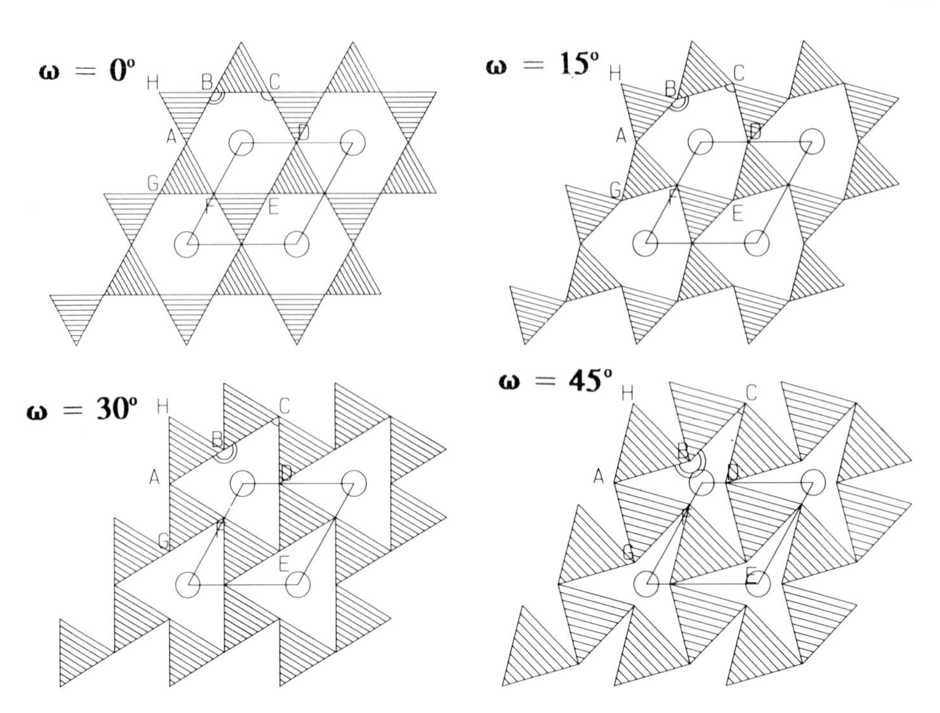

Figure 3 Projection of rhombohedral perovskite structure, viewed along threefold axis, for tilt angles, ω, = (a) 0° (b) 15° (c) 30° and (d) 45°. Shaded triangles represent faces of BO_6 octahedra, with A ions represented as circles. When $\omega > 30°$, a physically improbable structure results (d).

displacements, along the trigonal axis, are more favourable than antiparallel displacements.[4]

Orthorhombic perovskites, by virtue of their lower symmetry, present a wider range of polyhedral distortions, with a systematisation of these currently in progress.[5] The occurrence of significant distortions can be revealed from calculations of the polyhedral volume ratio, as defined in section 3.2.

The method used to quantify octahedral tilting ((c) above) depends on the symmetry of the perovskite. In rhomohedral systems, tilting takes place by rotation of octahedra about their threehold axes, as shown in Fig. 3 for tilt angles of 0, 15, 30 and 45°. The rhombohedral tilt angle, ω, can be inferred either from the deviation of either $\angle$ABC or $\angle$BCD from 120°. In the former case, $\omega_1 = (\angle ABC - 120°)/2$, and in the latter, $\omega_2 = (120° - \angle BCD)/2$. For adjacent triangular faces of equal area (Fig. 3), $\omega_1 = \omega_2$. However, in the general case, where the areas of adjacent triangular faces are unequal (due to octahedral distortion), ω is taken as $(\omega_1 + \omega_2)/2$.[4]

For tilting in perovskites of orthorhombic symmetry, the appropriate parametrisation requires six parameters (Fig. 4): s_1, s_2, s_3: separations of 3 pairs of opposite octahedral vertices (lengths of 3 perpendicular 'octahedral stalks'); $\theta_x, \theta_y, \theta_z$: tilt

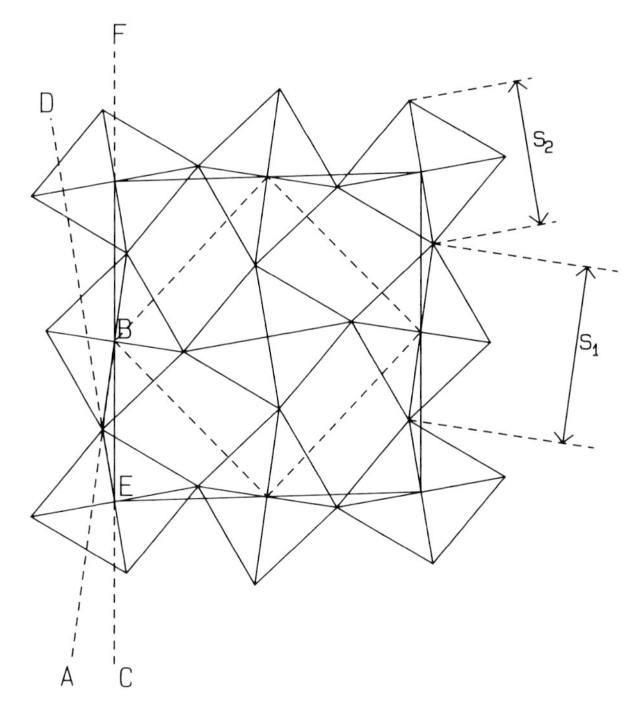

Figure 4 Definition of parameters s_1, s_2, θ_x, θ_y, used to quantify tilting in orthorhombic perovskites. (Parameters s_3 and θ_z refer to the perpendicular z axis, which is not shown.) θ_x corresponds to ∠ABC and θ_y to ∠DEF. Dotted cell corresponds to orthorhombic unit cell, with larger solid cell representing derived pseudotetragonal (i.e. $a = b$, $\gamma \neq 90°$) unit cell.

angles defined as angles between octahedral stalks and *pseudotetragonal axes* x, y and z.

3.2 *Use of the Polyhedral Volume Ratio,* $V_A:V_B$ *as a Structural Determinant*

Owing to the uniqueness of the perovskite structure, in that the AO_{12} and BO_6 polyhedra fill space, there exists a quantitative relationship between the volumes of the A and B ion polyhedra and the occurrence of tilting. Of particular relevance is the $V_A:V_B$ ratio. In the absence of octahedral tilting, this is exactly equal to 5, whereas in tilted perovskites, $V_A:V_B$ takes on a value less than five.

Calculation of $V_A:V_B$ is performed computationally[6,7] from the known unit cell parameters and atomic coordinates of the various crystal structures. Thus in barium titanate, where no octahedral tilting is to be found in either the paraelectric cubic phase, or in any of the three ferroelectric phases, $V_A:V_B$ is equal to 5.

A geometrical analysis reveals that, in rhombohedral systems, the following relationship holds:[4]

$$\frac{V_A}{V_B} = 6\cos^2\omega - 1 \quad (1)$$

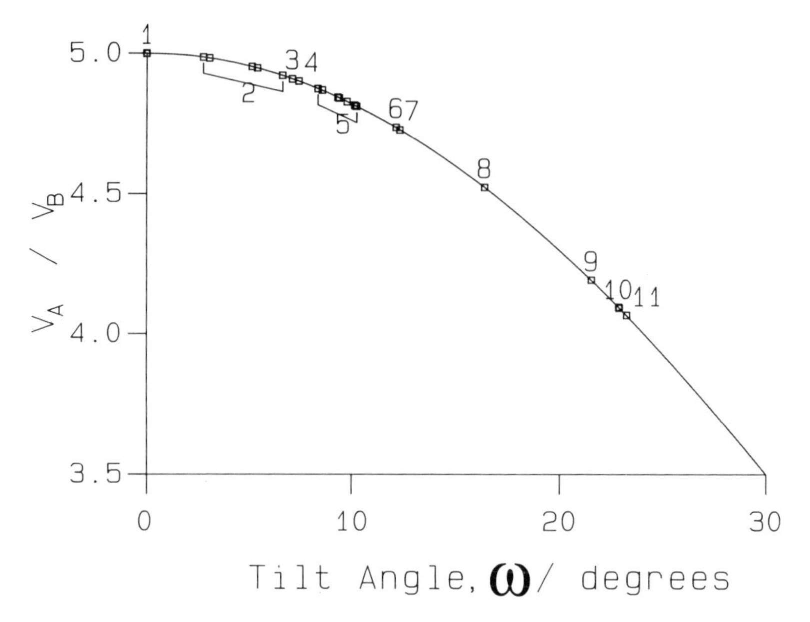

Figure 5 Variation of polyhedral volume ratio, $V_A : V_B$, with tilt angle ω in rhombohedral perovskites. Experimental points correspond to following structures: 1: $KNbO_3$, $LaAlO_3$, $PrAlO_3$, PZT in F_R(HT) phase; 2: PZT in F_R(LT) phase; 3: $LaCuO_3$; 4: $BaTbO_3$; 5: $LaCo_3$ at various temperatures, $NdAlO_3$; 6: $NaNbO_3$; 7: $BiFeO_3$; 8: $HgTiO_3$; 9: $LiReO_3$; 10: $LiUO_3$, $LiTaO_3$; 11: $LiNbO_3$.

In this equation, ω is the rhombohedral tilt angle, as defined in section 3.1. In orthorhombic perovskites, by comparison, the following approximate relationship holds:[5]

$$\frac{V_A}{V_B} = 6\cos^2\theta_m \cos\theta_z - 1 \tag{2}$$

where $\theta_m = (\theta_x + \theta_y)/2$. This relationship is more approximate than equation (1), since it has been assumed that the three octahedral stalks intersect at right-angles. This is not always the case. It is also appropriate to define a composite parameter, Φ, to represent *degree of tilt*:

$$\Phi = 1 - \cos^2\theta_m \cos\theta_z \tag{3}$$

Thus

$$\frac{V_A}{V_B} = 5 - 6\Phi \tag{4}$$

The variation of $V_A : V_B$ with ω for rhombohedral perovskites is given in Fig. 5, with an indication of the compositions giving rise to experimental points. Although the effective coordination number of the A ion can be considered to be less than 12

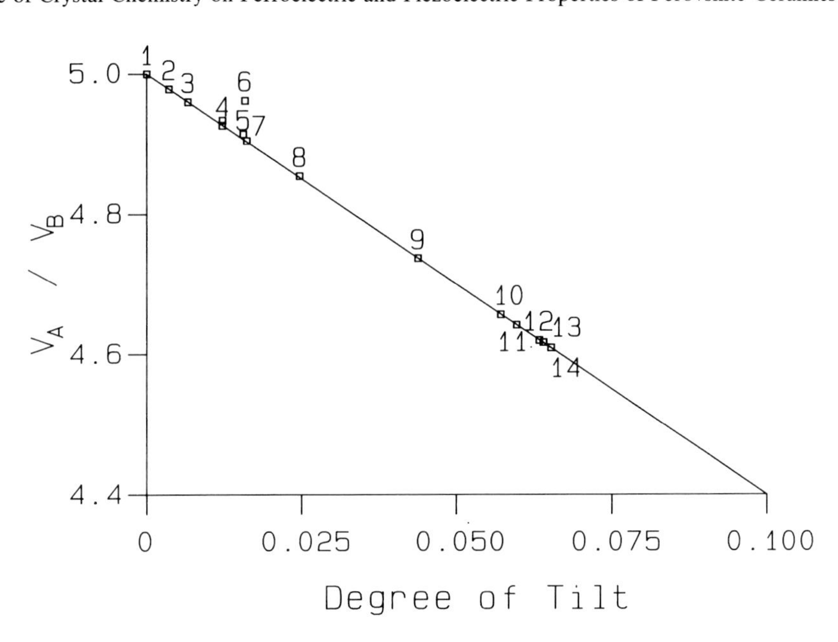

Figure 6 Variation of polyhedral volume ratio, $V_A:V_B$, with degree of tilt Φ in orthorhombic perovskites. Experimental points correspond to following structures: 1: $BaTiO_3$, $KNbO_3$; 2: $NaNbO_3$ (873 K); 3: $NaTaO_3$ (893 K); 4: $NaTaO_3$ (873 K), $SrZrO_3$ (1175 K); 5: $NaTaO_3$ (803 K); 6: $PbZrO_3$; 7: $NaNbO_3$ (813 K); 8: $SrZrO_3$ (1035 K); 9: $NaNbO_3$ (295 K); 10: $SrZrO_3$(r.t.); 11: K_3PtD_5; 12: Cs_3PtD_5; 13: $CaTiO_3$; 14: Rb_3PtD_5.

at larger tilt angles, a coordination number of 12 has been maintained for consistency in the calculation of $V_A:V_B$ values.[4] The proposed linear variation of $V_A:V_B$ with Φ in orthorhombic systems (Fig. 6) also shows good agreement with experiment, with the exception of the highly distorted $PbZrO_3$ structure.[8] In both cases, however, the parametrisation provides a coherent overview of a wide range of structural data.

3.3 *Use of the* $V_A:V_B$ *Ratio in Identifying Possible New Ferroelectric Compositions*

Whereas there is a direct relationship between polyhedral volume ratio and octahedral tilting, the connection between polyhedral volumes V_A, V_B and the occurrence of ferroelectricity is rather more subtle. In order to articulate the nature of this connection, reference must be made to the polyhedral volumes which would be adopted by the A and B ions if left to themselves, i.e. without the imposed constraint that the AO_{12} and BO_6 polyhedra should fill space.

The most general source of idealised ionic sizes is that of coordination number dependent ionic radii,[9] which find widespread use in ceramics. Thus the volume of a regular polyhedron will be equal to $N(r_{cation} + r_{anion})^3$, with N equal to $4/3 = 1.333$ for a BO_6 octahedron and $10/3\sqrt{2} = 2.357$ for an AO_{12} cuboctahedron, with cations and anions in contact.

It is argued that the off-centre displacement of the B ion in a ferroelectric perovskite like $BaTiO_3$ is a consequence of the TiO_6 octahedral volume being larger than that of an idealised, regular TiO_6 octahedron, since $V_B = V_A/5$, and barium is a large A site cation. Thus an off-centre displacement is interpreted as a consequence of short-range forces, with longer-range coulombic forces serving to promote ferroelectric ordering. This viewpoint is supported by the bond valence method,[10] and has been developed previously.[11]

These considerations may be used to predict possible new ferroelectric compositions, which are isostructural with known ferroelectric compositions, for example tetragonal $BaTiO_3$. These compositions will share the attribute that their BO_6 octahedral volumes will be larger than that of an idealised, regular octahedron. The method proceeds by calculating a value of the parameter N_{poly} for the BaO_{12} and TiO_6 polyhedra:

$$N_{poly} = \frac{V_{poly}}{(K_n r_c^{VI} + r_a)^3}$$

V_{poly} is a polyhedral volume derived from experimental crystallographic data, with r_a and r_c^{VI} tabulated anion and cation radii. K_n is a factor to convert from 6- to n-fold coordination. The r_c^{VI} values are thus considered to be standard indicators of ionic size. These N_{poly} values are subsequently used to calculate hypothetical polyhedral volumes of other cations (of known r_c^{VI}) in a geometrically similar coordination. In order for ferroelectricity to occur in a hypothetical composition, its calculated $V_A : V_B$ ratio must be close to 5, the value in barium titanate.

In assessing which of the hypothetical compositions lie closest in structure to barium titanate, the following 'goodness of fit' parameter is employed:

$$c_{fit} = MAX\left|\frac{[V_i]/[V_j]}{[V_i]_{calc}/[V_j]_{calc}} - 1\right| \quad (6)$$

Here, $[V_i]/[V_j]$ is the polyhedral volume ratio in the generating structure, i.e. $BaTiO_3$, with $[V_i]_{calc}/[V_j]_{calc}$ the volume ratio of a hypothetical composition. Results of the calculation are given in Table 1, quoting hypothetical compositions with lowest values of the parameter c_{fit}. The only composition which has been reported to date corresponds to potassium niobate, $KNbO_3$, which lies in 23rd position in Table 1, with a c_{fit} parameter of 0.0154. This lends tentative support to the methodology, since $KNbO_3$ undergoes the same sequence of phase transitions as barium titanate, with a cubic paraelectric phase and three ferroelectric phases, tetragonal, orthorhombic and rhombohedral. Clearly further experimental work is required to confirm the prediction of ferroelectricity in the other compositions quoted.

Caution must be exercised in applying this method, however, since specific electronic interactions can give rise to polyhedral geometries for which idealised ionic radii are not sufficiently sensitive. The lead ion, Pb^{2+}, for example, which is widespread in ferroelectric compositions, often forms directional, partly covalent Pb–O interactions, as a consequence of 6s–6p orbital mixing and the desirability of

Table 1
Compositions $A'_{0.5}A''_{0.5}B'_{0.5}B''_{0.5}O_3$ predicted to be isostructural with $BaTiO_3$

A′	A″	B′	B″	$(00)_{min}$/pm	c_{fit}
Sm(II)	Sm(II)	Se(IV)	Se(IV)	266.4	0.0016
{No(II), Np(II)}		Si(IV)	Si(IV)	252.5	0.0019
Au(I)	Au(I)	W(V)	W(V)	283.4	0.0019
Rb(I)	Rb(I)	Pu(V)	Pu(V)	298.5	0.0051
Ba(II)	Ba(II)	Co(III)	Mo(V)	281.3	0.0074
Ba(II)	Ba(II)	Rh(IV)	Rh(IV)	280.6	0.0075
Ba(II)	Ba(II)	Rh(IV)	Ti(IV)	280.6	0.0075
Ba(II)	Ba(II)	Ni(III)	{Sb(V), Tc(V)}	280.6	0.0075
Pb(II)	Pb(II)	Ni(IV)L	Ni(IV)L	263.0	0.0087
Rb(I)	Rb(I)	Np(V)	Np(V)	300.8	0.0089
Tl(I)	Tl(I)	Zr(IV)	Np(VI)	297.4	0.0104
Rb(I)	Rb(I)	Pu(V)	Np(V)	300.2	0.0139
Tl(I)	Tl(I)	Zr(IV)	U(VI)	297.4	0.0140
K(I)	{Au(I), K(I)}	W(V)	W(V)	283.4	0.0141
Ba(II)	Ba(II)	Pd(IV)	Pd(IV)	281.3	0.0148
Ba(II)	Ba(II)	Pd(IV)	Ti(IV)	281.3	0.0148
{six further compositions}					
K(I)	K(I)	Nb(V)	Nb(V)	284.7	0.0154

accommodating its lone pair away from coordinating oxygen ions. Another example would be electronic interactions giving rise to Jahn–Taller distortions in some transition metal/oxygen octahedra. However, in the majority of cases, which are concerned with closed shell metal ions, the method is a helpful indicator towards which chemical compositions are likely to be ferroelectric.

4. DISCUSSION

4.1 The PZT System Revisited

The phase diagram of the well-known $PbZrO_3$–$PbTiO_3$ system is shown in Fig. 7. Of particular interest is the *morphotropic phase boundary* (MPB) at the composition 52% $PbZrO_3$, since optimum piezoelectric properties are to be found near this boundary. As one proceeds from $PbTiO_3$ to $PbZrO_3$ (i.e. as x increases from 0 to 1), the mean B ion radius increases, according to

$$r_B = x\,r^{VI}(Zr^{4+}) + (1-x)\,r^{VI}(Ti^{4+}),\ \text{i.e.}$$

$$r_B/pm = 60.5 + 8.5x \qquad (7)$$

At the critical x value of 0.52, the structure switches over from a tetragonal to a rhombohedral distortion, i.e. from a structure in which the B ion is undersized with respect to its O_6 octahedron to one in which the B ion is oversized and

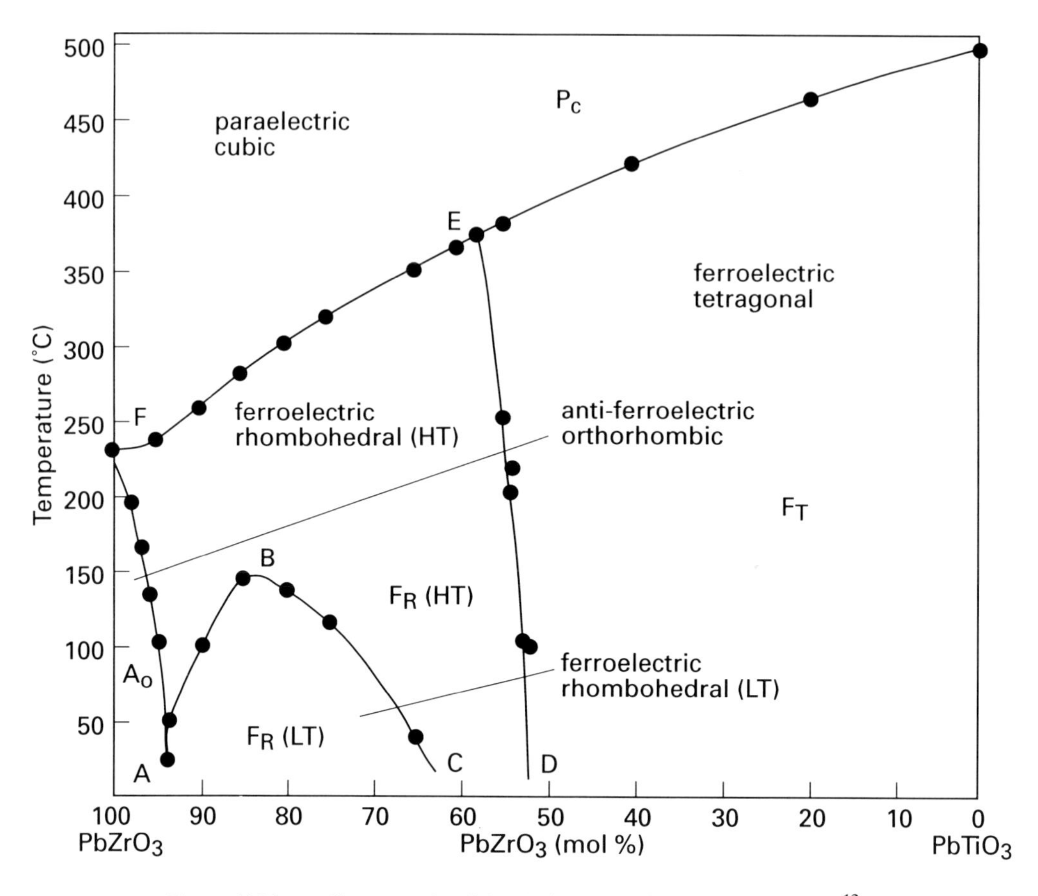

Figure 7 Phase diagram of solid solution $PbZrO_3$–$PbTiO_3$ system.[12]

accommodated by a rhombohedral distortion (see Section 3.1). The width of the stability field of the F_R(HT) phase is relatively small at room temperature, however, since the ability of the rhombohedral distortion to accommodate a larger B-ion is limited. At approximately $x = 0.64$, the F_R(LT) phase begins to be stabilised, in which there is octahedral tilting as well as distortion. With $V_A:V_B$ now less than 5, the BO_6 volume, V_B is greater than $V_A/5$. Thus tilting provides a mechanism by which relatively larger B ions can still be accommodated in a rhombohedral ferroelectric structure.

Of relevance also is the F_R(LT) to F_R(HT) transition at the Zr rich end of the phase diagram. For $x = 0.9$, this occurs at a temperature of approximately 75°C, and may be exploited in pyroelectric detectors. In structural terms, this temperature corresponds to that at which octahedral tilting is completely destabilised by thermal agitation. Since the tilting increases with decreasing $V_A:V_B$ ratio, it is anticipated that the temperature of this phase transition can be controlled by adjusting the A:B size ratio chemically. This bodes well for future device development.

4.2 Analysis of Aurivillius and Layered Perovskite Systems

Recent crystallographic data have been published on the ferroelectric phases of Aurivillius phases Bi_3TiNbO_9[13] and $Bi_4Ti_3O_{12}$.[14] Both of these show octahedral tilting about three axes, and can be analysed according to equations (2)–(4).[5] Significant octahedral tilting is also to be found in the layered (slab-like) perovskite $Sr_2Nb_2O_7$.[15] Each of these compositions has potential applications as a high-temperature ferroelectric or piezoelectric ceramic, by virtue of their high Curie points, T_c: 675°C in $Bi_4Ti_3O_{12}$, 940°C in Bi_3TiNbO_9 and 1340°C in $Sr_2Nb_2O_7$. An extension of this methodology to these systems is therefore envisaged.

4.3 Relevance to other Current Technological Problems

The current generation of solid oxide ceramic fuel cells frequently makes use of doped $LaMnO_3$ as a cathode and doped $LaCrO_3$ as an interconnect material, both of which are perovskites. The undoped compositions undergo an orthorhombic to rhomobedral phase transition upon heating above room temperature, at 387°C in $LaMnO_3$ and 280°C in $LaCrO_3$. It has been shown experimentally that doping with Sr^{2+} (larger than La^{3+}) stabilises the rhombohedral phase down to room temperature, whereas doping with Ca^{2+} (smaller than La^{3+}) has the converse effect of stabilising the orthorhombic phase to higher temperatures.[16] It appears, therefore, that in lanthanum based systems with tilted octahedra, orthorhombic phases are associated with smaller $V_A:V_B$ ratios than their rhombohedral counterparts. Thus this subject area is also likely to be a fruitful one for structural design of electroceramic materials.

5. CONCLUSION

A quantitative analysis of cationic displacements, distortions of cation coordination polyhedra and octahedral tilting has been used to rationalise the underlying ferroelectric properties of perovskite and perovskite related ceramics. Variations in these crystal structural features are reflected in the value of $V_A:V_B$, the polyhedral volume ratio. This ratio can also be used, in conjunction with standard ionic radii, to predict new ferroelectric compositions. The contribution of the analytical framework towards an understanding of the phase behaviour of the PZT system has been outlined. Its relevance to other areas of ceramic technology, e.g. solid oxide fuel cells, has also been discussed.

6. REFERENCES

1. J. W. Edwards, R. Speiser and H. L. Johnston: *J. Amer. Chem. Soc.*, 1951, **73**, 2934–2935.
2. J. Harada, T. Pedersen and Z. Barnea: *Acta Cryst.*, 1970, **A26**, 336–344.
3. R. J. Nelmes and W. F. Kuhs: *Solid State Comm.*, 1985, **54**, 721–723.
4. N. W. Thomas and A. Beitollahi: *Acta Cryst.*, 1994, **B50**, accepted for publication.

5. N. W. Thomas: to be submitted to *Acta Cryst.* **B**.
6. N. W. Thomas: *Acta Cryst.*, 1989, **B45**, 337–344.
7. N. W. Thomas: *Acta Cryst.*, 1991, **B47**, 180–191.
8. F. Jona, G. Shirane, F. Mazzi and R. Pepinsky: *Phys. Rev.*, 1957, **105**, 849–856.
9. R. D. Shannon: *Acta Cryst.*, 1976, **A32**, 751–767.
10. I. D. Brown: *Chem. Soc. Rev.*, 1978, **7**, 359–376.
11. N. W. Thomas: *Ferroelectrics*, 1989, **100**, 77–100.
12. B. Jaffe, W. R. Cook and H. Jaffe: '*Piezoelectric Ceramics*', Academic Press, London, 1971.
13. J. G. Thompson, A. D. Rae, R. L. Withers and D. C. Craig: *Acta Cryst.*, 1991, **B47**, 174–180.
14. A. D. Rae, J. G. Thompson, R. L. Withers and A. C. Willis: *Acta Cryst.*, 1990, **B46**, 474–487.
15. N. Ishizawa, F. Marumo, T. Kawamura and M. Kimura: *Acta Cryst.*, 1975, **B31**, 1912–1915.
16. N. Q. Minh: *J. Amer. Ceram. Soc.*, 1993, **76**, 563–588.

Influence of Structural Defects on Properties of Zirconium Titanate Based Microwave Ceramics

P. K. DAVIES and R. CHRISTOFFERSEN
Department of Materials Science and Engineering, University of Pennsylvania, Philadelphia, PA 19104-6272, USA

ABSTRACT
Pure and Sn doped zirconium titanate ceramics have been investigated to characterise their incommensurate crystal structures and evaluate the effect of cation ordering on their dielectric properties in the microwave region. Using electron diffraction and high-resolution transmission electron microscopy (HRTEM) the incommensurate ordered solid solutions in the zirconium titanate system were shown to consist of a random mixture of the commensurate structural units, or modules, of the $Zr_5Ti_7O_{24}$ and $ZrTiO_4$ end members. For the Ti-rich end member this unit has a ZrTiTi layered repeat and for $ZrTiO_4$ the module consists of two Zr-rich and Ti-rich layers with a ZrZrTiTi repeat. Limited interlayer cation diffusion inhibits the complete segregation of Zr and Ti, and individual (100) cation layers switch abruptly from a Zr-rich to Ti-rich occupancy. By investigating samples of $Zr_{1-x}Sn_xTiO_4$ with x ranging from 0.0 to 0.20 it was found that Sn substitutions reduce the correlation length of the Zr–Ti cation correlations and significantly increase the number of occupancy switching boundaries. It is proposed that the reduction in the driving energy for the coarsening of the ordered domains is due to the preferential segregation of Sn to the Zr–Ti domain boundary. Although the long-range cation ordering increases the dielectric loss of Sn-free ceramics by 30%, it results in a slight improvement in the loss properties of $(Zr_{0.91}Sn_{0.09})TiO_4$. These differences may be a further reflection of the effect of Sn in stabilising the Zr–Ti boundaries and reducing their contribution to the dielectric loss.

1. INTRODUCTION

Ceramic dielectric resonators have revolutionised microwave-based communications systems by reducing the size and cost of filter and oscillator components in systems ranging from cellular telephones to global positioning technologies. Microwave resonator ceramics are required to have a high dielectric constant ($k' = 30$–100), a low dielectric loss or high 'Q' ($Q = 1/\tan\delta = 5000$–$10\,000$) and a near-zero temperature coefficient of resonance frequency ($\tau_f = 0 \pm 3$ ppm). Several oxides have been found to meet these criteria and, for example, ceramics in the BaO–TiO_2 ($Ba_2Ti_9O_{20}$, $BaTi_4O_9$), ZrO_2–SnO_2–TiO_2 ($Zr_{1-x}Sn_xTiO_4$), BaO–Ta_2O_5–ZnO ($BaTa_{2/3}Zn_{1/3}O_3$) and BaO–PbO–Re_2O_3–TiO_2 systems have found widespread commercial application.[1,2] With the expanding use of these ceramics has come increased interest in their crystal structure, phase transformations and property-enhancing chemical additives, all of which are factors in influencing the dielectric properties. Crystallographic studies of these systems have shown that most of the prominent microwave ceramic systems exhibit some type of structural disorder. For the barium polytitanates the structural inhomogeneities arise from extensive intergrowths of different polytypes;[3,4] in barium zinc/magnesium tantalate perovskites[5,6] and zirconium titanate[7,8] they are due to the formation of phases with differing degrees of cationic order. In some cases alterations in the degree of structural order have a pronounced effect upon the dielectric response in the microwave region. For example, data for the dielectric properties of $BaTa_{2/3}Zn_{1/3}O_3$ show that by increasing the long-range Zn–Ta order the loss can be reduced by more than a factor of 10.[5,6]

Ceramics based on the solid solution $(Zr_{1-x}Sn_x)TiO_4$, with up to 20 at.% of the Zr replaced by Sn, have been in use as resonator materials since the 1970s.[1] For frequencies between 1 and 10 GHz, the Sn substitution progressively increases Q from 2000 to 5000 for $x = 0.0$, to 6000 to 10 000 for $x = 0.20$.[9,10] As well as being essential to attain a useful Q, the Sn substitution lowers τ_f to near zero and stabilises the resonator against long-term breakdown.[1,9] Despite the widespread use of these ceramics in commercial resonator systems, many aspects of their crystal chemistry and phase transformations have remained obscure and there has been no clear understanding of how the structural state of these phases affects their dielectric properties. In our work we have attempted to unravel some of the structural complexities of this system using high-resolution TEM, to modify their crystal chemical behavior through controlled chemical substitution and examine how changes in the degree of cationic order affect the dielectric properties in the microwave region.[7,11,12] In this paper we will summarise the results of our studies for the $Zr_{1-y}Ti_{1+y}O_4$ and $(Zr_{1-x}Sn_x)TiO_4$ systems.

The compositions of interest for microwave applications in the ZrO_2–TiO_2–SnO_2 system adopt the α-PbO_2 structure above 1200°C with a disordered arrangement of the cations on equivalent distorted octahedral sites (Fig. 1a).[13,14] In the ZrO_2–TiO_2 binary, single-phase $Zr_{1-y}Ti_{1+y}O_4$ solid solutions can be prepared for $0.0 \leq y \leq 0.167$, i.e. between the end members $Zr_5Ti_7O_{24}$ and $ZrTiO_4$.[15] Below 1200°C all the compositions in the system undergo a continuous phase transformation characterised by increasing order in the Zr–Ti distribution.[8,15] Ti-rich samples near $ZrTi_2O_6$ (e.g. $Zr_5Ti_7O_{24}$), adopt a 'fersmite-type' structure in which the larger Zr ions order onto every third octahedral layer along the a axis, Fig. 1b.[14] Because of associated shifts in the adjacent oxygen layers, the Zr-rich site is distorted to approximate 8-fold coordination in a fluorite-related layer. This commensurate phase has an a repeat exactly three times that of the disordered structure and can be described by a $Zr^{TiTi}Zr_{TiTi}$ occupancy sequence with alternating pairs of up-pointing (^{TiTi}) and down-pointing ($_{TiTi}$) Ti-octahedra.[14]

For compositions in the range $ZrTiO_4$ to near $Zr_5Ti_7O_{24}$ the structure corresponding to partial or complete Zr–Ti order is more problematical and the nominally ordered phase exhibits non-integral satellite reflections.[15] The satellite positions are consistent with the development of an incommensurate structural modulation along a, but the exact nature of this modulation and its associated Zr–Ti distribution was not clear prior to our studies using HRTEM.[7]

The substitution of Sn into zirconium titanate is known to inhibit the ordering transformation and because of the accompanying improvement in the dielectric properties it has been suggested that cation ordering may be detrimental in its effect on dielectric loss.[8] However, there are several unresolved questions pertaining to the mechanism by which Sn affects the structure. When Zr is replaced

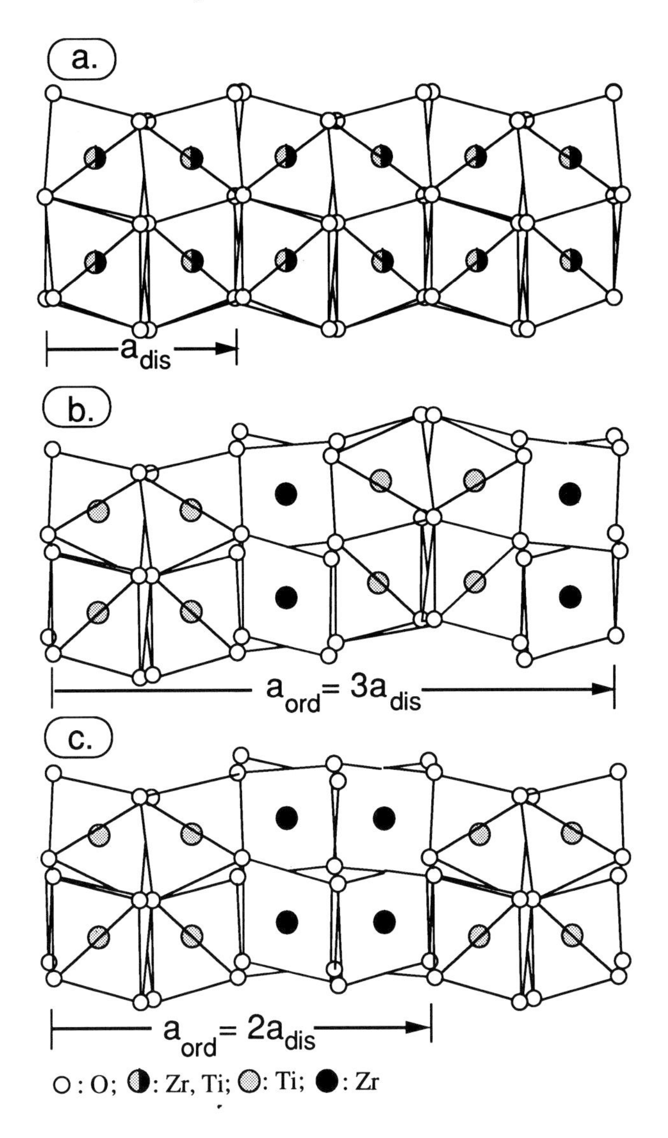

Figure 1 [010] projections of zirconium titanate structures (a) disordered α-PbO_2 structure (b) ordered $Zr_5Ti_7O_{24}$ (c) ordered $ZrTiO_4$.

by the smaller Sn ion the unit cell volumes of the resultant $Zr_{1-x}Sn_xTiO_4$ solid solutions show an anomalous increase with x, primarily due to a significant expansion along the b direction of the unit cell.[8,9] In addition, when samples with $x \geqslant 0.2$ are annealed for extended periods at temperatures known to induce order in $ZrTiO_4$, their X-ray patterns show no evidence for the contraction along b or the additional superlattice reflections that characterise the ordering transformation in the pure zirconium titanates.[8]

While these observations would imply that Sn stabilises a random arrangement of the cations in the zirconium titanate structure, investigations using extended X-ray absorption fine structure (EXAFS) spectroscopy and neutron powder diffraction point to a possibly more complicated situation.[16] Using neutron diffraction, Kudesia *et al.* confirmed that low-temperature annealed samples of $Zr_{0.8}Sn_{0.2}TiO_4$ can be refined within a disordered subcell, but their patterns also contained additional diffuse reflections that they felt may result from some type of correlated cation order.[16] In addition EXAFS measurements collected from the same samples showed the local coordination of Zr is similar to that adopted in the long-range ordered form of the pure zirconium titanates. Clearly many questions remain as to the exact state of order in the Sn substituted zirconium titanates. It was not known whether this substitution has an abrupt or gradual effect on the degree of order that can be developed for a given composition and other questions, such as the possibility that Sn substituted compositions have a fundamentally different cation ordering scheme, have also not been resolved.

In this paper we present a summary of our investigations of cation ordering in pure and Sn doped zirconium titanate using electron diffraction and high resolution transmission electron microscope (HRTEM) methods.[7,11] We also report data on the effect of the cation ordering on the microwave dielectric properties of ceramics in the $(Zr_{1-x}Sn_x)TiO_4$ solid solution.

2. HRTEM STUDIES OF COMMENSURATE AND INCOMMENSURATE PHASES IN $ZrTiO_4$–$Zr_5Ti_7O_{24}$ SYSTEM

The initial goal of our work on the pure zirconium titanate system was to determine the structural mechanisms by which ordering is accomplished in samples where the Zr:Ti ratio exceeds that of the commensurate $Zr_5Ti_7O_{24}$ phase. Of particular interest was the ordered structure of $ZrTiO_4$, the composition that is closest to that of commercial zirconium titanate dielectric ceramics. By examining crystals with a range of Zr:Ti ratios and thermal histories designed to maximise the attainment of Zr:Ti order, we found that although intermediate composition ordered zirconium titanates show incommensurate diffraction effects, their structures are not modulated in the true sense.[7] Instead they contain uniformly distributed commensurate structural units, or 'modules', that produce discrete non-integral diffraction spots by forming an 'interface modulated' structure.[17,18] In contrast, ordered phases near to or coincident with $ZrTiO_4$ exhibited a new commensurate structure that previously had not been reported for the zirconium titanate solid solution.

Figure 2 shows an [010] high-resolution image of a fragment of a grain with a chemically analysed composition $(Zr_{0.83}Ti_{1.17})O_4$. The diffraction pattern of this grain exhibits incommensurate satellites along a^* with $\alpha = 0.63$, where $r^* = ha^* + kb^* + lc^* \pm m\alpha a^*$ (for a commensurate supercell with a tripled *a*-repeat, $\alpha = 2/3$). The image contains regions whose overall contrast and sequence of Zr-rich and Ti-rich layers, $Zr^{TiTi}Z_{TiTi}$, is in good agreement with calculated

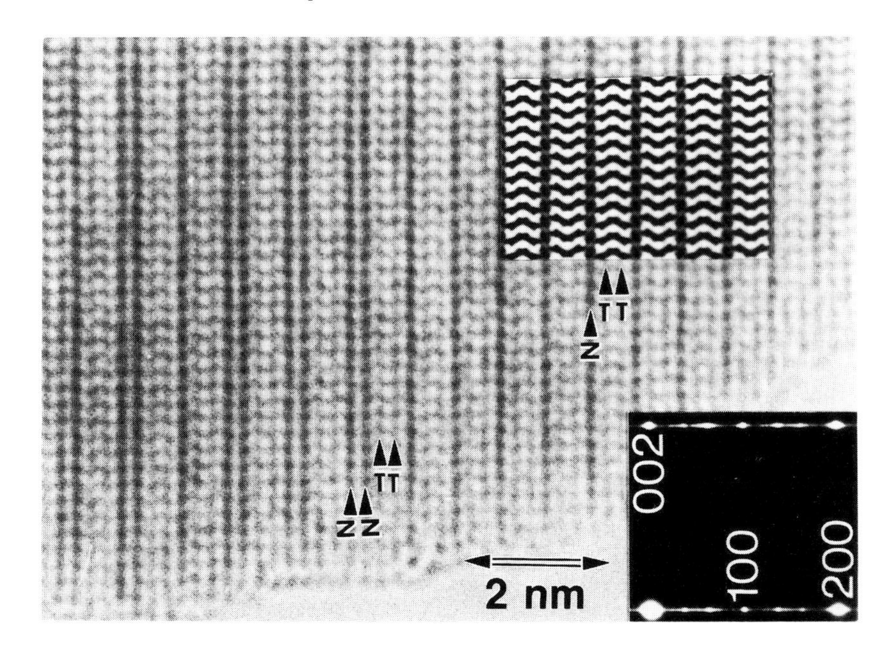

Figure 2 High-resolution image of $(Zr_{0.83}Ti_{1.17})O_4$. Zr-rich (Z) and Ti-rich (T) layers indicated by arrows. Calculated image for $Zr_5Ti_7O_{24}$ is inset.

images of commensurate $Zr_5Ti_7O_{24}$ (see Fig. 2). However, in many cases this sequence is interrupted by the insertion of extra Zr-rich layers, which form double-layer slabs of Zr. This leads to local regions with a ZrZrTiTi repeat. On a larger scale the structure of this sample is comprised of a more or less random sequence of integral numbers of the ZrZrTiTi units and half multiples of the $Zr^{TiTi}Zr_{TiTi}$ repeat unit (see Fig. 3). Images of other fragments confirmed that this mixing of ZrTiTi and ZrZrTiTi units is representative of the crystal as a whole.

Further details on the arrangement of the doubled Zr layers are visible in images collected along the [001] zone axis (see Fig. 4a). These show that the cation layers undergo abrupt (although in some cases gradational) changes in occupancy from a Zr-rich to Ti-rich composition, or vice-versa. Typically the switch converts one Zr layer of a ZrZrTiTi slab into a Ti layer, and this 'new' Ti layer is compensated by the conversion of an adjacent Ti layer to a Zr layer. In this manner the formation of triple- or single-Ti slabs are prevented and the total number of Zr and Ti layers is conserved. These 'switching boundaries' produce local alterations in the structural sequence, but do not change the relative fraction of the ZrZrTiTi and ZrTiTi modules.

By examining a series of samples, we found that the structures of all the compositions in the ordered zirconium titanate system can be described in terms of the random, uniform mixing of the ZrZrTiTi and ZrTiTi structural modules. The relative proportions of the two units was an approximately linear function of the Zr:Ti ratio. From this relationship it became apparent that the $ZrTiO_4$ end

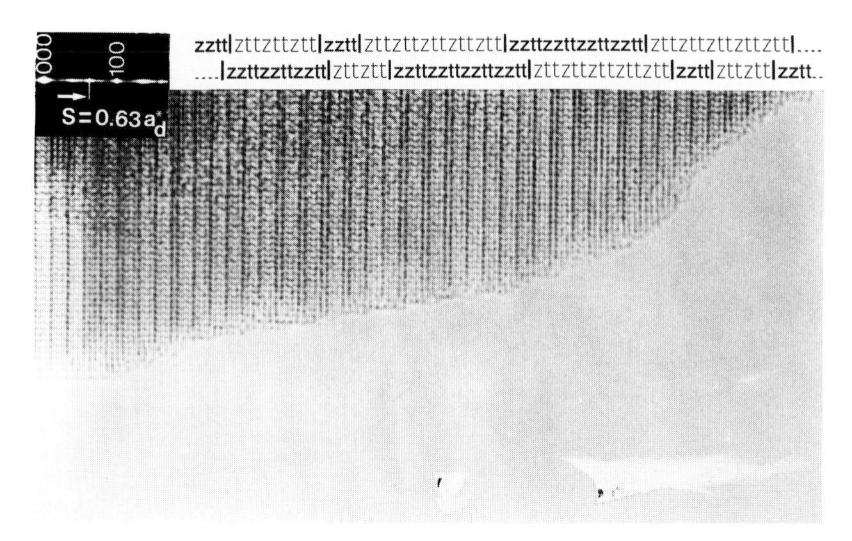

Figure 3 Lower magnification [010] image of $(Zr_{0.83}Ti_{1.17})O_4$. Darker rows correspond to Zr layers, insets show electron diffraction pattern and list of sequence of cation layers in the image.

member should have an uninterrupted $ZrZr^{TiTi}ZrZr_{TiTi}$ ordered sequence. HRTEM observations on ordered $ZrTiO_4$ powders [7] confirmed that most grains consisted predominately of ZrZrTiTi slabs and those that were entirely free of ZTT units had a commensurate $2x$ repeat along a with $\alpha = 0.5$. The ordered structure of $ZrTiO_4$ is shown in Fig. 1c.

To summarise, the incommensurate ordered solid solutions in the zirconium titanate system consist of a random mixture of the commensurate structural units, or modules, of the $Zr_5Ti_7O_{24}$ (ZrTiTi) and $ZrTiO_4$ (ZrZrTiTi) end members. Within each module Zr occupies an 8 coordinate site and Ti is in octahedral coordination. Although the individual modules are commensurate, the structure as a whole yields non-integral diffraction spots. The concept that random, uniform sequences of structural faults or modules can produce incommensurate satellite reflections is well established[17,19] and is fully detailed for this system in Ref. 7. The complete segregation of Zr and Ti into individual cation layers is apparently inhibited by the limited interdiffusion of the cations between layers which are separated by an approximately close-packed array of oxygen anions. The occupancy switching boundaries observed in [001] images are a direct consequence of this anisotropic diffusion and indicate that each (100) cation layer consists of two-dimensional Zr-rich and Ti-rich domains. The primary driving force for the cation ordering is the reduction in the number of Zr–Ti nearest neighbours within a given layer, particularly Zr and Ti nearest neighbours within the same chain of edge-sharing coordination sites. These linkages are energetically unfavourable because the larger size of the Zr ion and its preference for a higher local coordination induces concerted oxygen displacements that are incompatible with the octahedral

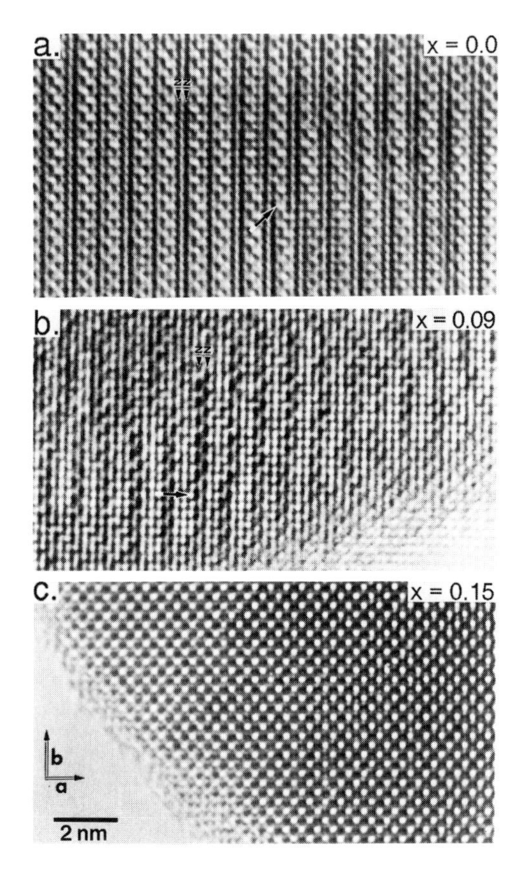

Figure 4 Series of [001] high-resolution images collected from slow cooled samples of (a) $ZrTiO_4$ (b) $Zr_{0.91}Sn_{0.09}TiO_4$ (c) $Zr_{0.85}Sn_{0.15}TiO_4$. Double Zr-layers (ZZ) are indicated in (a) and (b), arrows highlight locations of one of 'switching sites' (see text).

coordination preference of the neighbouring Ti site. However because of the kinetic limitations a certain number of these interfaces are unavoidable in this system and their significance will be discussed further below.

3. HRTEM STUDIES OF $(Zr_{1-x}Sn_x)$ TiO_4 SYSTEM

For the Sn-doped system we hoped to examine how the modulated structure of the ordered Sn-free zirconium titanates was altered by replacing Zr by Sn. In our studies using HRTEM,[11] which are summarised below, we found that the structure persists, albeit with some progressive changes, almost to the limits of Sn solubility in the $(Zr_{1-x}Sn_x)TiO_4$ solid solution. The primary effect of Sn was to gradually reduce the correlation length of the Zr–Ti ordering and increase the number of switching boundaries, these changes could be understood in terms of the role of Sn in mediating the interaction between Zr–Ti nearest neighbour cations.

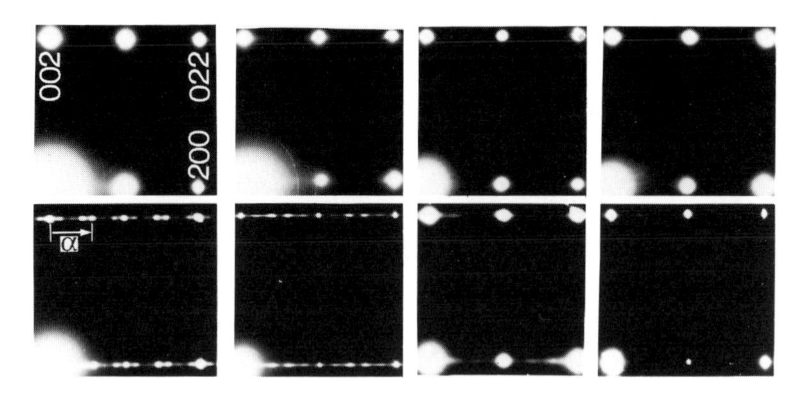

Figure 5 [010] electron diffraction patterns for (a) as synthesised and (b) slow cooled samples of $Zr_{1-x}Sn_xTiO_4$ with $x = 0.0$, 0.09, 0.15, and 0.20. Satellite spacing α is indicated by arrow.

A series of ordered and disordered $(Zr_{1-x}Sn_x)TiO_4$ ceramics with $x = 0.0$, 0.09, 0.15, and 0.20 were prepared using methods fully described in Ref. 11. The ordered materials were prepared by slow cooling from 1200 to 780°C at the rate of 1°C/hour. Figure 5 shows a representative series of [010] selected-area electron diffraction patterns for the samples of $Zr_{1-x}Sn_xTiO_4$ used in the present study. In their as-synthesised state all ceramic compositions from $x = 0.0$ to 0.20 show only diffuse continuous streaking along a^*, Fig. 5(a), with no sign of satellite reflections. These patterns are consistent with all samples having an equivalently high degree of long-range Zr–Ti disorder. However, the diffraction patterns for slow-cooled samples with $x = 0.0$ and 0.09 contain well-developed incommensurate satellites, which are somewhat less intense and more streaked along a^* for the $x = 0.09$ composition (Fig. 5b). The fact that the $x = 0.0$ composition developed an incommensurate satellite spacing rather than a commensurate one with a exactly doubled is not necessarily unusual, because in addition to the slow cooling treatment used in the present study, supplemental long-term isothermal annealing is sometimes required for the commensurate ordered structure to form.[7] The slow cooled samples with $x = 0.15$ and 0.20 have similar satellite reflections that are very weak and diffuse (Fig. 5b) and there is a fairly marked change with Sn content from the well-developed satellites in samples $x = 0.0$ and 0.09 to weak and diffuse satellites in compositions $x = 0.15$ and 0.20.

High-resolution imaging was carried out on the samples with compositions $x = 0.0$, 0.09 and 0.15; the $x = 0.20$ sample was not studied using high-resolution techniques because its diffraction pattern was very similar to that of the $x = 0.15$ sample. Imaging was performed along the [001] and [010] zone axes, directions that are parallel to the cation layers; the [001] orientation was of particular interest as this was the direction used to reveal the switches in the occupancy of individual cation layers in the Sn free system.

Figure 4 shows the [001] images for the slow cooled samples with $x = 0.0$, 0.09

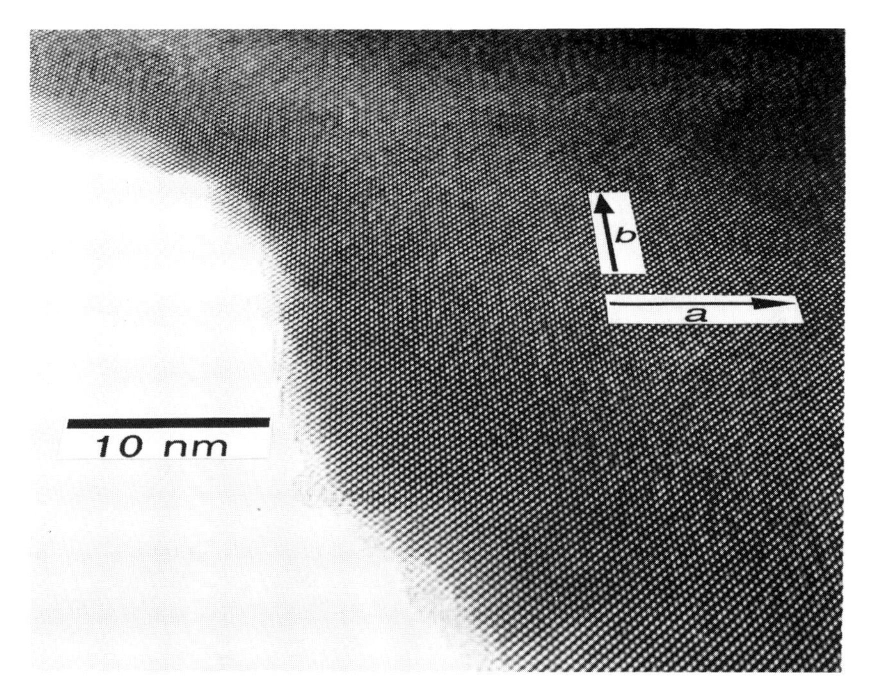

Figure 6 High-resolution image of $Zr_{0.85}Sn_{0.15}TiO_4$ collected along [001] and viewed at inclined angle. Weak additional contrast originates from 25–50 Å order domains.

and 0.15. The images have the same defocus value of −40 nm, which is close to Schertzer defocus and the atomic columns in the structure appear dark. For the $x = 0.0$ and 0.09 samples (Fig. 4a and b) the Zr-rich cation layers appear darker relative to the Ti-rich layers because of the stronger electron scattering cross-section of Zr relative to Ti. For the $x = 0.15$ sample there is significantly less contrast between the cation layers, and the image for this composition (Fig. 4c) shows no immediately obvious delineation between Zr-rich and Ti-rich layers. However, contrast is visible if the [001] images of the $x = 0.15$ sample are viewed at an inclined angle (Fig. 6). These images indicate that a visible degree of Zr–Ti segregation is present in the $x = 0.15$ sample, although the contrast in occupancy between the Zr-rich and Ti-rich layers, and the correlation length of the ordered regions is clearly much less than in the samples with a lower Sn content.

As observed for the pure zirconium titanates, the [001] image for the $x = 0.09$ sample reveals abrupt switches in occupancy of individual cation layers along the *b* axis direction (Fig. 4a, b). Again the switches convert one Zr-rich layer of a double Zr slab into a Ti-rich layer, and the 'new' Ti layer is compensated by a switch of opposite sense in a next adjacent layer. The two layers that switch occupancy are always separated by an intervening Ti-rich layer (Fig. 4a and b). Occupancy switches are also present in the cation layers in the $x = 0.15$ sample (Fig. 6), although they are less well defined because the contrast between the Zr-rich and Ti-rich layers is not as strong.

In comparing the images for the samples with $x = 0.0$ and 0.09 it was found that the $x = 0.09$ sample had a greater number of sites where the layers switched occupancy. By measuring the number of switches per unit area, we found that the average spacing between switches along the length of the layers was 425 Å for $x = 0.0$ and 180 Å for $x = 0.09$. For the $x = 0.15$ sample the switches were more difficult to resolve but on the average occurred every 25 to 50 Å. The $x = 0.15$ sample thus has a significantly shorter switch spacing than either the $x = 0.0$ and $x = 0.09$ compositions.

Switches in layer occupancy are a common feature in [001] images but occur less frequently in [010] images. In [010] images for the $x = 0.0$ and $x = 0.09$ compositions, individual cation layers retain the same occupancy for distances of several thousand Å, and typically only one or two switching sites are present in a given [010] image. This dependence of switch spacing on imaging orientation is related to the shape of Zr-rich and Ti-rich compositional domains on the (001) cation layers and is discussed in more detail below.

The two principal variations observed in slowly cooled samples as a function of increasing Sn content are a decrease in the 'switching length' along the layers in the *b* direction of the α–PbO_2-related cell and a decrease in the compositional contrast between adjacent Zr-rich and Ti-rich layers associated with the reduction in the correlation length of the cation order. For this discussion we focus on the difference in switching length between the samples. As shown in Fig. 7, each (100) cation layer consists of Zr-rich and Ti-rich two-dimensional domains and the boundaries between the domains project as occupancy switches in [001] and [010] high-resolution images. The distance between switches is considerably longer in [010] images than it is in [001] images (425–25 Å depending on the Sn content), indicating that the domains have a shape that is longer along the *c* axis than along the *b* axis.

As noted above, the Zr–Ti boundaries within a given layer are energetically unfavourable but are unavoidable because of the limited interdiffusion of cations between the layers. As the Zr-rich and Ti-rich domains develop with some initial size and shape, their stacking along the *a* axis apparently adjusts to maintain local compositional homogeneity and to avoid stacking sequences that contain single Ti-rich layers, e.g. ZrZrTiZrTiTi. Such sequences are unfavourable apparently because a Ti-rich layer can accommodate the oxygen distortions of the 8 fold Zr sites on one side, but not on both. Partly because of these constraints the boundaries between domains in a given layer will stack on top of the boundaries in other layers in such a way as to follow the 'switching rules' illustrated by Fig. 7. Because the boundary between coplanar domains is to some extent compositionally gradational, the region of crystal associated with switches of opposite sense in next-adjacent layers contains residual Zr–Ti nearest neighbours within the same layer (type A sites, Fig. 8), in addition to Ti sites that share oxygens on both sides with Zr sites (type B sites, Fig. 8).

We interpret the large decrease in domain size that occurs as a function of increasing Sn content to be linked to an effect of Sn on the driving force for coarsening the domains. Because this driving force is derived from nearest

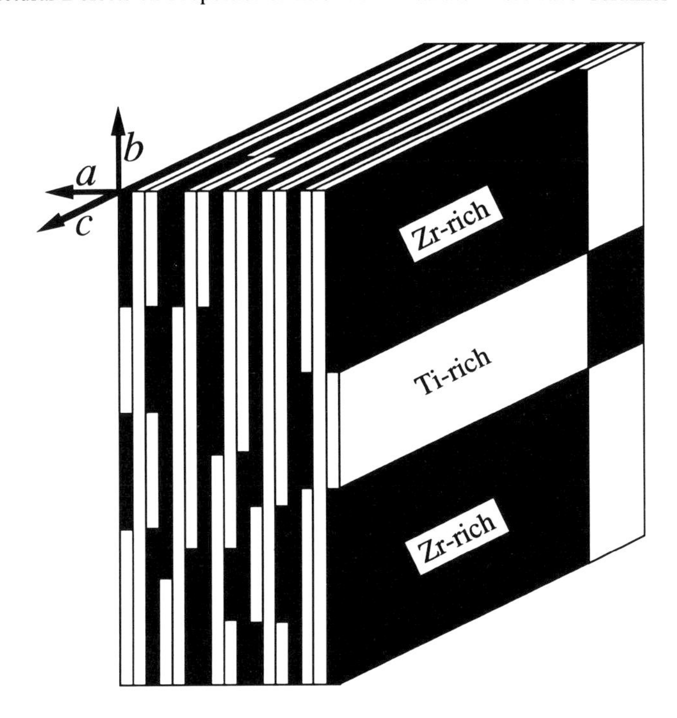

Figure 7 Schematic illustration of occupancy switches observed in zirconium titanates. Distance between switches is much longer along *c* than along *b*. Stacking sequence of Zr (dark) and Ti (white) layers along *a* axis adjusts to maintain compositional homogeneity and avoid sequences with single Ti layers.

neighbour cation interactions at sites associated with occupancy switches, we hypothesise that Sn affects these interactions and thereby reduces the driving energy for removing the sites. As the size of the Sn^{4+} ions (e.g. r_{Sn} = 0.69 Å in octahedral coordination)[20] lies between that of Ti (0.605 Å for octahedral coordination) and Zr (0.84 Å in 8 fold coordination) and Sn is known to adopt a range of coordination environments in different oxides, one possibility is that it replaces Zr and Ti in 'strained' coordination environments and thereby reduces their energy. Such sites would be either the type A sites in the switching layers themselves, or the type B sites in the Ti-rich layers between two switching layers, Fig. 8. Support for this hypothesis comes from the large change in the frequency of the occupancy switches with increasing Sn content.

The changes that we have observed in the length scale of the cation order, from the several hundred angstrom correlations in pure $ZrTiO_4$ to the 20–50 Å domains in $Zr_{0.85}Sn_{0.15}TiO_4$, are consistent with the previous studies of the system by X-ray and neutron diffraction.[8,16] For example, the neutron refinements[16] reported for $Zr_{0.8}Sn_{0.2}TiO_4$ gave a reasonable fit to a model with a disordered arrangement of the cations but the diffraction profiles also contained broad, weak additional reflections that presumably originate from the locally ordered domain structure observed in this study. The formation of these correlated domains, with a Zr

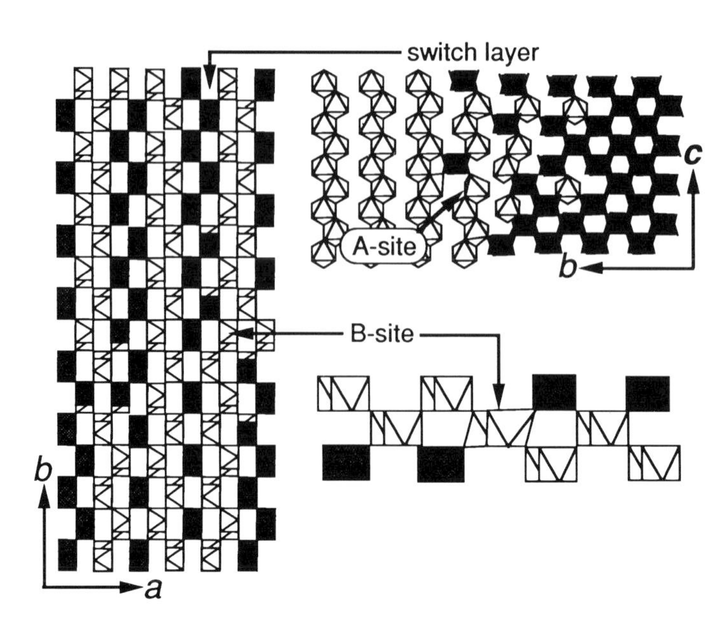

Figure 8 Schematic view of strained cation coordinations at Zr and Ti domain boundaries. Left, [001] projection of structure containing ordered Zr (dark) and Ti (white) layers. Upper right, strained coordination site within (100) switching layers — type A site; note large distortion of sites at Zr and Ti interface induced by shearing of oxygen ions surrounding Zr. Lower right, a type B site located in Ti layers between two switching layers; distortion originates from anion shear in adjacent Zr layers.

coordination greater than six, is also consistent with the investigations of the local environment of Zr using EXAFS.[16]

The role we propose for Sn in stabilising the Zr–Ti interfaces and reducing the length scale of the diffusional rearrangements required to satisfy the coordination preference of Zr, would lead to a situation where the high- and low-temperature forms of samples with large concentrations of Sn become virtually indistinguishable and different cooling conditions are no longer important.

4. EFFECT OF CATION ORDERING ON MICROWAVE DIELECTRIC PROPERTIES

To examine whether the cation ordering has any effect upon the microwave dielectric properties, we prepared a series of dense $(Zr_{1-x}Sn_x)TiO_4$ ceramics using sol-gel precursors for the composition $x = 0.0$ ($ZrTiO_4$), and mixed oxides processed by hot forging for compositions $x = 0.09$, 0.15, and 0.20. The dielectric constant, k', the dielectric loss, Q, and the temperature coefficient of resonant frequency, Tf, were measured at microwave frequencies using parallel plate and cavity methods. These measurements were carried out in collaboration with Dr T.

Table 1
Structure and Properties of Ordered and Disordered $Zr_{1-x}Sn_xTiO_4$

	Structural Properties						Dielectric Properties				
x	TEM	a(Å)	b(Å)	c(Å)	V(Å^3)	Density	K'	Q	f(GHz)	$Q \cdot f$	$\tau_{f(ppm)}$
0.0	Disordered	4.81	5.48	5.03	132.5	94%	42.4	3079	8.3	25556	58
0.0	Disordered	4.81	5.48	5.03	132.5	96%	43.8	3166	8.2	25961	59
0.0	Ordered	4.81	5.38	5.03	130.3	94%	38.3	2280	8.7	19836	74
0.0	Ordered	4.81	5.38	5.03	130.3	96%	39.7	2063	8.5	17535	79
0.09	Disordered	4.78	5.49	5.03	132.6	94%	38.6	3233	8.7	28127	24
0.09	Ordered	4.80	5.40	5.04	130.7	94%	35.9	3034	9.65	29278	31
0.09	Ordered	4.80	5.40	5.04	130.7	94%	36.2	3379	8.92	30141	30

Negas at Transtech Corporation, Maryland, USA. Quenched (disordered) and slow cooled (ordered) ceramics were prepared using methods and precursors that avoided the effects of extrinsic factors such as sample reduction, density differences, sintering aid additions and Sn loss. A complete description of the preparations can be found in Ref. 11.

For the high-temperature, disordered ceramics, $Q \cdot f$ increases significantly with increasing Sn content in a manner consistent with previous studies,[9,10] full details are provided in Ref. 11. Here we focus on the differences in the dielectric properties of ordered and disordered ceramics with $x = 0.0$ and 0.09 which are presented in Table 1. Relative to their high-temperature counterparts, the ordered $x = 0.0$ samples show a 30% decrease in $Q \cdot f$, a 10% lower value of k', and a 30% increase in τ_f. For the $x = 0.09$ ceramics the effects of slow cooling on k' and τ_f are of the same sense as those observed for the $x = 0.0$ samples, but the changes are less marked, a 7% decrease for k' and a 25% increase for τ_f. Of particular note for the $x = 0.09$ samples is the fact that $Q \cdot f$ does not decrease with ordering but instead increases slightly by an amount, 5.6%, which is just outside the overall limits of error of the Q measurements.

We believe that our HRTEM findings and other observations allow us to interpret the trends in the dielectric data at various levels. Starting with the effects of ordering on the dielectric constant, we find that ordering in the $x = 0.0$ samples results in about a 10% decrease in k', whereas for $x = 0.09$ k' decreases by 7%. The magnitude of this change in the relative permittivity of pure $ZrTiO_4$ is in general agreement with the lower-frequency measurements made by Ikawa *et al.*,[21] but somewhat larger than the microwave determinations reported by Hirano *et al.* on partially ordered samples.[10] The decrease in k' in both samples could be explained by a decrease in the total polarisability and/or an increase in the molar volume. Because the ordering transformation is accompanied by a decrease in the unit cell volume (see Table 1) in this case the lower dielectric constants must arise from a significant decrease in the total polarisability. Hirano has reported that the contribution of the electronic polarisation is unaffected by the degree of cation order,[10] therefore in agreement with his results we conclude that the lower permittivity of the long-range ordered structures is directly attributable to a reduction in the ionic polarisability.

As for dielectric loss, we had previously speculated that it should decrease upon Zr–Ti ordering, partly on the basis of comparison with other microwave dielectric ceramics such as the perovskites $Ba(M_{1/3}Ta_{2/3})O_3$ (M = Zn, Mg) in which the ordering of different-sized cations causes Q to increase substantially.[5,6] In both $Ba(M_{1/3}Ta_{2/3})O_3$ and zirconium titanate ordering has the similar effect of segregating cations of unlike size onto separate structural layers, thereby decreasing local structural strain. Because of this similarity we anticipated that the effect of ordering on Q would be similar in both compounds.[7] The loss properties reported in Table 1 do not support this analogy. Instead they show that the ordered form of $ZrTiO_4$ has a 30% lower $Q \cdot f$ than its disordered counterpart. However, for the Sn substituted ceramics with $x = 0.09$, the effect of ordering on loss is quite different and $Q \cdot f$ is slightly larger for the ordered sample by an amount, 5.6%, that is just outside the measurement error.

Because our processing routes were designed to ensure that we avoided the deleterious effects associated with the chemical reduction of titanium or the loss of Sn, the changes in loss properties summarised above could be due to changes in the grain boundary structure, or they may be tied to intrinsic changes in crystal structure such as cation ordering that occur within the grains themselves. Although they cannot be discounted completely, we believe we have taken sufficient measures to exclude extrinsic properties from principal consideration, and we suggest that the intrinsic structural changes induced by cation ordering are responsible for the changes in loss properties that we observe. Considering first Sn free $ZrTiO_4$ it should be noted that the presence of aperiodic structural discontinuities in the form of the switching sites distinguishes this phase from other microwave ceramics such as $Ba(Zn_{1/3}Ta_{2/3})O_3$ and $Ba(Mg_{1/3}Ta_{2/3})O_3$ that order with perfectly periodic superstructures. Therefore we suspect that in $ZrTiO_4$ it is these defects that adversely affect the dielectric loss. For the Sn substituted ceramic $(Zr_{0.91}Sn_{0.09})TiO_4$, which has a much greater number of switching sites but better loss properties upon ordering, this hypothesis is consistent only if the loss qualities of the switches are reduced or eliminated, possibly as suggested above by the preferential segregation of Sn to or near the switches. In this case, the hypothesised effect of Sn in reducing the structural strain and energetics of the Zr–Ti interfaces, may also be related to an effect on the contribution of these sites to the dielectric loss of the ordered phase.

Although no direct measurements were made to determine the effect of slow cooling on the dielectric properties of samples with higher Sn contents, following the arguments presented above we would expect that at sufficiently high levels of Sn doping the dielectric response would be insensitive to the cooling protocol, given that all other extrinsic factors remain constant. This expectation seems to be supported by the measurements made by Hirano on $Zr_{0.8}Sn_{0.2}TiO_4$,[10] where ceramics subjected to different thermal treatments showed a maximum deviation in their dielectric constant of only 2%, and no change, within the limits of error, in their loss properties at 10 GHz.

5. SUMMARY AND CONCLUSIONS

Most of the prominent microwave ceramic systems exhibit some type of structural disorder. For the zirconium titanates examined in this work the structural inhomogeneities arise from the formation of ordered 'modular solid solutions'. The incommensurate ordered phases in the $ZrTiO_4$-$Zr_5Ti_7O_{24}$ system result from the uniform mixing of two commensurate structural units, one with a ZrZrTiTi repeat, the other with a ZrTiTi sequence. Within each unit Zr occupies an 8 coordinate site and Ti is in octahedral coordination. Because of limited cation diffusion perpendicular to the close-packed anion layers, individual (100) cation layers are observed to switch abruptly from Zr-rich to Ti-rich occupancy. These linkages are energetically unfavourable because the larger size of Zr and its preference for higher local coordination requires concerted oxygen displacements that are incompatible with the octahedral preference of the Ti ions. The primary effect of the substitution of Zr by Sn was found to be a reduction in the correlation length of the Zr–Ti long-range order and an associated increase in the number of occupancy switching boundaries. We propose that this is due to the preferential segregation and resultant stabilisation of Sn at the Zr–Ti interface. Measurements of the effect of ordering on the microwave dielectric properties lend some support to this hypothesis. Whereas the loss properties of Sn free ordered samples show a significant deterioration compared to their disordered counterparts, samples containing Sn do not suffer any reduction in their loss properties upon ordering and in fact exhibit a small increase in Q.

ACKNOWLEDGEMENTS

The authors wish to thank R. S. Roth, A. McHale and R. Kudesia for access to their data and samples throughout this study, D. Ricketts-Foot for technical support, T. Negas and S. Bell at Transtech Corporation for conducting the dielectric measurements and Y. Zhang and X. Wei for assistance in preparation of ceramic samples. This work was supported by the National Science Foundation under Grant No. DMR 92-00800. The electron microscopy facility is supported by the National Science Foundation, MRL program, under Grant No. DMR 91-20688.

REFERENCES

1. K. Wakino, K. Minai and H. Tamura: *J. Am. Ceram. Soc.*, 1984, **67** (4), 278–281.
2. T. Negas, G. Yeager, S. Bell and R. Amren: *Chemistry of Electronic Ceramic Materials*, P. K. Davies and R. S. Roth eds, 21–38, NIST Special Publication 804, Washington, DC, 1990.

3. P. K. Davies and R. S. Roth: *J. Solid State Chem.*, 1987, **71**, 490–502.
4. P. K. Davies and R. S. Roth: *J. Solid State Chem.*, 1987, **71**, 503–512.
5. K. Matsumoto, T. Hiuga, K. Takada and H. Ichimura: Proc. 6th IEEE International Symp. on Application of Ferroelectrics, 118–121, Institute of Electrical and Electronic Engineers, New York, 1986.
6. S. Kawashima, M. Nishida, I. Ueda and H. Ouchi: *J. Am. Ceram. Soc.*, 1983, **66** (6), 421–423.
7. R. Christoffersen and P. K. Davies: *J. Am. Ceram. Soc.*, 1992, **75** (3), 563–569.
8. A. E. McHale and R. S. Roth: *J. Am. Ceram. Soc.*, 1983, **66** (2), C18–20.
9. G. Wolfram and H. E. Gobel: *Mater. Res. Bull.*, 1981, **16** (11), 1455–1463.
10. S. Hirano, T. Hayashi and A. Hattori: *J. Am. Ceram. Soc.*, 1991, **74** (6), 1320–1324.
11. R. Christoffersen, P. K. Davies, X. Wei and T. Negas: *J. Am. Ceram. Soc.* (in press).
12. Y. Zhang and P. K. Davies: *J. Am. Ceram. Soc.*, 1994, **77** (3), 743–748.
13. R. E. Newnham: *J. Am. Chem. Soc.*, 1967, **50**, 216.
14. P. Bordet, A. E. McHale, A. Santoro and R. S. Roth: *J. Solid State Chem.*, 1986, **64**, 30–46.
15. A. E. McHale and R. S. Roth: *J. Am. Ceram. Soc.*, 1986, **69**, 827–32.
16. R. Kudesia, R. L. Snyder, R. A. Condrate Sr and A. E. McHale: *J. Phys. Chem.*, 1993, **54**, 671–684.
17. K. Fujiwara: *J. Phys. Soc. Jpn.*, 1957, **12** (1), 7–13.
18. S. Amelinckx: in *Modulated Structures*, J. B. Cohen, M. B. Salamon and B. J. Wuensch eds, Proc. American Institute of Physics Conference, 102–113, Kailua Kona, Hawaii, American Institute of Physics, New York, 1979.
19. J. M. Cowley: *Diffraction Physics*, 384–387, Elsevier, Barking, Essex, UK, 1984.
20. R. D. Shannon: *Acta Crystallogr.*, 1976, Sect. A, **32** (9), 751–767.
21. H. Ikawa, H. Narita and O. Fukunaga: in *Materials and Processes for Microelectronic Systems*, Ceramic Transactions, Vol. 15, 143–152, K. M. Nair, R. Pohanka and R. C. Buchanan eds, American Ceramic Society, Westerville, OH, 1990.

Simulation of the Dielectric Function of $Pb(Mg_{1/3}Nb_{2/3})O_3$ from the Superparaelectric Model

A. BELL and A. GLAZOUNOV
Laboratoire de Céramique, EPFL, CH-1015 Lausanne, Switzerland

ABSTRACT
A recently proposed method for calculating the dielectric permittivity according to the superparaelectric model of relaxors is outlined. The model assumes that a superparaelectric comprises small coherently polarising regions, which obey the Landau–Ginzburg–Devonshire theory of ferroelectrics. For sufficiently small regions the direction of polarisation can fluctuate with thermal energies and it is assumed that the permittivity due to the reorientation of such regions under small applied fields follows the Debye equations. Employing statistical mechanics to determine the static permittivity and Devonshire theory to estimate the 'high frequency' permittivity, calculations are carried out for a fictitious relaxor based on $Pb(Zr_{0.7}Ti_{0.3})O_3$. Qualitative comparison of the results with experimental data suggests that there is a distribution of sizes of the polar regions, with an increase in the size and distribution width with decreasing temperature.
From high-temperature permittivity and field-induced polarisation data, the relevant free energy coefficients and polar region size of $Pb(Mg_{1/3}Nb_{2/3})O_3$ are estimated, allowing an approximate simulation of the temperature and frequency dependence of permittivity in the superparaelectric region. For a temperature-independent polar region size of 3.2 nm, the calculated permittivity of PMN over the temperature range 200–500 K is similar to that seen experimentally.

1. INTRODUCTION

Relaxor dielectrics[1] are of significant technological and scientific interest. The class comprises a large number of complex perovskites of the type $A(B'_xB''_{1-x})O_3$, some perovskite solid solutions and a number of tungsten bronze structure oxides. They all exhibit a broad peak in the real part of the dielectric permittivity as a function of temperature, with a characteristic frequency dependence of the real and imaginary parts as shown in Fig. 1. The large permittivities of such materials are of interest for use in multilayer ceramic capacitors, whilst the large field induced strains observed close to the permittivity maximum are employed in electrostrictive actuators. However, in both cases, the range of applications is limited by the strong temperature dependence of the permittivity and the saturation polarisation respectively. Although much scientific effort is presently being directed towards the elucidation of the nature of the low-temperature state of relaxors, and in particular that of $Pb(Mg_{1/3}Nb_{2/3})O_3$ [PMN], any model which can provide an account of the temperature dependence of the permittivity around the peak may be of great technological interest.

Relaxor behaviour appears to be associated with a transition from a macroscopic paraelectric to a ferroelectric phase at a temperature below that of the peak in

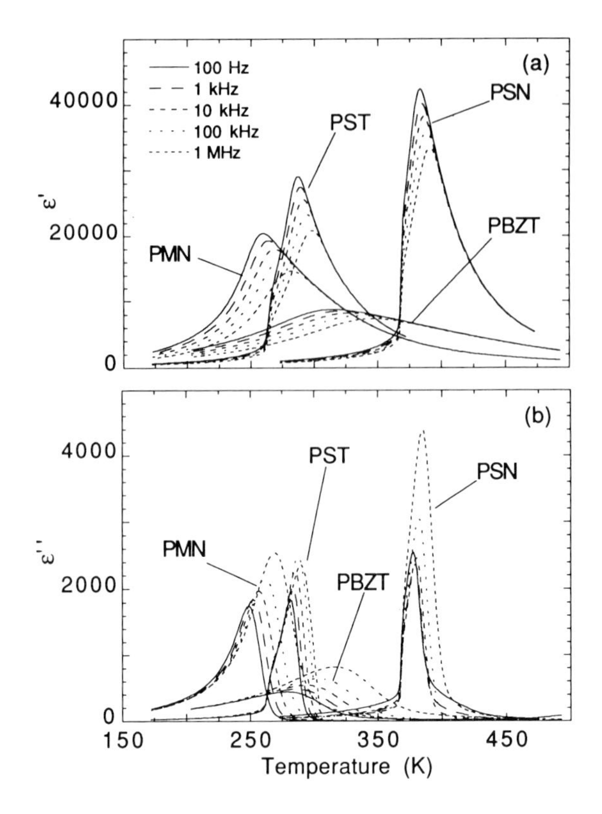

Figure 1 (a) real and (b) imaginary parts of permittivity for $Pb(Mg_{1/3}Nb_{2/3})O_3$ [PMN], $Pb(Sc_{1/2}Ta_{1/2})O_3$ [PST], $Pb(Sc_{1/2}Nb_{1/2})O_3$ [PSN] and $(Pb_{0.7}Ba_{0.3})(Zr_{0.7}Ti_{0.3})O_3$ [PBZT] as function of temperature.

permittivity. In PMN the ferroelectric phase is induced by the application of an electric field, the transition temperature being dependent upon the applied field strength;[2] zero field de-poling occurs on heating at 213 K, well below the peak in the permittivity at 270 K. In contrast, in $Pb(Sc_{1/2}Ta_{1/2})O_3$ [PST] a spontaneous, zero field, transition to a ferroelectric state, the appearance of which has been shown to be critically dependent upon the Pb-vacancy concentration, occurs at 263 K, with the dielectric maximum at approximately 287 K.[3]

Transition electron microscopy of PMN[4,5] has revealed a partitioning of the structure into regions with different degrees of B site cation ordering on the nanometer scale. X-ray and neutron diffraction studies[6] have shown that high-temperature PMN is macroscopically cubic, but with significant atomic shifts around the ideal perovskite structure. On cooling, correlated polar clusters with ⟨111⟩ distortions develop leading to the formation, at low temperature, of polar nanodomains with an estimated diameter of 10 nm at 5 K. This indicates that in the absence of applied fields the structure changes from a high temperature paraelectric through to a microscopically polar, but macroscopically isotropic state.

Initial interpretations of the dielectric anomaly focused on that of a diffuse ferroelectric phase transition,[7] however more recent work suggests that the dielectric anomaly and frequency dispersion are due to the slowing down of superparaelectric moments.[1] An ideal superparaelectric might be considered to be an ensemble of small polar regions, referred to here as clusters, the sizes of which are characterised by the spatial coherence of their spontaneous polarisation. Where the size of the clusters is sufficiently small, the direction of the polarisation of each cluster can fluctuate with thermal energies. On reducing the temperature of the crystal, the frequency of the fluctuations will decrease. Such a situation would be expected to give rise to a classical Debye-like relaxation in the dielectric properties with a temperature-dependent relaxation frequency. The observed transition in PST might correspond to a spontaneous alignment of superparaelectric clusters to form a macroscopic ferroelectric state. On the other hand the moments in PMN appear to slow down into a glassy state.

Although the peak in permittivity is thought to be associated with the dynamics of polar clusters rather than with a diffuse ferroelectric phase transition, there is still evidence for what might be interpreted as an underlying ferroelectric-like transition. At temperatures above 500–600 K, the real part of permittivity of PMN follows Curie–Weiss behaviour,[8] with a Curie temperature of 398 K, almost 100 K above the permittivity maximum. The departure from Curie–Weiss behaviour below 600 K is taken as evidence in support of polar clusters and in this respect is consistent with the observations of Burns[9] who showed that local polarisation fluctuations exist up to 600 K.

The true nature of relaxors and the origin of the peak in permittivity are still a matter of some debate. Although there is much evidence in favour of a superparaelectric/dipolar glass model, a mechanism for the slowing down of the polar clusters is still very much a topic of discussion, with dipolar interactions[2] and random fields being cited as possible progenitors.[10] However, there have been few attempts to model fully the effects of the proposed mechanisms on the dielectric properties.

Recently a method was proposed for simulation of the dielectric properties of a superparaelectric.[11] A basic assumption of the calculations is that the clusters may be treated as independent, classical ferroelectrics, the main justification for this assumption being the high-temperature Curie–Weiss behaviour. The approach employs the Landau–Ginzburg–Devonshire (LGD) formalism for ferroelectrics which although often regarded as inaccurate far from the Curie temperature, has been shown to be successful when applied to the PZT system over a wide temperature range.[12] The Debye model of dielectric relaxation is assumed to apply to the ensemble. The calculations carried out for the fictional superparaelectrics based upon $Pb(Zr_{0.7}Ti_{0.3})O_3$ are summarised below and new work, on the application of the model to provide an approximate simulation of PMN, is described.

2. THEORY

An ideal superparaelectric is defined as an ensemble of non-interacting polar regions, in which each region, or cluster, behaves as an independent, classical ferroelectric. One might consider the concept of a ferroelectric crystal which includes features, such as crystalline disorder, which limit the coherence of the spontaneous polarisation. The density of these features determines the size of a coherently polarising volume, λ, which is characterised by a local, temperature-dependent polarisation, P_s. The reorientation of the polarisation vector of a single region, between the variants allowed by the crystal symmetry, is a thermally activated process with activation energy E_a. In terms of an energy density, G_a, the activation energy is dependent on the region size ($E_a = G_a\lambda^3$). For E_a of the order of kT, the polarisation vector fluctuates in direction with frequency, f_r, given by:

$$f_r = f_0 \exp\left(\frac{-G_a\lambda^3}{\mathrm{kT}}\right) \tag{1}$$

A temperature-dependent dielectric relaxation is observed at frequencies below f_0, with the dielectric function being given by the Debye equations:

$$\varepsilon' = \varepsilon_\infty + \frac{\varepsilon_s - \varepsilon_\infty}{1 + (f/f_r)^2} \quad \text{and} \quad \varepsilon'' = \frac{(\varepsilon_s - \varepsilon_\infty)(f/f_r)}{1 + (f/f_r)^2} \tag{2}$$

in which ε_s is the static permittivity and ε_∞ the permittivity for $f > f_r$. Both the static and high-frequency permittivities might be expected to be temperature dependent. ε_∞ represents changes in the magnitude of the polarisation vector under applied fields at frequencies higher than the relaxation frequency; i.e. that permittivity which would be exhibited by a corresponding 'macroscopic' ferroelectric. The static permittivity, ε_s, represents the change in polarisation due to reorientation of the polarisation vector under static applied fields.

Considering the case of a superparaelectric perovskite with low temperature rhombohedral symmetry and ferroelectric transition of second order it can be shown that the total polarisation as a function of applied field is given by:

$$P_t = P_s \tanh\left(\frac{EP_s\lambda^3}{3\,\mathrm{kT}}\right) + \varepsilon_\infty E \tag{3}$$

and hence the static permittivity follows a modified Curie law:

$$\varepsilon_s = \frac{P_s^2\lambda^3}{3\,\mathrm{kT}} + \varepsilon_\infty . \tag{4}$$

The polarisation, high-frequency permittivity and anisotropy energy density may all be found in terms of the Landau–Ginzburg–Devonshire theory of ferroelectrics.[13] The elastic Gibbs free energy of a ferroelectric, with reference to the unpolarised state, is expressed in terms of a Taylor series expansion in polarisation. For simplicity, the series is terminated here at terms in P^4, which is sufficient for second-order transitions, but not necessarily accurate.

$$\Delta G_1 = \alpha_1(P_1^2+P_2^2+P_3^2)+\alpha_{11}(P_1^4+P_2^4+P_3^4)+\alpha_{12}(P_1^2P_2^2+P_2^2P_3^2+P_3^2P_1^2) \quad (5)$$

where $P_{i=1,2,3}$ are the components of polarisation parallel to the pseudo-cubic axes and α_1 is temperature dependent: $\alpha_1 = \alpha_1'(T-T_0)$. For rhombohedral symmetry the expression reduces to:

$$\Delta G_1 = 3\alpha_1 P_r^2 + 3(\alpha_{11}+\alpha_{12})P_r^4 \quad (6)$$

The stable states are found by solving $d\Delta G_1/dP_r = 0$, to give the well-known result for the temperature dependence of the polarisation:

$$P_s^2 = \frac{-3\alpha_1'(T-T_0)}{2(\alpha_{11}+\alpha_{12})},\ (T<T_0). \quad (7)$$

For macroscopic crystals, the stable state polarisation is taken to be that corresponding to the minimum in ΔG_1. However for an ensemble of polar clusters, if their size is such that $\Delta G_1\lambda^3$ is of the order of kT, a distribution of states about the minimum energy should be considered. This would imply a corresponding distribution in the magnitudes of polarisation and permittivity, ε_∞. The effects of the distribution are most significant when the potential well is shallow, that is, close to T_0. Averaging of the polarisation is carried out by integrating over possible values:

$$\bar{P}_s = \frac{\displaystyle\int_0^\infty P_s \exp\left(\frac{-G_1(P_s)\lambda^3}{\mathrm{kT}}\right)dP_s}{\displaystyle\int_0^\infty \exp\left(\frac{-G_1(P_s)\lambda^3}{\mathrm{kT}}\right)dP_s} \quad (8)$$

where $G_1(P_s)$ is given by Equation (6) with $P_s^2 = 3P_r^2$, $(P_1 = P_2 = P_3)$. For computational brevity averaging is carried out for polarisation in the $\langle 111\rangle$ direction only.

Following normal practice for macroscopic ferroelectrics, the value of ε_∞ may be found from the second differential of the free energy with respect to polarisation $1/\varepsilon_{ij} = \eta_{ij} = \partial^2\Delta G_1/\partial P_i\partial P_j$, i.e. the dielectric stiffness. However, as in the calculation of polarisation, it is necessary to take into account the effect of thermal fluctuations on the magnitude of polarisation, necessitating an averaging of the stiffness over probable values:

$$\bar{\eta} = \frac{\displaystyle\int \eta(P_s)\exp\left(\frac{-G_1(P_s)\lambda^3}{\mathrm{kT}}\right)dP_s}{\displaystyle\int \exp\left(\frac{-G_1(P_s)\lambda^3}{\mathrm{kT}}\right)dP_s} \quad (9)$$

The anisotropy energy density, G_a, can be calculated from the free energy expansion as being the lowest energy path for reorientation of the polarisation between two ⟨111⟩ directions. The activation energy for orientation from one rhombohedral orientation to another is the difference between the energy of the rhombohedral state and that corresponding to the next lowest polarisation state i.e. the orthorhombic orientation. This can be shown to be:

$$G_a = \alpha_1^2 \frac{2\alpha_{11} - \alpha_{12}}{4(\alpha_{11} + \alpha_{12})(2\alpha_{11} + \alpha_{12})}. \tag{10}$$

3. FICTITIOUS SUPERPARAELECTRIC: PZT 70/30

Calculations were made for $Pb(Zr_{0.7}Ti_{0.3})O_3$ [PZT 70/30], which is a perovskite with a second-order transition from a paraelectric to a rhombohedral ferroelectric phase at 605 K, but is not regarded as a relaxor. A full discussion of the results is given in Ref. 9. The effect on the polarisation and permittivity of the finite size

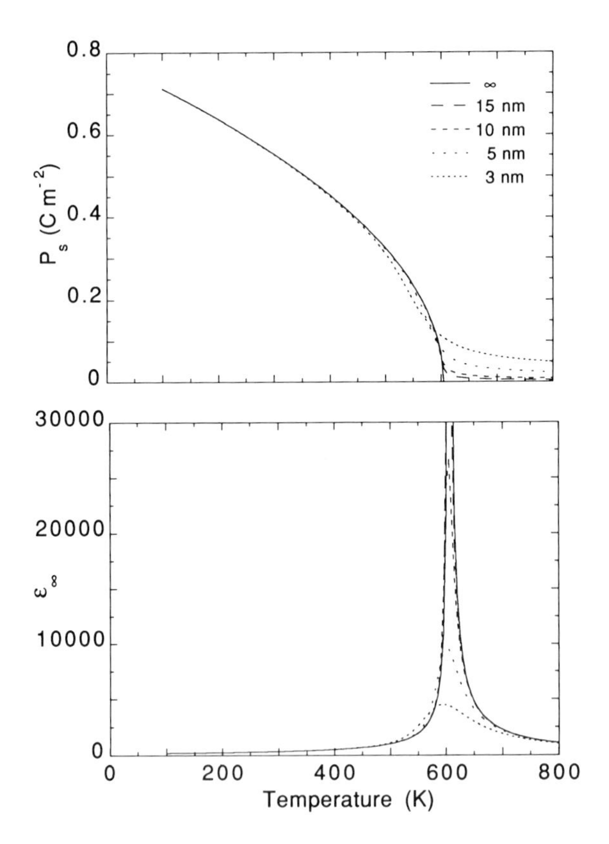

Figure 2 (a) polarisation and (b) high-frequency permittivity of PZT 70/30 as a function of temperature for various cluster sizes.

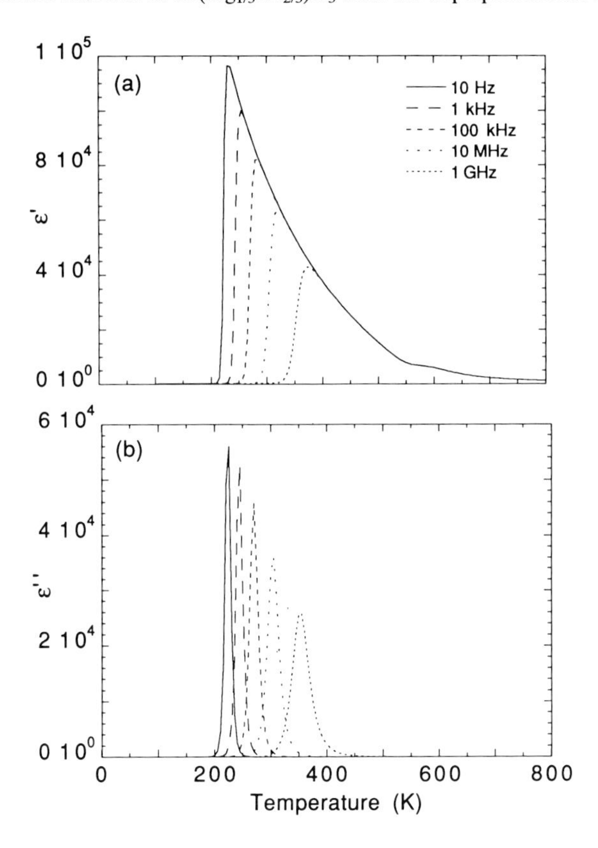

Figure 3 (a) real and (b) imaginary parts of permittivity as function of temperature for PZT 70/30 with cluster size of 3 nm.

averaging over the distribution of energy states is most apparent for small cluster sizes (<10 nm) near to T_0 (Fig. 2). There is a suppression of the permittivity at T_0 and non-zero polarisation fluctuations up to temperatures much greater than T_0. In general, relaxation peaks are exhibited for $\lambda < 15$ nm, with both the temperature and magnitude of the permittivity maximum decreasing with decreasing cluster size. As an example the complex permittivity is shown for $\lambda = 3$ nm in Fig. 3. The peak in the real part is due purely to the combined effects of the Curie law divergence of the static permittivity and the slowing down of the superparaelectric moments, with only a minor contribution from the underlying local ferroelectric transition at 605 K.

Although the calculated real part of permittivity exhibits frequency dispersion on the low-temperature side of the peak, the behaviour differs somewhat from that exhibited by known relaxors. Moreover, the calculated imaginary part of the permittivity is contrary to that found experimentally; that is, the maximum ε'' increases as a function of frequency, with no tendency towards frequency independence at lower temperatures. These disparities have been attributed to the

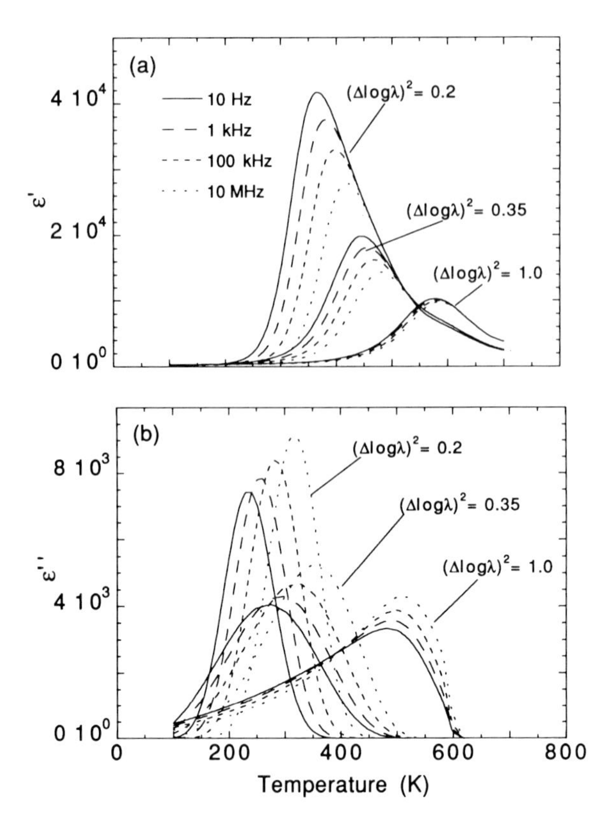

Figure 4 (a) real and (b) imaginary parts of permittivity as function of temperature for PZT 70/30 with cluster size of 3 nm and various widths of size distribution.

use of unique values of T_0 and λ in the calculations, hence calculations were also carried out for distributions of cluster size.

Figure 4 shows the real and imaginary parts of permittivity for log Gaussian distributions of cluster size, with $\bar{\lambda} = 3\,\text{nm}$ and distribution widths given by $(\Delta \log \lambda)^2 = 0.2$, 0.35 and 1. The imaginary part of permittivity shows evidence of the broadening of the relaxation time spectrum with decreasing temperature and increasing $\Delta \log \lambda$. However with the parameters used here, the peak in ε'' for the largest value of $\Delta \log \lambda$ becomes extremely broad in the temperature domain when compared with experimental data. It has therefore been suggested that the broadening of the relaxation frequency spectrum is stronger than that given by a temperature-independent distribution of cluster sizes, and that the cluster size distribution itself is temperature dependent.

A number of simulations were made which suggested that a divergence at a non-zero temperature of both the cluster size and distribution width might describe better the situation in real systems. Figure 5 shows the real and imaginary parts of

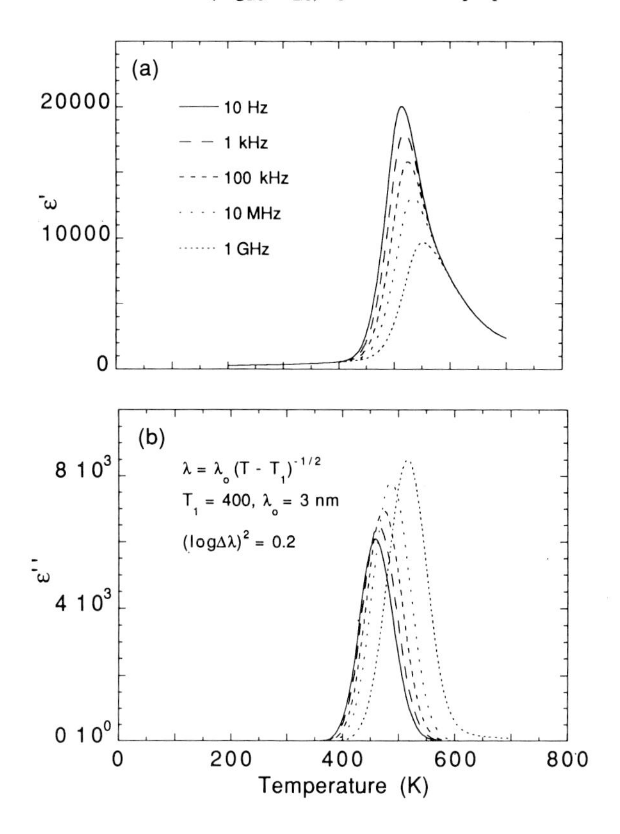

Figure 5 (a) real and (b) imaginary parts of permittivity as function of temperature for PZT 70/30 with temperature-dependent cluster size.

permittivity for log Gaussian distributions of λ, in which $\bar{\lambda}(T) = \lambda_0(T - T_1)^{-1/2}$, where $T_1 = 400$ K and λ_0 is such that $\bar{\lambda}(T_0) = 3$ nm; f_0 is 10^{10} Hz. The form of the temperature dependence, reflecting the behaviour of the ferroelectric correlation length in Landau theory, was chosen as an illustration only and not as a physical hypothesis.

The requirement of cluster growth to achieve sensible results for the above model has been interpreted as evidence for significant interactions between the polar clusters. A limited description of the consequences of one form of coupling was demonstrated, indicating the possibility of a ferroelectric ground state.[11] The model, signifying dipolar cluster coupling. The onset of long-range ferroelectric order at a temperature which is dependent upon the mean field coupling strength was demonstrated, indicating the possibility of a ferroelectric ground state.[11]. The question of whether the ground state is kinetically attainable is not easily answered by the present model as it does not easily lend itself to the calculation of dynamic properties.

4. AN APPROXIMATE PMN SIMULATION

Although a number of conclusions can be drawn from the simulation of a 'fictitious' superparaelectric, more conclusive judgements concerning the accuracy of the model can only be made if it is applied to a real material. An attempt has been made to simulate the dielectric properties of PMN using the type of calculation outlined above. At least 7 parameters are involved in the calculation. Attempts to 'fit' to existing experimental data using all 7 as variable parameters would be neither instructive nor conclusive, therefore as far as is possible the parameters should be estimated from other experimental data. The minimum list of parameters required for the simulation is: α_1', T_0, α_{11}, α_{12}, f_0, $\bar{\lambda}(T)$ and $\Delta\lambda$. However, as only an approximate simulation is sought in order to judge the applicability of the approach, a number of further simplifications may be made. The estimation of the free energy parameters is facilitated by the assumption of zero coupling between the components of the local polarisation, that is, $\alpha_{12} = 0$. In that case the value of α_1' and T_0 may be estimated from the high-temperature Curie–Weiss extrapolation, whilst α_{11} may be estimated from fitting $P_s(T)$ to Landau theory predictions. Data for $P_s(T)$ and $\bar{\lambda}(T)$ may be obtained by fitting field-induced polarisation measurements to Equation (3), letting $\lambda = \bar{\lambda}$. For complete rigour such data should be obtained from measurements made on single crystals. However, for a first attempt, the more readily obtainable data for polycrystalline material may be used, bearing in mind that there are geometrical factors to be included if one wishes to relate polycrystaline measurements to true single-crystal values.

4.1 Experimental

Single-phase PMN powder was produced by a columbite route,[14] employing a basic magnesium carbonate as the magnesium oxide precursor as described by Butcher and Daglish.[15] Pressed pellets were sintered at 1225°C for 2 h using a PMN atmosphere powder and achieving a density of approximately 98% of the theoretical. Electrodes of either gold or platinum (for the high-temperature measurements) were sputter deposited on the polished surfaces of the pellets. Capacitance and dielectric loss were measured using a HP 4284A LCR meter, over five decades of frequency (100 Hz–1 MHz), and over the temperature range 150–900 K. Polarisation field characteristics were measured using a virtual Sawyer–Tower circuit over the temperature range 175–375 K at 10 K intervals.

4.2 Estimation of the Simulation Parameters

The real and imaginary parts of permittivity for the temperature range 150–500 K are as shown in Fig. 1. Figure 6 shows the dielectric stiffness (reciprocal relative permittivity, ε_r) as a function of temperature. At temperatures above 700 K the effects of conductivity become apparent in the lower frequency permittivity, whilst at temperatures below 475 K there are significant departures from linearity of the

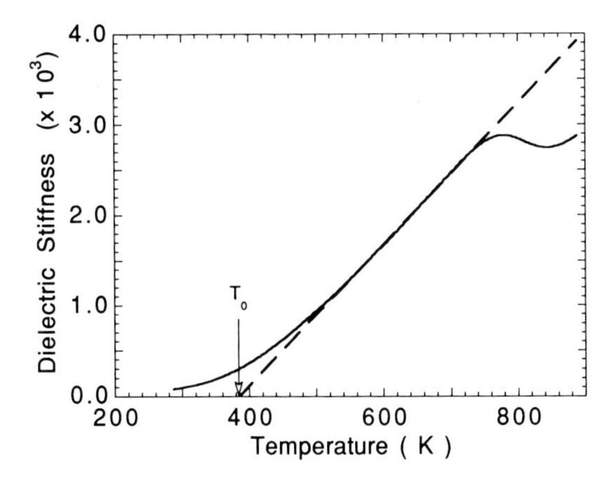

Figure 6 Curie–Weiss plot for PMN.

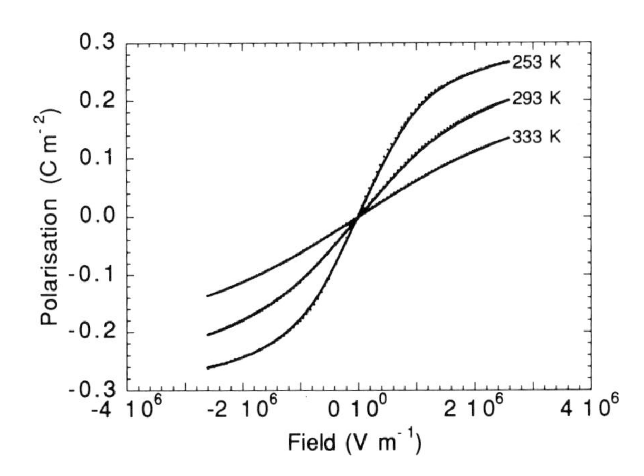

Figure 7 Polarisation of PMN as function of applied electric field at 3 different temperatures.

dielectric stiffness. By linear regression a fit to $1/\varepsilon_r = 7.796 \times 10^{-6}\ (T - 385)$ was obtained, thus $T_0 = 385$ K and $\alpha_1' = 1.468 \times 10^5$.

Examples of the polarisation field characteristics are shown in Fig. 7. Equation 3 can be fitted to this data for temperatures above the onset of significant hysteresis (i.e. $T \geqslant 250$ K). However with three fitting parameters P_s, λ and ε_∞, unambiguous fitting is not possible for unsaturated data. To reduce the number of fitting parameters, the value of ε_∞ for $T < T_0$ can be deduced from: $\varepsilon_\infty = 1/(12\varepsilon_0\alpha_1'$ $(T - T_0))$. However, as shown from the simulations for PZT 70/30 (Fig. 3), this is only true for temperatures away from T_0: $T_0 + 50 < T < T_0 - 50$. Hence fitting to Equation 3, using derived values for ε_∞ was carried out over the temperature range

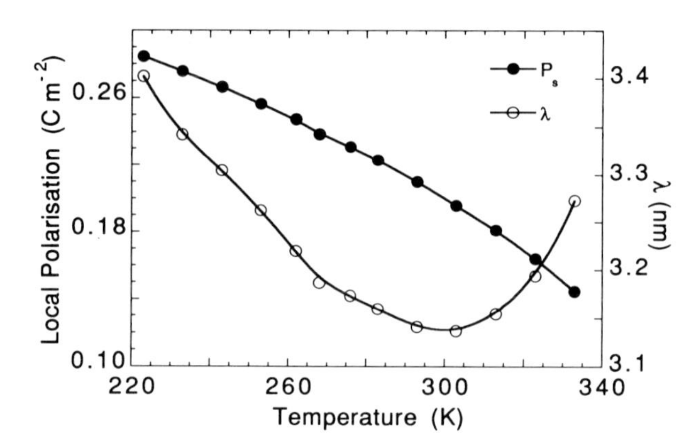

Figure 8 Experimentally derived values of local polarisation and cluster size of PMN as function of temperature.

255–340 K only. The fitted values of P_s and $\bar{\lambda}$ are shown in Fig. 8 as a function of temperature.

The local polarisation, P_s, decreases as a function of increasing temperature. A good fit is obtained to the equation $P_s = 0.0325\ (363 - T)^{0.44}$ reminiscent of that predicted by LGD, but with an exponent of 0.44 rather than 0.50 and a Curie temperature, T_0, = 363 K rather than the 385 K predicted from the high-temperature stiffness extrapolation. Such inconsistencies are not surprising given the departures from true LGD predicted by the finite size constraints of Equation 8 and illustrated in Fig. 2 for PZT 70/30. Although the implications of the observed departure from LGD have yet to be resolved, a value for α_{11} can be estimated from the $P(T)$ data in the range 2 to 5×10^8. Hence an approximate value of α_{11} of 3.5×10^8 has been used for the simulations.

It can be seen that the derived value of mean cluster size, $\bar{\lambda}$, increases with decreasing temperature for $T < 295$ K. The apparent increase for $T > 295$ K is not fully understood. This may be a consequence of aspects of the procedure for fitting to Equation 3, particularly concerning the accuracy of the assumed values of ε_∞. Alternatively it could be a real effect with a continuing divergence of the cluster size towards T_0. This aspect is worthy of further investigation. Nevertheless, for simplicity, initial simulations were carried out with a temperature-independent value for $\bar{\lambda}$ of 3.2 nm which is not inconsistent with interpretations of TEM images.[5]

It is less clear how the value for f_0, the attempt frequency in the Arrhenius equation for the Debye relaxation frequency, should be estimated. The pre-exponent in Vogel–Fulcher analyses[16] suggests a frequency of 10^{12} Hz, however the high-frequency measurements of Elissade *et al.*,[17] which revealed a high frequency relaxation in PMN at below 5×10^8 Hz, are more indicative of a lower attempt frequency. The value used for the present simulations is 5×10^{10} Hz.

Six of the seven parameters necessary to carry out a simulation have thus been

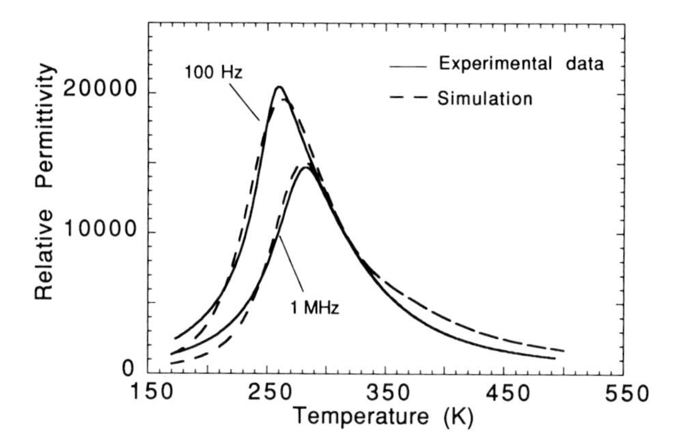

Figure 9 Example of simulated real part of permittivity of PMN at 100 Hz and 1 MHz compared with experimental data.

extimated from experimental evidence. The remaining parameter is that of the distribution of cluster sizes. As in the case of PZT 70/30 a log Gaussian distribution was employed and the distribution width was set somewhat arbitrarily as $(\Delta \log \lambda)^2 = 0.15$.

5. RESULTS AND DISCUSSION

Only a small number of simulations has been run within the limits of the parameters α_{11} and f_0 referred to above. All produced a permittivity (real part) characteristic which resembles that measured for PMN. An example is shown in Fig. 9 for the specific parameter values detailed in the previous section. For all simulations the peak permittivity at 100 Hz was $20\,000 \pm 2000$, while the temperature of the peak permittivity varied from 240 to 300 K. The latter is notably sensitive to the value of α_{11}, but not critically so. In general therefore, for the range of estimates of the simulation parameters, behaviour similar to that of PMN was always observed. With respect to the result in Fig. 9, the most significant deviations from the experimentally observed behaviour occur at high temperatures (>350 K) and low temperatures (<240 K). The former is due to an overestimate of the contribution due to polar fluctuations. As the cluster size in this temperature range has not been estimated from experimental data, it might be assumed that there is longer-range coherence than that assumed for the calculation. At low temperatures, the calculation underestimates the permittivity. The present model does not take into account the nature of the low-temperature state. If indeed PMN is a dipolar glass with a ferroelectric ground state, there may be as yet undefined permittivity contributions from this state. Moreover, it is likely that the high-frequency permittivity, ε_∞, towards which the frequency-dependent permittivity is assumed to converge, does not follow the form predicted by LGD which is implicit in this model.

Despite these discrepancies, the model appears to be particularly successful in describing the permittivity in what is assumed to be the superparaelectric region. Although more work must be undertaken in examining the output of the model with respect to changes in input parameters, and hence assess its integrity, it might be postulated from this first result that cluster coarsening, and hence cluster–cluster interactions, are not essential for the apparent freezing of the moments in PMN. This is contrary to previous understanding, and at present it is difficult to reconcile with the long-range order induced by field cooling.

6. CONCLUSIONS

A mathematical model of the dielectric behaviour of an ideal superparaelectric has been proposed, based upon the LGD theory of ferroelectrics. From experimentally derived estimates of the parameters of the model, simulations of the relative permittivity of PMN have yielded results which bear a strong resemblance to the observed properties. Discrepancies at high and low temperature may be attributed to the relatively low level of sophistication of the model, particularly with respect to the assumption of temperature-independent cluster size. The results suggest that cluster interactions do not play a significant role in the freezing of the superparaelectric moments in PMN. The slowing down is a consequence of the combination of the decrease in thermal energy of the crystal with a Landau-like increase of the local polarisation.

ACKNOWLEDGEMENTS

The authors thank their colleagues at EPFL, particularly F. Chu, M. Daglish and Prof. N. Setter, for stimulating discussions during the course of this work. The financial support of the Fonds National Suisse de la Recherche Scientifique is gratefully acknowledged.

REFERENCES

1. L. E. Cross: *Ferroelectrics*, 1987, **76**, 241.
2. D. Viehland, J. F. Li, S. J. Jang, L. E. Cross and M. Wuttig: *Phys. Rev. B*, 1992, **46**, 8013.
3. F. Chu, A. K. Tagantsev and N. Setter: *J. Appl. Phys.*, 1993, **74**, 5129.
4. J. Chen, H. Chan and M. Harmer: *J. Amer. Cer. Soc.*, 1989, **79**, 593.
5. C. Randall and A. Bhalla: *J. Mat. Sci.*, 1990, **29**, 5.
6. N. De Mathan, E. Husson, G. Calvarin, J. R. Gavarri, A. W. Hewat and A. Morell: *J. Phys.: Condens. Matter*, 1991, **3**, 8159.
7. G. A. Smolenskii: *J. Phys. Soc. Jpn.*, 1970, **28**, 26.
8. D. Viehland, S. J. Jang, L. E. Cross and M. Wuttig: *Phys. Rev. B*, 1992, **46**, 8003.

9. G. Burns and F. Dacol: *Solid State Commun.*, 1983, **48**, 853.
10. V. Westphla, W. Kleeman and M. D. Glinchuk: *Phys. Rev. Lett.*, 1992, **68**, 847.
11. A. J. Bell: *J. Phys.: Condens. Matter*, 1993, **5**, 8773.
12. M. J. Haun, E. Furman, S. J. Jang and L. E. Cross: *Ferroelectrics*, 1989, **99**, 13.
13. A. F. Devonshire: *Adv. Phys.*, 1954, **3**, 85.
14. S. L. Swartz and T. R. Shrout: *Mat. Res. Bull.*, 1982, **17**, 1245.
15. S. J. Butcher and M. Daglish: *3rd Euro-Ceramics*, 1993, **2**, 121.
16. D. Viehland, S. J. Jang, L. E. Cross and M. Wuttig: *J. Appl. Phys.*, 1992, **68**, 2916.
17. C. Elissade, J. Ravez and P. Gaucher: *Mat. Sci. Eng.*, 1992, **B13**, 327.

Chemical Synthesis and Processing of Bismuth Titanate ($Bi_4Ti_3O_{12}$) Electroceramics in Thin-Layer Form by a Sol-Gel Method

LINQING MA, CHRISTOPHER M. BECK,* and DAVID A. PAYNE

Department of Materials Science and Engineering, Materials Research Laboratory, and the Beckman Institute for Advanced Science and Technology, University of Illinois at Urbana-Champaign, Urbana, IL 61801, USA

**School of Materials, University of Leeds, Leeds, LS2 9JT, UK*

ABSTRACT

Data are reported for the solution deposition of bismuth titanate ($Bi_4Ti_3O_{12}$) thin layers on silicon. The method is based upon a 2-methoxyethanol (2-MOE) solution of metal alkoxides. Bismuth 2-methoxyethoxide ($Bi(OCH_2CH_2OCH_3)_3$) was found to be a suitable precursor and was used with titanium 2-methoxyethoxide ($Ti(OCH_2CH_2OCH_3)_4$) for the sol-gel processing of $Bi_4Ti_3O_{12}$ thin layers. Precursor solutions were characterized by spectroscopic methods and the formation of Bi-Ti bimetallic alkoxide precursor(s) in solution was observed. Xerogels crystallised into the perovskite phase at temperatures as low as 550°C, as determined by thermal analyses, FT-IR, and X-ray powder diffraction studies. By using a 0.05 M polymeric sol with a hydrolysis ratio of 6:1, and by heating above the crystallisation temperature (550°C) between successive deposition of layers, with a final heat-treatment at temperatures between 625 to 700°C, thin layers were densified on platinised substrates in the perovskite structure. Integrated thin-layer capacitors (0.3 μm) had a dielectric constant K' of ~200 with a tan δ ~0.03 when measured at 25°C, 1 kHz and 100 mV_{ac}.

1. INTRODUCTION

Novel methods for the synthesis and processing of ceramic materials have evolved in recent years which avoid the use of powders and make use of chemical solution methods.[1] One such method for the powderless processing of ceramics is polymeric sol-gel processing. In this paper we report the sol-gel processing of thin-layer bismuth titanate ($Bi_4Ti_3O_{12}$) electroceramics. The chemical method allows for the densification and crystallisation of ceramics at sufficiently reduced temperatures that they are compatible with integrated semiconductor technology.

$Bi_4Ti_3O_{12}$ is a layered-structure compound which was first reported by Aurivillius.[2] The structure consists of three perovskite '$BiTiO_3$' units between two Bi_2O_2 layers,[3,4] and the unit cell is monoclinic at room temperature. A polar (m)–nonpolar ($4/mmm$) transformation occurs on heating at 675°C (T_c).[5,6] In the polar state, the spontaneous polarisation (P_s) lies in the a–c plane, with the major component along a ($P_{s,a} = 0.5\,C\,m^{-2}$, $P_{s,c} = 0.04\,C\,m^{-2}$).[7,8,9] The room temperature values of dielectric constant, K, are reported to be $K_a = 120$, $K_b = 205$, and $K_c = 140$.[6] The switchable polarisation and field-induced birefringence is of interest for electro-optic devices. Other potential applications include pyroelectric detectors, ferroelectric memories and integrated capacitors.[10]

Recent developments in the application of the sol-gel method for the deposition of thin-layer dielectrics at reduced crystallisation temperatures (400–700°C) have generated interest in integrating these materials with silicon devices. By mixing at the molecular level, reactions proceed at lower temperatures compared with conventional mixed-oxide powder processing. In addition, sol-gel processing also provides possibilities for precise compositional control, and homogeneity and stoichiometry. However, even though $Bi_4Ti_3O_{12}$ has been deposited by physical methods, such as RF sputtering,[11] pulsed laser deposition,[12] and plasma sputtering,[13] only recently has the sol-gel method been used in the preparation of bismuth titanate thin-layer ceramics.[14,15,16,17,18,19]

Previous investigators of sol-gel derived $Bi_4Ti_3O_{12}$ thin-layers used solid bismuth nitrate hydrate ($Bi(NO_3)_3 \cdot 5H_2O$)[14,15] or bismuth acetate ($Bi(OAc)_3$)[16,18,19] as starting materials together with liquid titanium alkoxides ($Ti(OR)_4$). The choice of these compounds also requires the use of a large amount of acetic acid (CH_3COOH,HOAc) in addition to 2-methoxyethanol ($CH_3OCH_2CH_2OH$,2-MOE) as solvents, and heating of the precursor solution is necessary, in order to dissolve the alcohol-insoluble bismuth starting materials. As a consequence of limited solubilities of these bismuth compounds, and the unavoidable esterification reaction between 2-MOE and HOAc at elevated temperatures which produces water, reproducibility in precursor solution preparation was often difficult to achieve.[19] In this work bismuth 2-methoxyethoxide ($Bi(OCH_2CH_2OCH_3)_3$) was chosen as the bismuth starting material because of its high solubility in 2-MOE. Titanium 2-methoxyethoxide ($Ti(OCH_2CH_2OCH_3)_4$) was used as the titanium precursor in the sol-gel processing of bismuth titanate thin layers. No additives were used, nor heat was necessary, to dissolve the precursor compounds in solution. Here we report the preparation and characterisation of precursor solutions, crystallisation behaviour of xerogels, processing conditions for thin-layer deposition, development of ceramic microstructures on heat treatment, and dielectric properties of bismuth titanate thin layers.

2. EXPERIMENTAL PROCEDURES

2.1 Precursor Solution Synthesis and Characterisation

Due to the moisture-sensitive nature of the metal alkoxides used in this work, they were stored in oven-dried glassware and handled in an argon-filled glove box or by using standard Schlenk-line procedures. $Bi(OCH_2CH_2OCH_3)_3$ was prepared from bismuth trichloride ($BiCl_3$, Aldrich, Milwaukee, WI) according to a literature method.[20] $Ti(OCH_2CH_2OCH_3)_4$ was prepared from titanium tetraisopropoxide ($Ti(O\text{-}Pr^i)_4$, Aldrich) by an alcohol exchange reaction.[21]

Pre-hydrolysis precursor solutions with concentrations from 0.05 to 0.20 M were prepared by dissolving $Bi(OCH_2CH_2OCH_3)_3$ and $Ti(OCH_2CH_2OCH_3)_4$ in 2-MOE followed by stirring for several hours at room temperature. An excess of 10% $Bi(OCH_2CH_2OCH_3)_3$ was used to compensate for possible Bi loss during thermal processing at temperatures above 700°C. For spectroscopic and mass spectrometric characterisation of the species formed in the precursor solution, the solvent was removed in vacuo, and the oily residue was used net or dissolved in a suitable non-coordinating solvent (e.g., chloroform). ^1H nuclear magnetic resonance (NMR) spectra were obtained on a GE QE-300 spectrometer at 300 MHz. Infra-red (IR) spectra were acquired with a Perkin-Elmer 1600 FT-IR spectrometer. Chemical ionisation mass spectrometry (CI-MS) was carried out on a VG 70-VSE instrument with methane as the chemical ionisation reagent.

2.2 Precursor Solution Hydrolysis and Gelation

For the preparation of polymeric sols and xerogels, a hydrolysis solution was first prepared by dissolving an appropriate amount of water in 2-MOE, which was then combined in equal volume with a precursor solution, resulting in 0.025 to 0.10 M polymeric sols. R_w ratios (defined as the molar ratio of water to a hypothetical '$Bi_4Ti_3(OCH_2CH_2OCH_3)_{24}$' bimetallic alkoxide) between 0 and 6 were varied for the hydrolysis and condensation of polymeric sols used in the spin-casting of thin layers, in order to study the effects of partial hydrolysis; and a R_w of 24 was used to initiate gelation, which was complete within 10 min, for the thermal processing of standard powders. The polymeric sols were aged for 24 hours after the addition of water and before thin-layer deposition. Wet gels were dried in an oven at 145°C to form xerogels, and the thermal decomposition and crystallisation behaviour was studied by differential thermal analysis (DTA) on a Du Pont 1200°C analyser, thermogravimetric analysis (TGA) on a 951 Du Pont Thermogravimetric analyser, IR, and X-ray diffraction (XRD) on a Rigaku D/Max IIIA diffractometer with Cu K_α radiation.

2.3 Thin-layer Deposition and Testing

Platinised silicon substrates ($Pt/Ti/SiO_2/Si$) were used in the deposition of bismuth titanate thin layers. A published multilayer spin-casting procedure was followed.[22] Some of the thermal processing conditions (including temperature, heating rate and dwell time) for each layer, and for the final crystallisation step, were varied to study their effects on perovskite phase development and dielectric properties.

For dielectric property measurements, gold electrodes were sputter-deposited through a 100 mesh (200 μm opening) transmission electron microscope (TEM) grid on top of the crystallised thin layers. Dielectric properties were measured on an HP 4284A LCR meter. Microstructures were determined on a Hitachi S-800 scanning electron microscope (SEM).

3. RESULTS AND DISCUSSION

3.1 Precursor Solutions

No apparent colour change was observed when bismuth and titanium 2-methoxyethoxides were dissolved together in 2-MOE and the solutions remained clear and colourless after stirring for several hours at room temperature. However, white precipitates formed from all clear solutions when they were heated. After partial hydrolysis, the 0.10 M polymeric sol also gave precipitates when aged longer than 24 h even when the flask was tightly capped. Therefore, care was taken with the extreme heat and moisture sensitivity of $Bi(OCH_2CH_2OCH_3)_3$ during solution preparation.

^1H NMR spectra were recorded at room temperature for $Bi(OCH_2CH_2OCH_3)_3$, $Ti(OCH_2CH_2OCH_3)_4$, and the reaction mixture of the two, and they are shown in Figure 1. Three groups of resonances were observed for the three types of H atoms in the 2-methoxyethoxy ligands. An obvious change in chemical shift of the H_a resonance (see figure for designation) in the $Bi(OCH_2CH_2OCH_3)_3$ and $Ti(OCH_2CH_2OCH_3)_4$ mixture from the positions of the individual components strongly indicates the formation of new compounds in solution. IR spectral changes in the C-O (1200–900 cm^{-1}) and M-O (<650 cm^{-1}) stretching regions provide additional support for the occurrence of a reaction.

A more convincing piece of evidence for the formation of Bi-Ti bimetallic species in solution was offered by mass spectrometry. In the CI-MS spectrum shown in Figure 2, in addition to the $[M\text{-}OCH_2CH_2OCH_3]^+$ peaks at 273 and 359 amu for $Ti(OCH_2CH_2OCH_3)_4$ and $Bi(OCH_2CH_2OCH_3)_3$ respectively, a peak at 707 amu was assigned to the '$[BiTi(OCH_2CH_2OCH_3)_6]^+$' fragment. No molecular ion could be assigned, as the peaks over 800 amu all had very low intensities, probably due to the easiness of multiple fragmentation of these ions.

^1H NMR spectra of partially hydrolysed sols showed significant broadening of the three resonances, suggesting the formation of higher molecular weight species in solution. This confirmed the polymeric nature of the precursor sols used for spin casting.

3.2 Gels and Powders

Xerogels dried at 145°C contained only 1.26 wt% C and 0.88 wt% H, as determined by elemental analysis. This result indicated an efficient removal of organic groups in the sol-gel process, which was expected because only hydrolysis-prone 2-methoxyethoxy groups were present in the precursors. The DTA characteristics for the xerogel are shown in Figure 3. Decomposition of most of the organic groups occurred below 300°C, as indicated by the two most intense exothermic peaks. This also agrees well with the presence of low temperature thermally degradable 2-methoxyethoxy ligands as the only organic species. TGA data, shown in Figure 4, reveal that the majority of the weight loss occurred below 300°C, and because of the low-percentage of C and H in the xerogel, the total

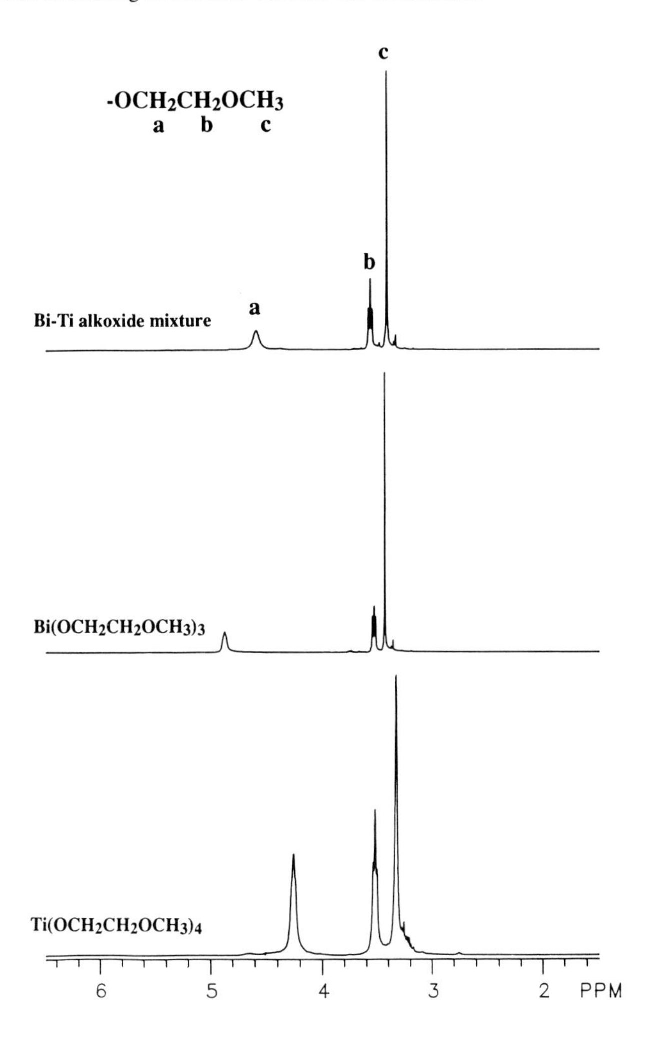

Figure 1 ^{1}H NMR spectra (20°C, $CDCl_3$) of $Ti(OCH_2CH_2OCH_3)_4$, $Bi(OCH_2CH_2OCH_3)_3$ and their reaction mixture.

percentage of weight loss was also low (~10%). A FT-IR study further showed that very little organic groups remained in the powder heat-treated at 400°C for 30 min.

Crystallisation of the xerogel into the perovskite bismuth titanate phase was monitored by XRD as well as by FT-IR. A sequence of room temperature XRD patterns as a function of heat-treatment temperature is shown in Figure 5. The onset temperature for crystallisation was below 550°C, which agrees well with DTA results (525°C), and by 750°C a fully crystalline monoclinic phase was identified at room temperature. In addition, two absorption bands at 816 and 576 cm^{-1} appeared in the FT-IR spectrum of the 550°C heat-treated specimen and grew

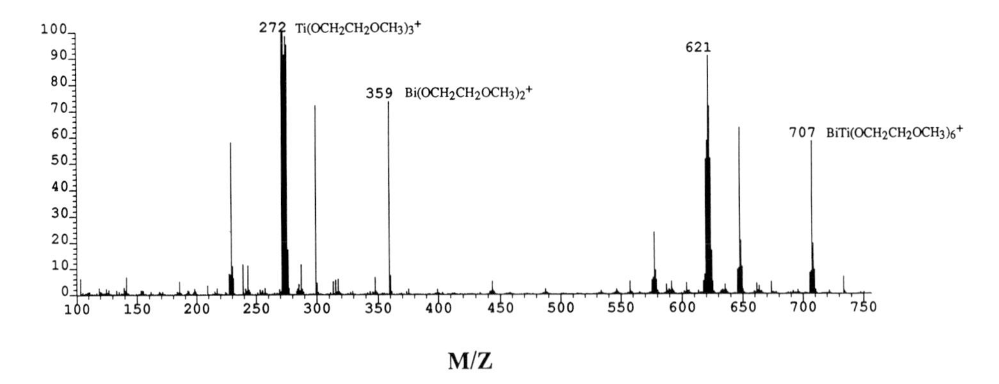

Figure 2 CI-MS spectrum of the $Bi(OCH_2CH_2OCH_3)_3$-$Ti(OCH_2CH_2OCH_3)_4$ reaction product.

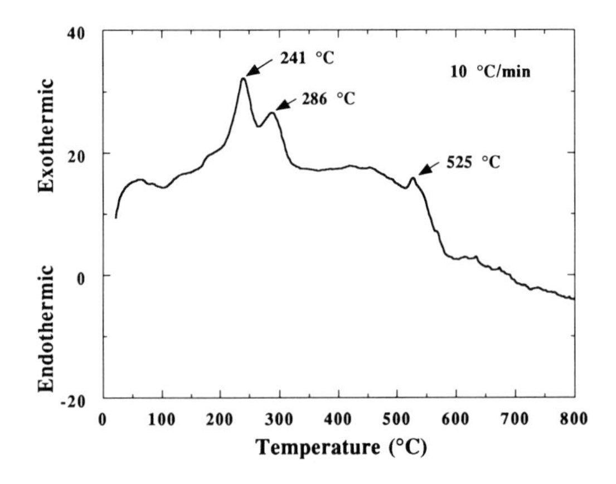

Figure 3 DTA data for the xerogel precursor to $Bi_4Ti_3O_{12}$.

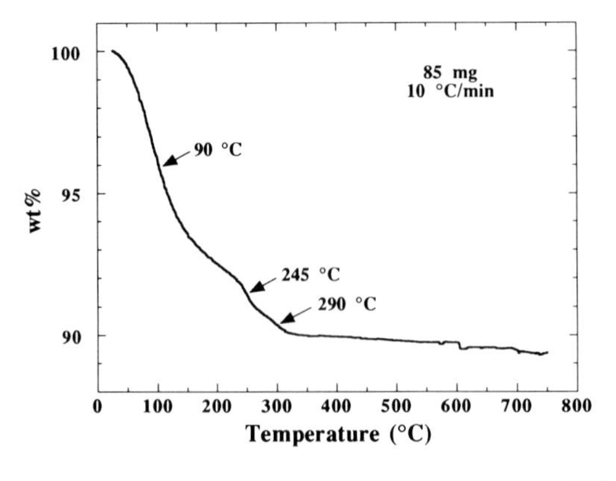

Figure 4 TGA data for the xerogel precursor to $Bi_4Ti_3O_{12}$.

sharper as the temperature was increased to 850°C. These two bands have been assigned to TiO_6 octahedral stretching.[23]

3.3 *Thin-Layer Studies*

A 0.05 M polymeric sol was used for thin-layer deposition since it gave the highest 'solid loading' without precipitation. Typically, an 8-layer deposition led to a 0.25 to 0.30 μm-thick layer. All thin layers with a final heat treatment above 500°C contained only the perovskite $Bi_4Ti_3O_{12}$ phase without the detection of a pyrochlore $Bi_2Ti_2O_7$ phase by XRD, when precursor solutions with 10% excess $Bi(OCH_2CH_2OCH_3)_3$ were used in the starting composition. Some of the thin layers displayed a certain amount of preferred orientation along the (00l) direction, as determined by increased diffraction intensities in the XRD data, when the final heat treatment was carried out at 700°C with a slow (200°C h^{-1}) heating rate. A hold time of 1 h was determined to be adequate to develop thin layers with required crystallinity and dense microstructures.

More detailed studies were centred around the investigation of the effects of the following processing variables: precursor solution hydrolysis ratio, deposition conditions, including the use of a Bi_2O_3 top coating, and thermal processing conditions, on the dielectric properties of $Bi_4Ti_3O_{12}$ thin layers. The effect of each parameter is discussed separately below.

A significant effect of precursor solution hydrolysis ratio, R_w, on the dielectric properties of sol-gel-derived thin layers was determined for the lead barium titanate ((Pb,Ba)TiO_3) system.[24] A higher R_w ratio resulted in a higher dielectric constant. This was attributed to the more complete elimination of organic ligands in the precursors through hydrolysis and evaporation of alcohols during drying, as well as the possible formation of more extended network structures by polycon-

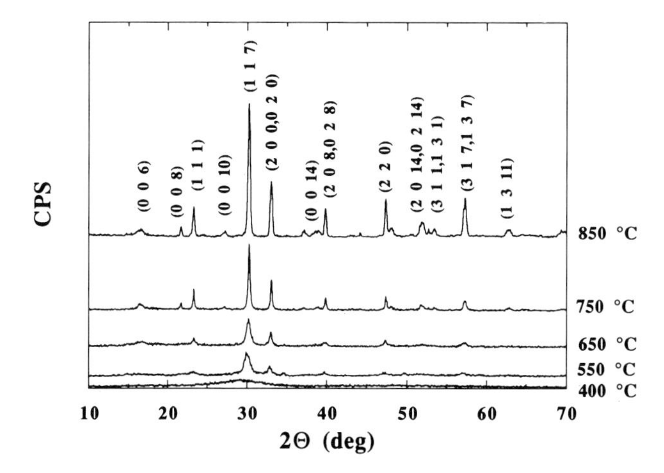

Figure 5 XRD data for a $Bi_4Ti_3O_{12}$ xerogel as a function of heat-treatment temperature.

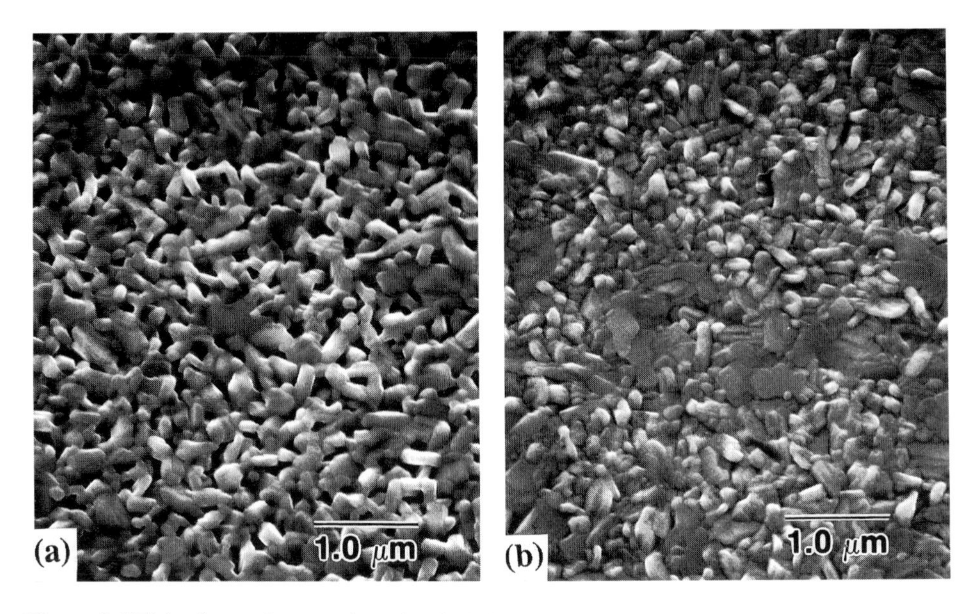

Figure 6 SEM photomicrographs of $Bi_4Ti_3O_{12}$ thin layers. (a) heat-treated at 300°C between each layer deposition; (b) heat-treated at 675°C between each layer deposition.

densation reactions. Both mechanisms would contribute to the densification of thin layers. In the present study, R_w was varied from 0 to 6, and only the thin layers from a precursor solution with $R_w = 6$ were found to be insulating, and were able to support an electric field (20 V/μm). We propose that a reduced amount of organic decomposition was responsible for fewer defects which yielded an improvement in properties in thin-layer electroceramics.

Heat treatments at temperatures greater than 300°C between successive depositions have been reported to have a beneficial effect on the development of microstructure and dielectric properties for $BaTiO_3$ thin layers, possibly because of seeded densification and crystallisation.[25] In this study, SEM photomicrographs (Figure 6) illustrate that layers heated at 300°C between each deposition had more porosity than layers heated to 675°C between depositions, after both were heat treated at a final temperature of 675°C, and the former material did not support an electric field. DTA data showed that elimination of organic species from partially hydrolysed ($R_w = 6$) precursors was achieved at ~340°C and crystallisation occurred between 500 and 550°C. Hence, when heating at 300°C between each layer deposition, the specimen may still retain some organic residues, which on final heat treatment, would be released, possibly forming porosity within the specimen. On the other hand, heating the specimen above 500°C between each deposition should remove all organic residue and also crystallise each layer, which could aid crystallisation in each successive step and contribute to the formation of a more dense microstructure.

Previous work by Tani *et al.* reported that the application of a cover coat of lead

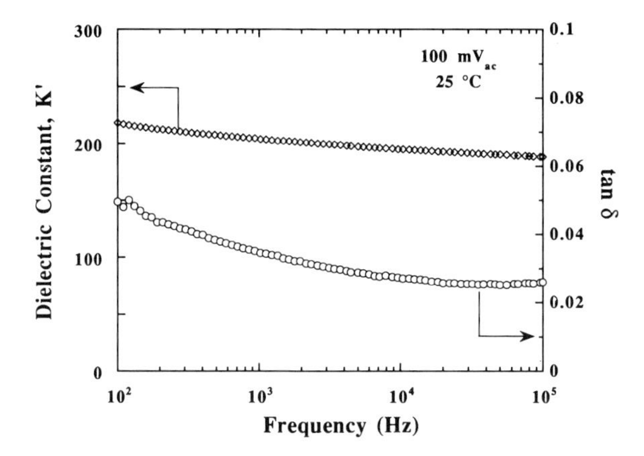

Figure 7 Dielectric data (K' and $\tan\delta$) for a $Bi_4Ti_3O_{12}$ integrated capacitor on silicon as a function of frequency.

oxide (PbO) to lanthanum modified lead zirconate titanate (PLZT) promoted the formation of uniform microstructures of single-phase perovskite material with improved dielectric and ferroelectric properties by reducing the loss of PbO from the surface of the layers on heat treatment.[26] In the present work a surface layer of Bi_2O_3 was found to be beneficial to the microstructure development for certain $Bi_4Ti_3O_{12}$ thin layer specimens, especially for those which were heat treated only at 300°C between each layer deposition. In fact, a semi-insulating specimen could be rendered insulating after the application of a Bi_2O_3 cover coat.

The effect of final heat-treatment temperature on microstructure development is noteworthy. The surface of an as-deposited amorphous layer was smooth and featureless. Crystallisation and grain growth occurred on further heat treatment. At 500°C, the grains had an average size of 0.1 μm with grain growth occurring at temperatures between 500 to 625°C. Layered-structure grains (0.3–0.5 μm) developed in the ceramic microstructure. The surface of a specimen which was heat-treated at 750°C appeared rough and contained some porosity, which was attributed to the loss of Bi_2O_3 (m.p. = 817°C) from the surface. Similar microstructural features were reported for $Bi(OAc)_3$-$Ti(OPr^i)_4$-derived thin layers.[16] Bismuth loss could also contribute to A-site vacancies which could render the specimen semi-insulating, in addition to the formation of pores. Such specimens had resistivities less than $10^{10}\,\Omega\cdot$cm.

Finally, the specimens which were insulating and could support an electric field, had low values of $\tan\delta$ (~0.03) and with good values of K' (~200), when measured at 1 kHz and 25°C, and are comparable with values reported in the literature for polycrystalline $Bi_4Ti_3O_{12}$ thin layers.[15,18] Corresponding data are shown in Figure 7. The effect of temperature on K' and $\tan\delta$ was also studied and there appeared to be little variation in K' or $\tan\delta$ with temperature as would be expected for a material with a high Curie temperature (T_c). The material could find applications in temperature-stable devices. The ferroelectric, piezoelectric, and pyroelectric properties will be reported elsewhere.

4. CONCLUSIONS

Bismuth titanate thin layers were prepared from 2-MOE precursor solutions containing $Bi(OCH_2CH_2CH_3)_3$ and $Ti(OCH_2CH_2CH_3)_4$. Under suitable conditions, i.e., hydrolysis ratio of $R_w = 6$, heat treatment after each layer deposition in the temperature range of ~625–700°C and with a final heat treatment below 750°C, insulators were formed which could be used for integrated dielectrics. A surface coating of bismuth oxide appeared to be beneficial in certain circumstances. The integrated capacitors had a temperature-stable dielectric constant of ~200 with low loss ($\tan\delta$ ~0.03). The capacitors may find applications in DRAMs and MCMs.

5. ACKNOWLEDGEMENTS

The research was supported by the US Department of Energy under Contract DMS-DEFG02-91ER45439. We acknowledge the use of facilities in the Center for Microanalysis of Materials in the Materials Research Laboratory, and in the Molecular Spectroscopy Laboratory and the Microanalytical Laboratory of the School of Chemical Sciences, at the University of Illinois. We thank Mr J. E. Stewart for invaluable aid in the initial experiments. Technical assistance by Mr C. D. E. Lakeman, Mr J. R. Roubik, Mr M. H. Frey and Dr Z. Xu is also acknowledged.

REFERENCES

1. L. L. Hench and J. K. West: *Chemical Processing of Advanced Materials*, 1992, John Wiley and Sons, New York, USA.
2. B. Aurivillius: *Arkiv. Kemi.*, 1950, **1**, 499–512.
3. J. F. Dorrian, R. E. Newhnam, D. K. Smith and M. I. Kay: *Ferroelectrics*, 1971, **3**, 17–27.
4. R. L. Withers, J. G. Thompson and A. D. Rae: *J. Solid-State Chem.*, 1991, **94**, 404–417.
5. E. C. Subbarao: *Phys. Rev.*, 1961, **122**, 804–807.
6. A. Fouskova and L. E. Cross: *J. Appl. Phys.*, 1970, **41**, 2834–2838.
7. S. E. Cummins and L. E. Cross: *J. Appl. Phys.*, 1968, **39**, 2268–2274.
8. S. Ehara, K. Muramatsu, M. Shimazu, J. Tanaka, M. Tsukioka, Y. Mori, T. Hattori and H. Tamura: *Jpn. J. Appl. Phys.*, 1981, **20**, 877–881.
9. Y. Masuda, H. Masumoto, A. Baba, T. Goto, M. Minakata and T. Hirai: *Jpn. J. Appl. Phys., Part 1*, 1992, **31**, 3108–3112.
10. T. S. Kalkur, J. Kulkarni, Y. C. Lu, M. Rowe, W. Han and L. Kammerdiner: *Ferroelectrics*, 1991, **116**, 135–146.
11. W. J. Takei, N. P. Formigoni and M. H. Francombe: *Appl. Phys. Lett.*, 1969, **15**, 256–258.
12. H. Buhay, S. Sinharoy, W. H. Kasner, M. H. Francombe, D. R. Lampe and E. Stepke: *Appl. Phys. Lett.*, 1991, **58**, 1470–1472.

13. Y. Masuda, H. Masumoto, A. Baba, T. Goto, M. Minakata and T. Hirai: *Jpn. J. Appl. Phys., Part 1*, 1991, **30**, 2212–2215.
14. P. C. Joshi, A. Mansingh, M. N. Kamalasanan and S. Chandra: *Appl. Phys. Lett.*, 1991, **59**, 2389–2390.
15. P. C. Joshi and S. B. Krupanidhi: *J. Appl. Phys.*, 1992, **72**, 5827–5833.
16. E. Dayalan, C. H. Peng and S. B. Desu: in *Ferroelectric Films*, Editors A. S. Bhalla and K. M. Nair (Am. Ceram. Soc., Westerville, OH, USA, 1992), pp. 279–291.
17. N. Tohge, Y. Fukuda and T. Minami: *Jpn. J. App. Phys., Part 1*, 1992, **31**, 4016–4017.
18. M. Toyoda, T. Hamaji, K. Tomono and D. A. Payne: *Jpn. J. Appl. Phys., Part 1*, 1993, **32**, 4158–4162.
19. M. Toyoda and D. A. Payne: *Mater. Lett.*, 1993, **18**, 84–88.
20. M.-C. Massini, R. Papiernik and L. G. Hubert-Pfalzgraf: *Polyhedron*, 1991, **10**, 437–445.
21. S. D. Ramamurthi and D. A. Payne: *J. Am. Ceram. Soc.*, 1990, **73**, 2547–2551.
22. L. F. Francis and D. A. Payne: *J. Am. Ceram. Soc.*, 1991, **74**, 3000–3010.
23. O. Yamaguchi, N. Maruyama and K. Hirota: *Br. Ceram. Trans. J.*, 1991, **90**, 111–113.
24. J. R. Roubik, Z. Xu and D. A. Payne: Manuscript in preparation.
25. Z. Xu, M. H. Frey and D. A. Payne: in *Better Ceramics Through Chemistry V*, Editors M. J. Hampden-Smith, W. G. Klemperer and C. J. Brinker (Mater. Res. Soc., Pittsburgh, PA, USA, 1992), pp. 339–344.
26. T. Tani and D. A. Payne: *J. Am. Ceram. Soc.*, 1994, in press.

Sol-Gel Ferroelectric PZT Thin Films for Non-Volatile Memory Applications

J. S. OBHI, A. PATEL and D. A. TOSSELL
GEC–Marconi Materials Technology Ltd, Caswell, Towcester, Northants, NN12 8EQ, UK

ABSTRACT
PZT thin films have been prepared via a sol-gel route using standard butoxide and acetate precursors. The solution compositions were modified by the addition of acetylacetone in 2 methoxyethanol, which has the effect of changing the solution complex and therefore its deposition and drying characteristics. Varying amounts of excess lead were included in some cases. The films were prepared by spin coating onto Pt/Ti or RuO_2 electroded silicon with an intermediate barrier layer of either boron phosphate silica glass (BPSG) or thermal silicon oxide, and the resulting samples subjected to a range of thermal annealing conditions in an oxidising atmosphere. Several methods have been investigated including two regimes using rapid thermal annealing (RTA) at 650°C in O_2 and slow annealing at 450°C in air. Using Pt/Ti electrodes both methods yielded highly (111) oriented and crack-free perovskite films with no evidence of other phases or orientations. With the RuO_2 electrodes, however, only the RTA route yielded the perovskite phase of PZT and was highly (110) oriented. Surface electrode dots have been evaporated or sputtered onto the film to allow electrical measurements. The ferroelectric and fatigue properties of these films have been assessed with respect to non-volatile memory applications. Although RuO_2 electrodes required a more severe thermal regime they gave noticeably better ferroelectric and fatigue results over the Pt electrode system.

1. INTRODUCTION

Interest in the area of ferroelectric materials for non-volatile, high speed random access memories (FRAMs) has increased in recent years.[1–3] Among the plethora of available perovskite ferroelectrics, the lead zirconate titanate (PZT) class is receiving considerable attention.[4] This is due to the fact that it has become possible to prepare high-quality and uniform films by either sol-gel processing or OMCVD,[5] allowing integration with existing silicon technology. A further benefit of a thin film format is that the switching voltages required are decreased to standard logic levels of approximately 3 V–5 V.

For successful implementation in FRAM-type applications, it is desirable to have material properties which include relatively large polarisations and low coercive fields. There exists a morphotropic phase boundary at 53 mol.% Zr:47 mol.% Ti which consists of both rhombohedral and tetragonal phases, exhibiting maximum remnant polarisations and dielectric constants in bulk, and therefore this study has concentrated on this composition. In addition, these PZT films should be capable of enduring a large number of switching reversals without extensively compromising their ferroelectric properties. Finally, the processing required for the production of PZT must not have any deleterious effects on underlying circuits. This places restrictions on the extent to which samples can be subject to high temperature processing, making a relatively low-temperature or a minimal timescale high-temperature route highly desirable. It is the effects on the ferroelectric behaviour (including durability) of a low thermal anneal route at 450°C and a higher, rapid thermal anneal (RTA) at 650°C on PZT films prepared via the sol-gel technique that is the subject of this paper. Results are discussed in the light of currently available evidence, and the viability of a low-temperature route assessed in terms of switching and fatigue properties.

The use of metal electrodes such as platinum in conjunction with PZT thin films has been widespread, but with several associated disadvantages. Generally, PZT films suffer from fatigue on metals, and this has been related to Schottky barrier heights at the electrode PZT interface.[6] Further, the use of platinum complicates processing considerations, especially etching which is required for VLSI applications. This has resulted in a recent shift towards transition metal oxides as a viable alternative, primarily due to their low resistivities and good thermal stability. The oxides are also compatible with conventional dry etching methods, leading to easy assimilation into processing schemes.

2. EXPERIMENTAL PROCEDURES

2.1 PZT Film Preparation

Films for this study were deposited via the sol-gel technique involving the dissolution and complexation of relevant metal organic compounds in an organic solvent. The compounds are reacted and partially polymerised in solution. The basic reaction steps are complexation, hydrolysis and polycondensation. The resulting sol is then applied as a uniform layer onto the substrate of interest by a standard spin coating process. Thermal treatment in an oxidising atmosphere then converts the deposited layer to the oxide phase. Due to the relatively large range of available metal/organic starting materials, it is relatively easy to vary compositions and obtain new and complex oxide materials.

Preparation of the solution has been described in detail elsewhere[7] and this method was used to prepare samples of sol with 53:47 Zr:Ti molar ratios. In addition, 10% excess lead was included. The composition of the solutions was further modified by adding some actylacetone in 2 methoxyethanol. This has the effect of altering the solution complex and therefore its deposition and drying characteristics.

Films were deposited by spin coating the solutions onto electroded 4 inch Si wafers and the deposited layers were dried on a hotplate using a proprietary technique developed at GMMT. This was continued until an overall film thickness of ~0.3 μm was achieved. Finally, samples were either baked for ~30 minutes at 450°C in air, or RTA treated for 30 seconds at 650°C in O_2.

2.2 *Electrodes*

2.2.1 *Metal electrodes*

The 4 inch Si wafer was coated with 50 Å Ti and 1000 Å of Pt via electron beam evaporation and annealed at 700°C for 2 h in air. This metal layer served as a bottom electrode, after a corner of the PZT film had been wet etched to allow electrical contact.

Two different top electrodes were prepared, Cr/Au and Pt, by evaporation onto the PZT film surface via either a shadow mask consisting of laser drilled holes of 500 μm and 200 μm diameter, or one with 300 μm diameter holes. A dependence of ferroelectric hysteresis on the particular metal used as the top electrode was observed[7] and has been linked to the difference in the work functions of the two metals used.

2.2.2 *Metal oxide electrodes*

Silicon wafers having thermal silicon oxide planarisation layers were coated with a bilayer consisting of 20 nm ruthenium and 100 nm of ruthenium oxide, by direct and reactive RF magnetron sputtering of a 99.95% ruthenium target. The sheet resistance of the resulting plane electrode was measured to be approximately 6–7 Ω/square and varied by ~12% over the central 70 mm of a 100 mm wafer. A top electrode consisting of ruthenium oxide was deposited via a shadow mask as described above.

2.3 *Physical Analysis*

All samples were examined by X-ray crystallography (XRD) in order to determine the extent of crystallisation and the phase. The surface was also examined by optical microscopy, and if necessary by scanning electron microscopy (SEM) to obtain information on the surface morphology of the samples — an important factor, as it has bearing on the yield of working electrodes.

2.4 *Electrical Characterisation*

The ferroelectric and fatigue properties of the PZT films were measured using the Radiant Technologies RT66A ferroelectric tester. The hysteretic behaviour was assessed by application of a single triangular waveform of period 500 ms, using the RT66A in virtual ground mode. Peak voltages ranging from 2 to 20 V were applied and the polarisations at peak voltage (P_s), polarisation at zero voltage (P_r), coercive field (i.e., field at zero polarisation, E_c) and effective dielectric constant, determined during the hysteresis acquisition.

Pulse polarisation, measurements with the RT66A systems was carried out by the application of a series of 5 triangular pulses of 2 ms width to the sample and a 1 second delay between each pulse. The pulse train consisted of one negative followed by two positive pulses and finally two negative pulses. The values of the switched and non-switched polarisations obtained during pulses 2 and 3 respectively were used to characterise the fatigue properties, after each cycling period.

Fatigue analysis using the RT66A tester was limited to applied frequencies of ~14 kHz, if the internally generated pulses were used. However, in order to acquire fatigue information within a reasonable period of time, an external signal generator (Topward 8115) was input via a Radiant Technologies high-voltage interface in series with the RT66A to cycle the samples studied. This has the effect of reducing the output resistance from ~1.6 kΩ to 50 Ω. The waveform applied consisted of a simple square wave (of variable amplitude) with the pulse width being dependent on the frequency applied. Thus, with an applied frequency of 200 kHz, the pulse width was 5 μs. The frequencies were varied from 1 kHz to an upper limit of 0.5 MHz owing to the RC time constant of the samples. The amplitude of the waveforms was varied between ±3 V and ±10 V. In addition, a comparison of hysteresis measurements was made before and after fatigue.

3. RESULTS AND DISCUSSION

3.1 Platinum Electrode System

The integrity of the films obtained via the routes described above determines the extent to which they can be used. An assessment of this useability can be made by measuring the yield of working capacitor structures, i.e. those that have not shorted. A major influence on this is the thermal history of the sample, and therefore various anneal routes and temperatures were attempted.

Recent work has concentrated on a slightly modified route using a final RTA anneal for 30 s at 650°C in O_2, replacing the 30 min anneal at 450°C. This gave a better thermal uniformity across 100 mm compared to that of the hotplate used. The ferroelectric results were comparable to those obtained at 450°C and are discussed below. Although this RTA step has a higher temperature of 650°C, it ensures that all post-PZT deposition processing occurs at a lower temperature and therefore the likelihood of any PZT disruption due to back end processing temperatures is low. Furthermore, the X-ray diffraction profile appeared to be relatively unaffected by the anneal change yielding a highly (111) oriented film. An example is shown in Fig. 1a. This should be compared with that obtained at 450°C in air (Fig. 1b).

The thermal annealing route included a bake at 450°C after the first deposited layer. This had the effect of assisting the formation of perovskite at lower temperatures giving the film the predominantly (111) orientation (Fig. 1) observed as a result of pseudo-epitaxial matching to the underlying (111) oriented platinum. This is supported by TEM studies which indicated the films to be highly (111) oriented with grains (~150 nm in size) exhibiting an approximately columnar structure through the film from the substrate.[7]

Previous work[8] has emphasised the potential advantage of such a structure for

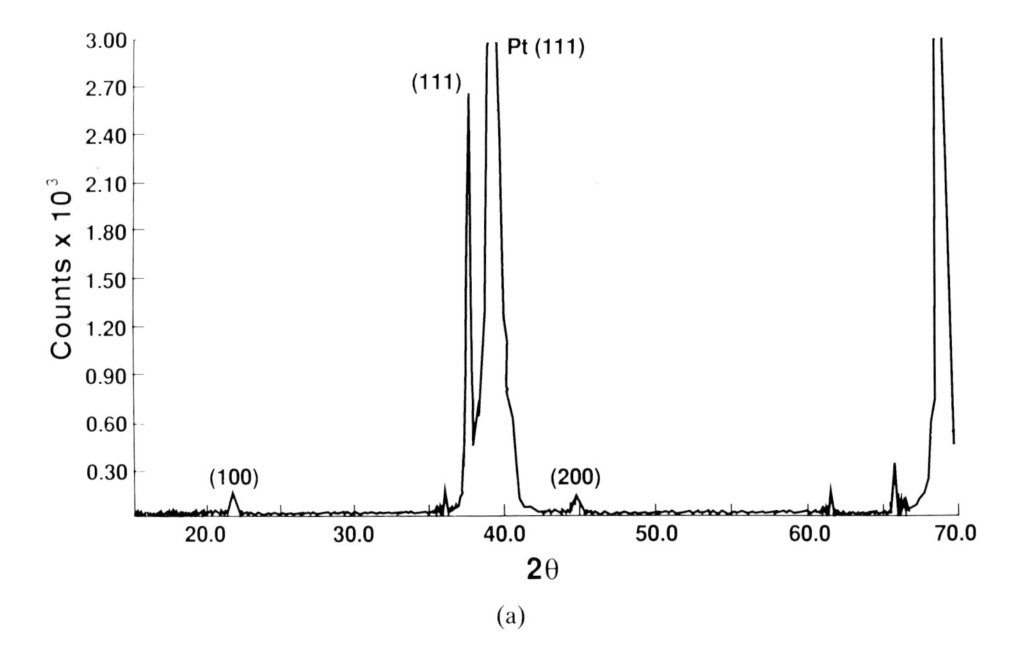

(a)

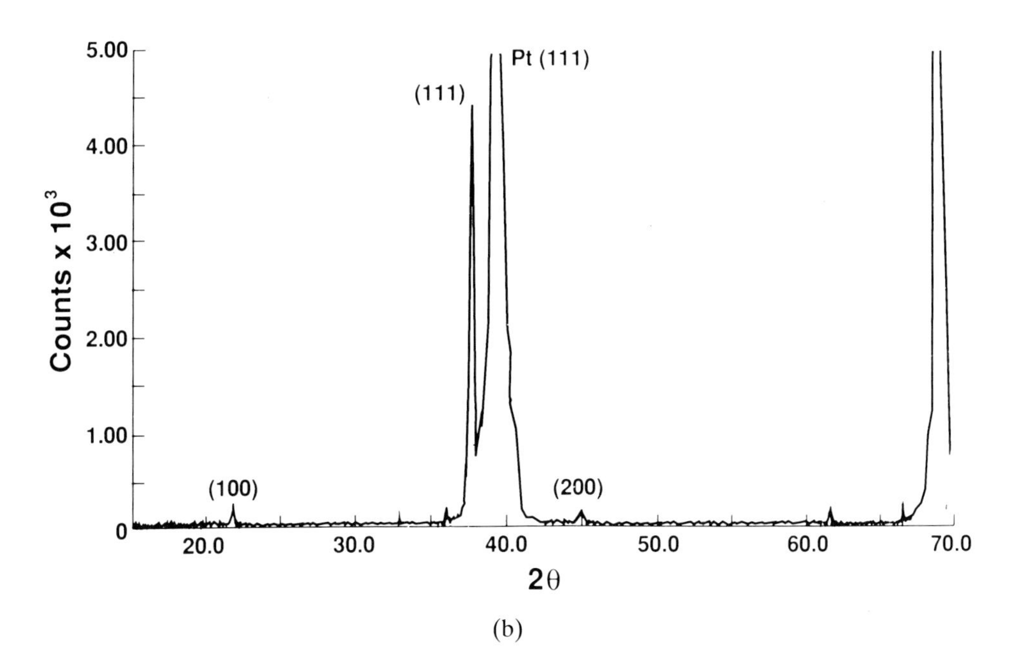

(b)

Figure 1 Comparison of resulting X-ray diffraction profiles obtained after (a) 450°C processing with final 650°C RTA for 30 s (b) 450°C processing only for 30 min.

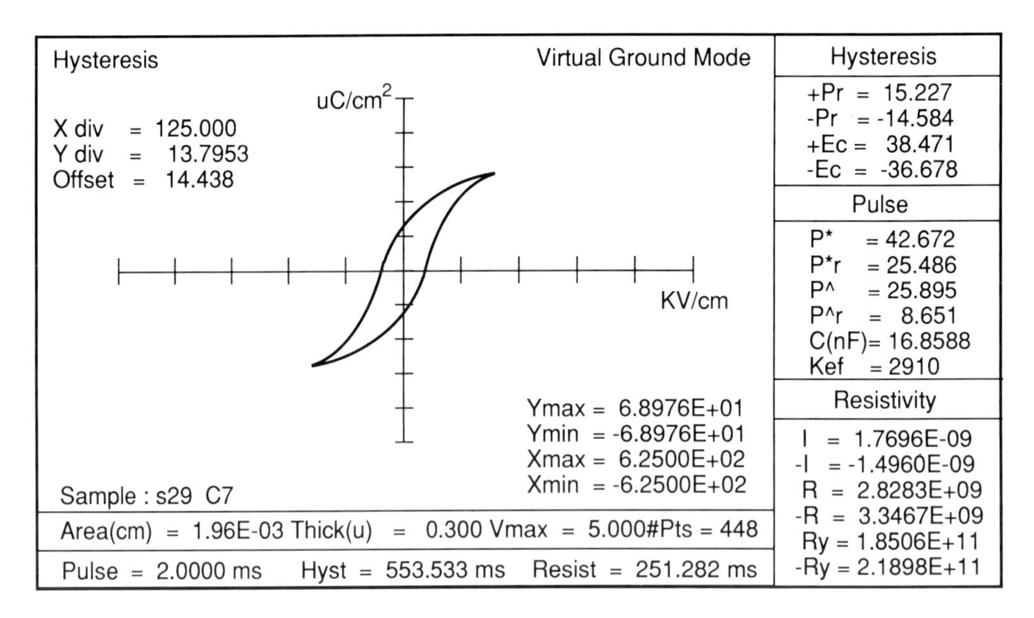

Figure 2 Typical ferroelectric hysteresis behaviour of samples subjected to RTA treatment route.

switching applications where speed is important, because it reduces the contribution of the slower switching 90° domains prevalent in a ceramic-type structure.

The ferroelectric hysteresis behaviour was monitored as described earlier, and good, symmetrical hysteresis loops with switched polarisations of ≈30–48 $\mu\ \mathrm{cm}^{-2}$ and coercive fields of ≈3–5 V $\mu\mathrm{m}^{-1}$, were obtained with 5 V switching, as shown in Fig. 2.

3.2 Ruthenium Oxide Electrode System

Perovskite PZT growth on a planar bottom electrode consisting of RuO_2 deposited as described above appeared to be more difficult. Use of the low-temperature route employed for the Pt electrode case merely resulted in the formation of the pyrochlore phase. An increase in temperature was required in order to convert this to perovskite. In particular, perovskite formation only occurred when each bake at 450°C in the PZT deposition cycle was replaced by an RTA treatment at 650°C for 30 s in O_2 atmosphere. This resulted in the formation of highly oriented (110) PZT as determined by X-ray diffraction (Fig. 3a). SEM analysis indicates that these samples consisted of a dense rosette-type PZT structure with rosettes 3–5 μm in diameter as shown in Fig. 3b.

The hysteretic behaviour, however, appeared to be better than in the case of Pt electrodes, exhibiting good, symmetrical hysteresis loops with switched polarisations of 60–70 $\mu\mathrm{C\ cm}^{-2}$ and coercive fields of 1.5–3 V $\mu\mathrm{m}^{-1}$ with 3 V switching,

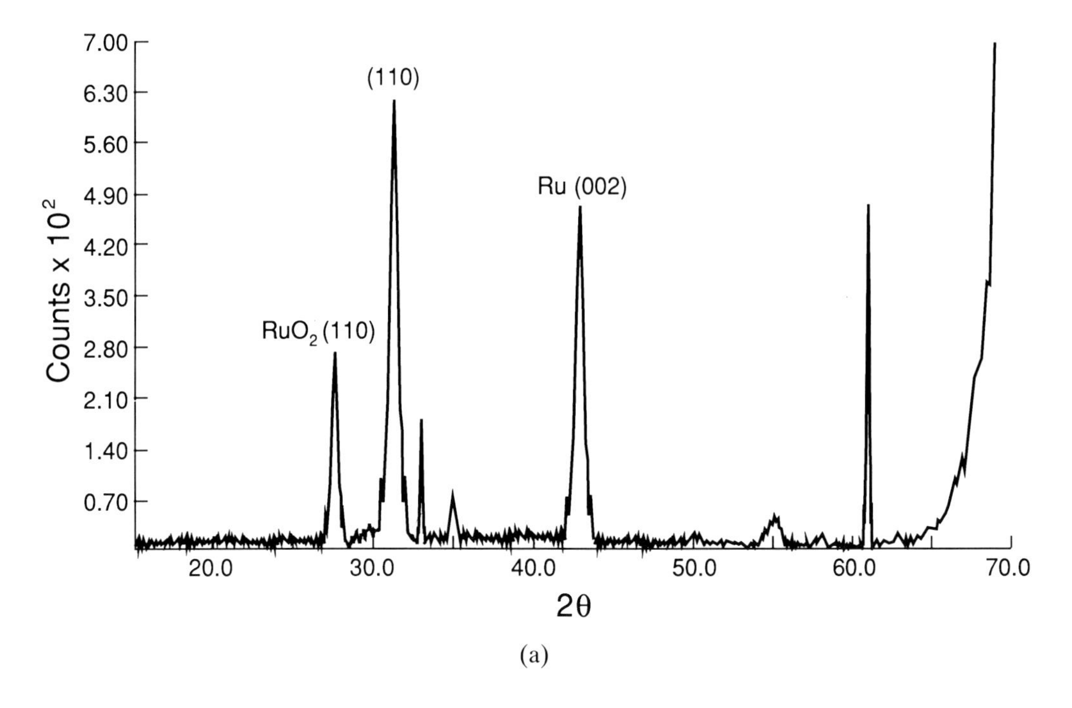

(a)

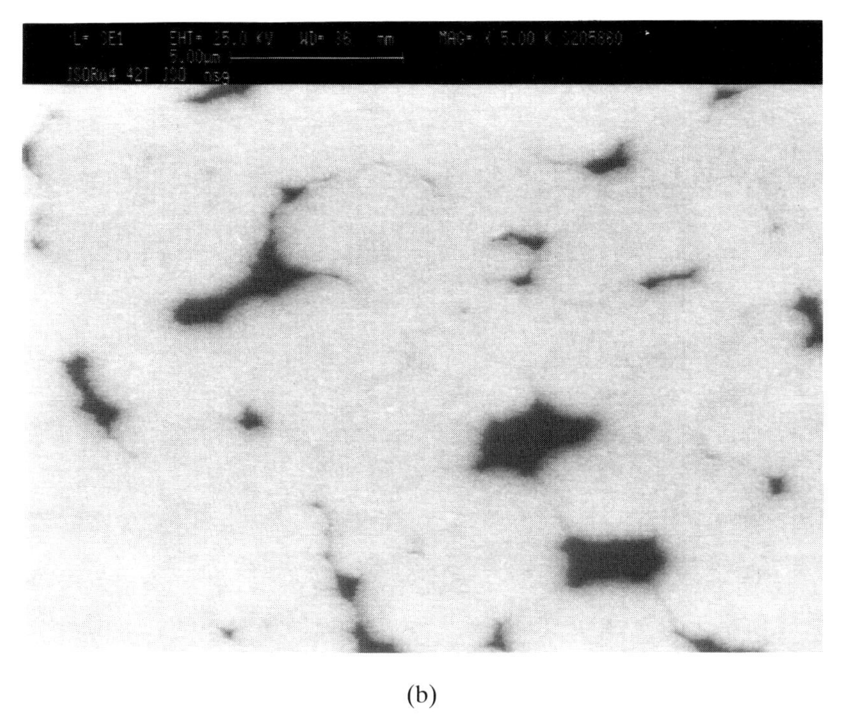

(b)

Figure 3 (a) X-ray diffraction trace of PZT on RuO_2 electrode system and (b) resulting surface structure as indicated by SEM.

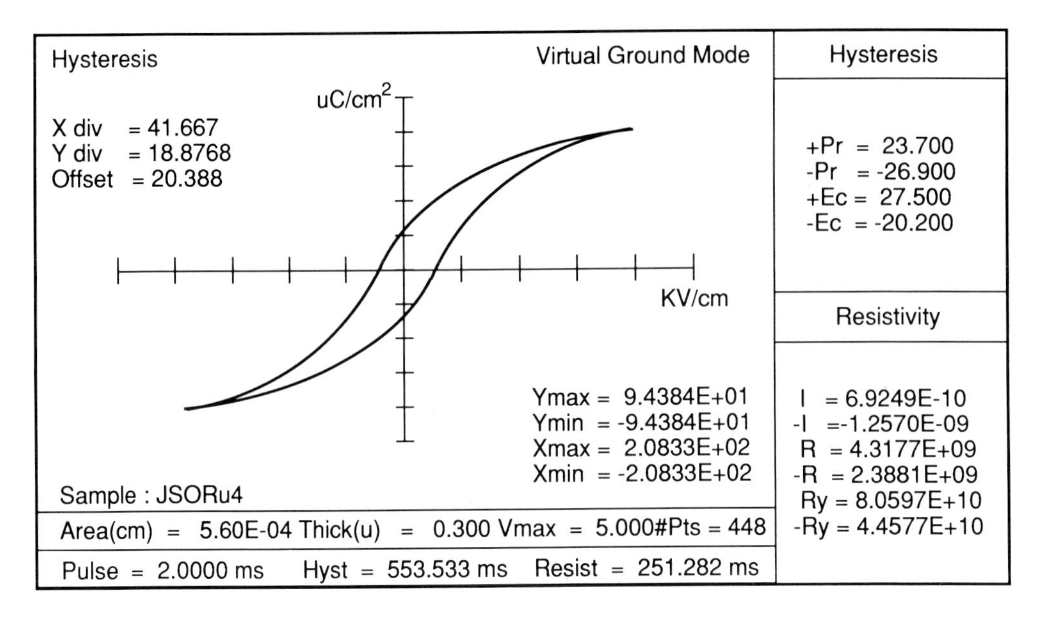

Figure 4 Hysteretic behaviour of PZT in conjunction with RuO_2 electrodes.

despite having a resistivity an order of magnitude lower than in the case of Pt. An example of the hysteresis is shown in Fig. 4.

3.3 Switching Measurements

It was stated above that grain morphology plays an important role in the switching characteristics of the material. The polarisation characteristics of the samples were therefore assessed as a function of applied pulse width after fatiguing for $\sim 10^6$ cycles. The particular set-up used has been described in detail elsewhere;[9] the pulse sequence used in this instance consisted of double bipolar pulses. The delay between the leading edge of two consecutive pulses of the same polarity was 6 μs, the delay between pulses of the opposite polarity was 13 μs and the pulse width was varied between 100 ns and 2 μs. The difference between the switched and non-switched polarisations is plotted as a function of pulse width in Fig. 5 for a capacitor area of 5000 μm^2.

The increase in polarisation with increased pulse width can be attributed to a distribution of switching times, and corresponds to the behaviour expected from a ceramic-like microstructure with at least two grains across the film.[8] There is evidence to suggest that this strong dependence on the pulse width is also a consequence of the particular thermal annealing route used. Larsen *et al.*[10] and Klee *et al.*[8] showed that the use of low-temperature intermediate firing is insufficient to produce the highly crystalline, columnar morphology necessary for fast switching characteristics and, hence, low dependency on pulse width. It was shown in a previous study[7] that the grain structure, although columnar, was

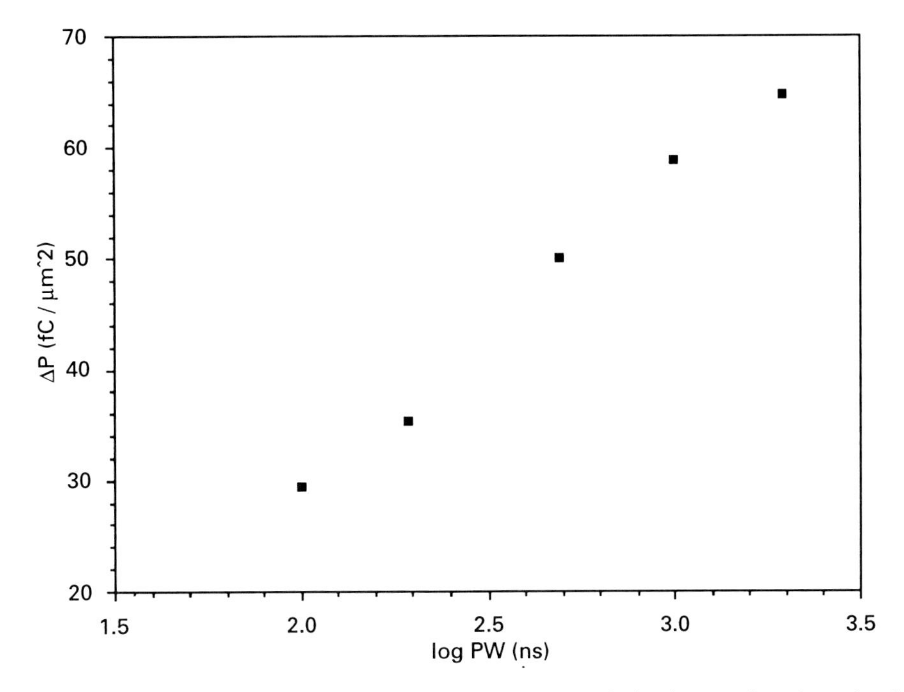

Figure 5 Difference between switched and non-switched polarisations as function of pulse width after $\approx 6 \times 10^6$ cycles using 10 V at 20 kHz.

sufficiently disrupted by the intermediate 450°C bakes that the behaviour under the measurements described above mimics that of ceramic-like structures. It must be emphasised, however, that these measurements were only performed on samples that had undergone the low-temperature route, i.e. those not exposed to temperatures higher than 450°C, and it is therefore unclear how the samples that were RTA treated would behave under such conditions.

3.4 Fatigue

Performance of the films under repetitive stress was assessed as described above, using the RT66A tester. The behaviour of samples under these conditions was found to depend on the particular ferroelectric stack structure under test. This can broadly be categorised as the Pt or the RuO_2 system and may depend on the structural differences described earlier.

3.4.1 Platinum electrodes

With an applied signal of amplitude 5 V, the difference between the switched and non-switched polarisations reduced to 50% of its initial value after $\sim 2 \times 10^6$ cycles.[7] It was further shown that if the applied signal was increased to 10 V the number of cycles achieved increased to 7×10^8 and was attributed to a poling

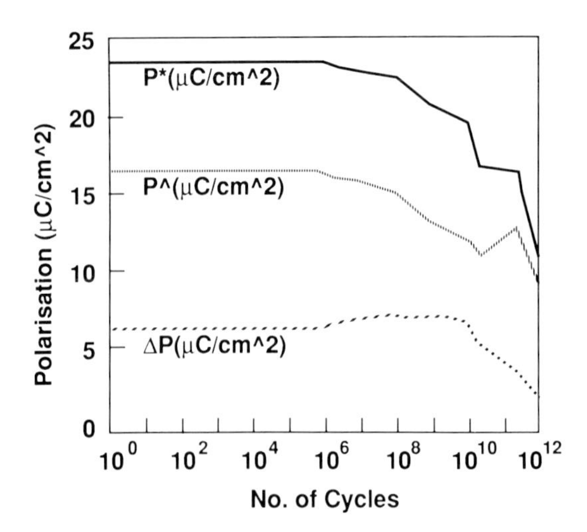

Figure 6 Fatigue behaviour of PZT with platinum electrode system.

mechanism at the higher field, as a greater quantity of the grains would be switched.

The ferroelectric studies carried out on the RTA treated samples showed evidence of improved performance as discussed above. Under fatigue test conditions therefore, it would seem likely that the durability will have improved as well. Indeed, when an applied signal amplitude of 3 V at 500 kHz was used the onset of fatigue was greatly delayed. In fact the sample underwent 9.6×10^{11} cycles, at which time the difference between the switched and non-switched polarisations was $\approx 2\ \mu C\ cm^{-2}$ before finally breaking down, as shown in Fig. 6. The temporary rise in polarisations observed around 10^{11} cycles is attributable to a slight recovery which was inadvertently permitted to occur as a result of a power outage. This event highlights one of the problems associated with fatigue measurements — the extended periods of time required without interruptions in order that the sample can undergo a sufficient number of field reversals. Resistivity was always observed to decrease with the onset of fatigue, in some cases by 3 orders of magnitude.

3.4.2 Ruthenium oxide electrodes

The PZT film which exhibited improved ferroelectric hysteresis behaviour as outlined earlier was tested under the same fatigue conditions as for the Pt case: the sample was cycled with a 3 V signal amplitude applied at 500 kHz. Despite the initial sample resistivity being an order of magnitude lower than the Pt case the fatigue properties were vastly improved. The sample was cycled for 1×10^{12} cycles and the switched and non-switched polarisations monitored. The difference between these two parameters remained at approximately $10\ \mu C\ cm^{-2}$ throughout the test and is shown in Fig. 7. Furthermore, the resistivity was observed to drop by only an order of magnitude after the fatigue test. It is noticeable that the profile

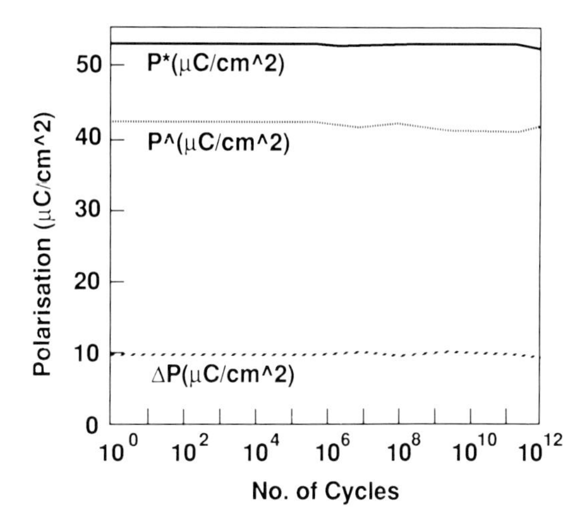

Figure 7 Fatigue behaviour of PZT with RuO_2 electrode system.

is flat until approximately 10^{11} cycles whereupon a slight decay is observed. Extrapolation of this decay implies that the difference between the switched and non-switched polarisations will decrease to $2\,\mu C\,cm^{-2}$ after approximately 10^{30} cycles.

These observations may be understood to be as a consequence of the fatigue mechanism involved as proposed in Ref. 11, where it is argued that the onset of fatigue is due to O_2 deficient sites growing dendritically with time from the electrode–ferroelectric interface under stress, accompanied by an increase in conductivity and actual field. Eventually, the sample will break down, which was seen to occur in the case of platinum. The use of conducting oxide electrodes however, is thought to reduce the Schottky barrier height,[6] owing to the better lattice match between the generally rutile metal oxide and the perovskite PZT, and therefore does not allow this growth of O_2 vacancies or defect density at the interface.[12] Indeed, the apparent importance of the PZT–electrode interface properties as a key element in device performance over the bulk PZT properties is supported by work demonstrating that the fatigue is relatively independent of PZT stoichiometry.[13]

4. CONCLUSIONS

This study indicated that good perovskite thin films of 53/47 PZT could be obtained on 4 inch metallised Si wafers. Highly oriented perovskite films with sharp XRD peaks were obtained at temperatures as low as 450°C. The ferroelectric properties were investigated and good hysteresis loops with switched polarisations and coercive fields of $\approx 35\,\mu C\,cm^{-2}$ and 3–5 V μm^{-1} respectively were obtained when the platinum electrode system was used. It was shown that when the electrode system was changed to ruthenium oxide the switched polarisations increased to 60–70 $\mu C\ cm^{-2}$ and the coercive fields decreased to 1.5–3 $V\ \mu m^{-1}$.

Fatigue behaviour tested using 3 V switching amplitude for the applied square wave with the RuO_2 system showed greatly improved performance over that of Pt, with no real degradation being observed even after 10^{12} cycles. This supports theories suggesting that the fatigue is dominated by the electrode–PZT interface rather than the bulk effects as previously thought. Although the low-temperature route used in conjunction with the platinum electrodes gave (111) crystalline PZT films, exhibiting good hysteretic behaviour with good remnant polarisations and coercive fields, it is apparent that their use for applications where fatigue and continuous switching are important is dubious and the use of RuO_2 electrodes would give a clear advantage, especially as the processing implications of using RuO_2 are now being realised. These include good diffusion barrier properties, low resistivity, high thermal stability and chemical compatibility.[6] However, the low-temperature route may be advantageous in regimes of integrated sensors where the good ferroelectric properties coupled with the low thermal loading can be employed successfully.

5. ACKNOWLEDGEMENTS

The financial support of GEC Plessey Semiconductors, Lincoln, and the European Commission under the ESPRIT 6137 FELMAS programme, is gratefully acknowledged. Thanks also go to all the members of the consortium for many useful discussions.

6. REFERENCES

1. E. R. Myers and A. Kingon: in Proc. of MRS Spring Meeting, April 1990.
2. Proc. 3rd Int. Symp. on Integrated Ferroelectrics, April 1991.
3. J. F. Scott and C. A. Paz de Araujo: *Science*, 1989, **246**, 1400–1405.
4. *Integrated Ferroelectrics*, 1992, **1** (1).
5. G. J. M. Dormans, *et al.*: MRS symposium Proc., **243**; private communication.
6. D. P. Vijay and S. B. Desu: *J. Electrochem. Soc.*, 1993, **140** (9), 2640.
7. J. S. Obhi and A. Patel: presented at ISIF5, Colorado Springs, April 1993, to appear at Integrated Ferroelectrics.
8. M. Klee, U. Mackens and A. de Veirman: in Proc. 2nd Int. Symp. on Domain Structure of Ferroelectrics and Related Materials, Nantes, July, 1992.
9. P. K. Larsen, *et al.*: *Appl. Phys. Lett.*, 1991, **59**, 611.
10. P. K. Larsen, *et al.*: Proc. 8th Int. Symp. on Applied Ferroelectrics, Greenville, SC, August–September, 1992.
11. H. M. Duiker, *et al.*: *J. Appl. Phys.*, 1990, **68** (11), 5783.
12. S. D. Bernstein, *et al.*: *J. Mater. Res.*, 1993, **8** (1), 12.
13. S. D. Bernstein, *et al.*: in *Ferroelectric Thin Films II*, A. I. Kingon, E. R. Myers and B. A. Tuttle eds, MRS Symp. Proc., **243**, 373, Pittsburgh, PA, 1992.

Effect of Thermal Processing Conditions on the Structure and Properties of Sol-Gel Derived PZT Thin Layers

C. D. E. LAKEMAN, D. J. GUISTOLISE, T. TANI and D. A. PAYNE

Department of Materials Science and Engineering, Materials Research Laboratory and Beckman Institute, University of Illinois at Urbana-Champaign, Urbana, IL 61801, USA

ABSTRACT

A number of competing processes occur during the heat treatment of sol-gel derived electroceramic materials. They include: pyrolysis of residual organic species, densification of the gel network, crystallisation of the ceramic microstructure, and grain growth. In the $Pb(Zr, Ti)O_3$ system, lead loss can also occur at high temperatures by evaporative processes which leads to the formation of a PbO-deficient surface layer, non-stoichiometry and additional phase formation. Ceramic/electrode interfacial reactions are also possible at higher temperatures. In our work, we report the effect of thermal processing conditions on the densification and crystallisation of sol-gel derived $Pb(Zr_{0.53}Ti_{0.47})O_3$ (PZT 53/47) thin layers as determined by X-ray diffraction. The results are correlated with dielectric properties. It was determined that heating rate and firing temperature had a significant effect on properties. Fast heating rates lead to an increase in the onset temperature for shrinkage, a decrease in overall shrinkage during heating and an increase in the crystallisation temperature. Based upon observed structural features arising from extended times at the firing temperature, a novel processing route was followed to avoid the formation of a PbO-deficient surface layer. Details of the processing route are reported together with improved properties for sol-gel derived PZT.

INTRODUCTION

Sol-gel processing of ferroelectric PZT thin layers has been extensively studied for potential applications in non-volatile RAM elements and integrated sensor devices.[1] Research has focused, in varying degrees, on the effects of solution chemistry,[2] substrate materials,[3] and thermal processing conditions[4] on microstructure development, phase assemblage, and resultant properties. However, property values reported in the literature have varied considerably[5] due to a lack of control over the many process variables involved.

It has recently been shown that control of solution chemistry in the sol-gel processing of PZT 53/47 thin layers deposited on platinised silicon substrates, results in reproducibly high dielectric constant (K') values, and improved ferroelectric polarisation (P) reversal characteristics with applied electric field (E) strength.[2] This was attributed to the increased solution homogeneity resulting in more compositionally homogeneous thin layers. Since the crystallisation of perovskite and pyrochlore phases from PZT gels is compositionally dependent, more homogeneous thin layers lead to a more uniform crystallisation into the perovskite phase. However, as is the case with much of the published data, property values are still lower than those of bulk materials.

The effects of controlling the firing atmosphere, in particular the lead oxide partial pressure, p_{PbO}, is well known in the processing of bulk lead-based perovskite ceramics.[6] In the preparation of thin layers, to compensate for the volatility of PbO, it has been shown that the addition of up to 15 mol.% excess PbO in solution resulted in thin layers with the highest perovskite yield, and with the highest dielectric constant.[7] In addition, recent work from this laboratory has shown that the loss of PbO may be suppressed by the deposition of a coating of the Pb precursor on top of the gel layer prior to firing, to compensate for the loss of volatile components.[8]

Another method that has been proposed to avoid the formation of undesirable second phases is rapid thermal processing (RTP). By kinetically limiting the nucleation of second phases, improvements in measured properties of PZT layers have been reported.[9] However, few details have been reported on the relationships between the heating rate and the densification and crystallisation of sol-gel derived PZT. Sintering of sol-gel derived materials is frequently described in terms of viscous flow, in which the energy gained by the reduction of surface area of a porous body is expended at a rate proportional to the square of the strain rate.[10] However, such models assume either isothermal sintering, or, in a few constant heating rate experiments, a simple dependence of gel viscosity on temperature. Since the viscosity of gels evolves constantly during heating, these models have had limited success in predicting their densification behaviour. In this paper, we describe various processes which occur during firing for sol-gel derived $Pb(Zr_{0.53}Ti_{0.47})O_3$ (PZT 53/47) thin layers on platinised silicon substrates, including, the effects of different heating rates on the shrinkage, and perovskite phase formation, and interfacial reactions.

EXPERIMENTAL

The preparation of precursor solutions was carried out by a method similar to that of Budd *et al.*,[11] and is described in detail elsewhere.[2,8] Amorphous PZT 53/47 thin layers were deposited onto platinised silicon substrates by spin coating prehydrolysed alkoxide precursors, with each layer being dried at 300°C on a hotplate before subsequent depositions to build up the desired thickness. For the heating rate study, the layers were then heated at 50, 500, or 5000°C/min to temperatures between 350 and 700°C using a Research Incorporated Micristar 828D/E RTP system (Fig. 1). The thickness of each coating was measured before and after firing on a Dektak 3030 profilometer, and the crystallisation behaviour was examined by X-ray diffraction (XRD, Rigaku D-Max IIIA). In order to evaluate the effects of high-temperature reactions on the properties of PZT thin layers, two series of specimens were prepared, with different thicknesses, by repeating the deposition procedure. For one series of specimens a top coating of the PbO precursor was deposited before firing. Thermal processing was carried out at 140°C/min to 700°C,

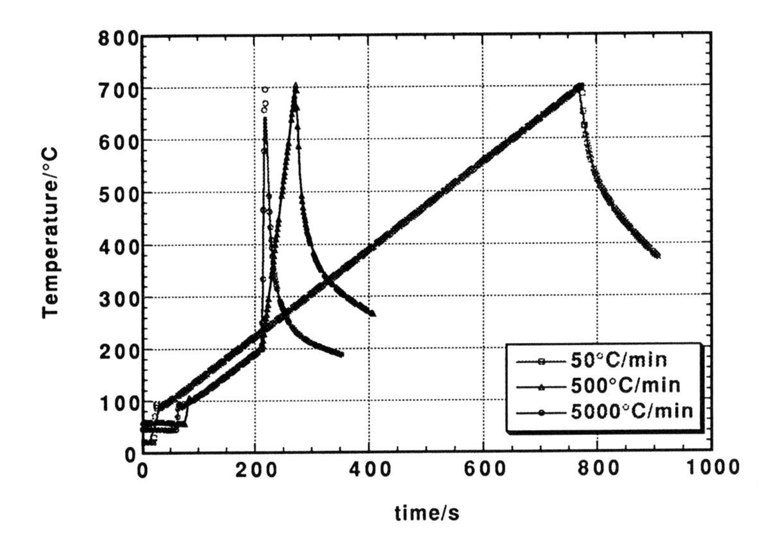

Figure 1 Rapid thermal processing.

and held at temperature for 30 min. After firing, gold counter electrodes were sputter deposited through 100 mesh ($200 \times 200\ \mu m^2$) grids, and the specimens were heated to 300°C for 2 min. under vacuum to remove any adsorbed moisture and surface hydroxyl groups. Dielectric properties were measured using a Hewlett Packard 4284 LCR meter interfaced with a Hewlett Packard Vectra computer for data acquisition.

RESULTS AND DISCUSSION

Figure 2 compares the shrinkage of PZT thin layers as a function of temperature for different heating rates up to ~700°C. Until 450°C, there was little difference in the shrinkage of specimens fired at the three heating rates, however, at higher temperatures, it became apparent that at faster heating rates, the onset of the major portion of the shrinkage was shifted to higher temperatures. In addition, for very fast heating rates (5000°C/min), the overall shrinkage was significantly less than for slower heating rates.

Figure 3 illustrates the variation in dielectric constant values with firing temperature, which reflects the crystallisation behaviour, as determined by XRD. The data indicate that crystallisation into the perovskite phase was shifted to higher temperatures with faster heating rates. This is in agreement with a number of reports for various sol-gel derived materials,[12–14] and with theoretical models, e.g. Matsusita *et al.*[15] Interestingly, as the firing temperature increases after crystallisation, the apparent dielectric constant decreases. This may be a result of PbO volatility, interfacial reactions, or a combination of both.

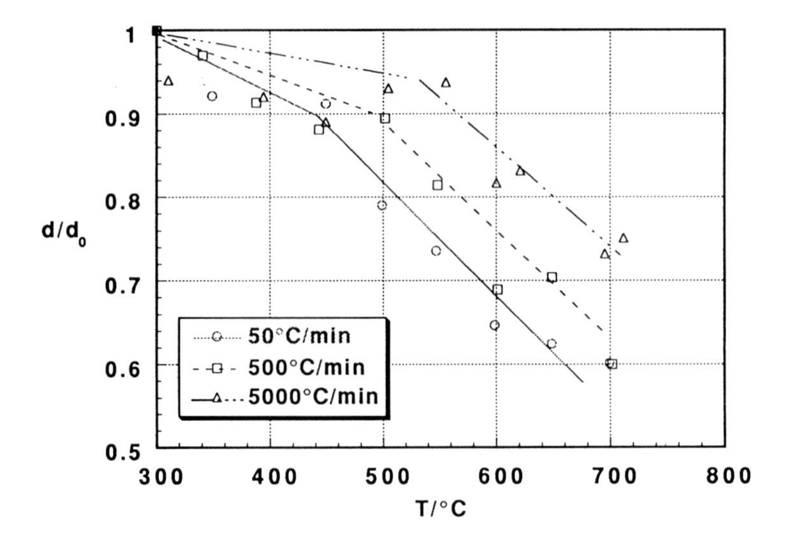

Figure 2 Shrinkage of rapidly thermally processed PZT thin layers.

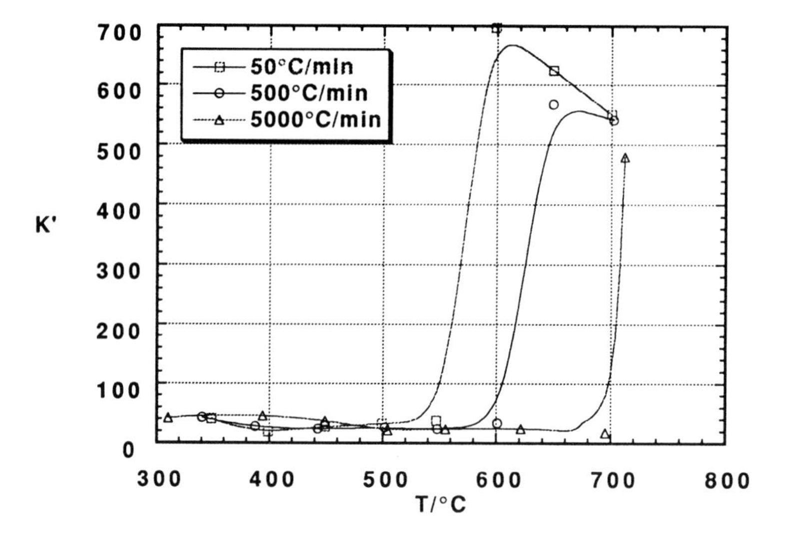

Figure 3 Calculated values of room temperature dielectric constant as a function of RTP conditions.

As discussed earlier, it is difficult to model the sintering behaviour of sol-gel derived materials due to the complexity of the temperature-dependent behaviour of the viscosity, however, from these results, it is apparent that the conditions under which sol-gel PZT thin layers are fired significantly influence their final properties due to induced structural variations. For example, too rapid a heating rate may result in crystallisation of the gel before densification has been completed,

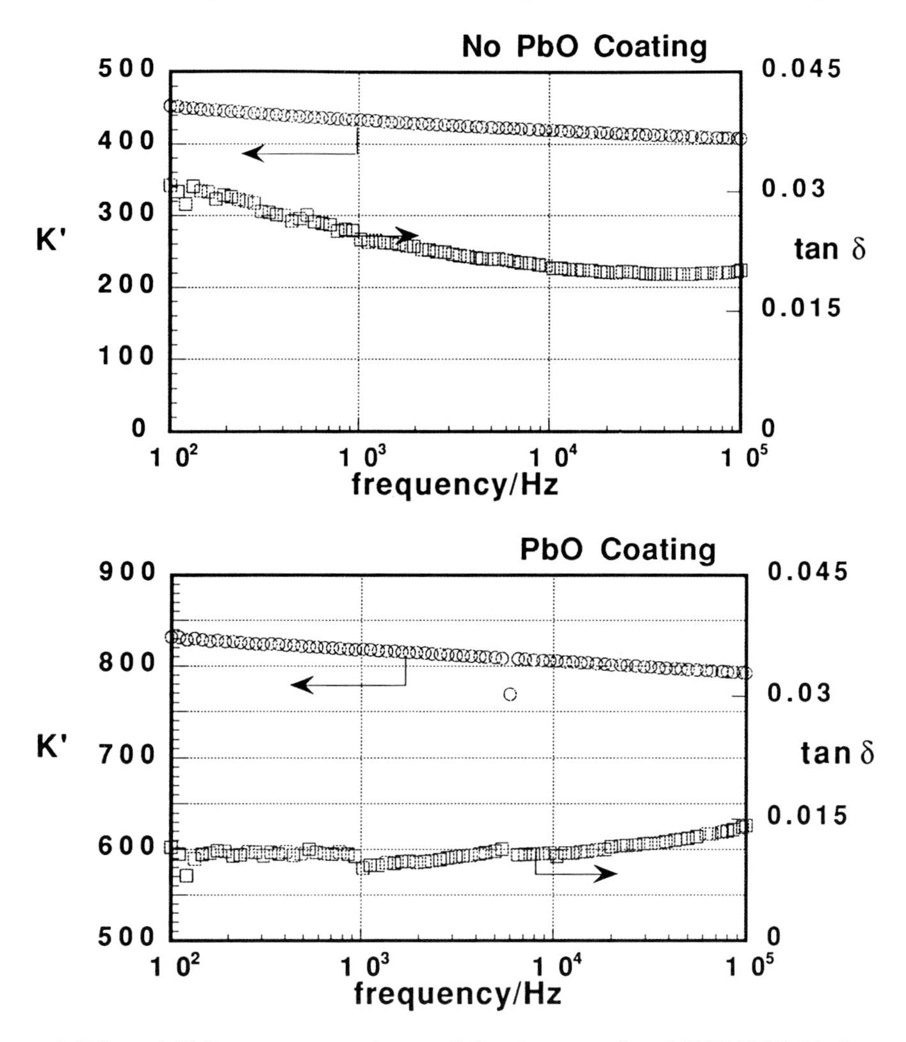

Figure 4 Effect of PbO precursor coating on dielectric properties of PZT 53/47 thin layers.

thereby entrapping porosity within the structure, giving an apparently lower value for the dielectric constant. Thus the use of RTP needs further study.

During extended treatments at high temperatures (30 min. at the firing temperature), loss of PbO can result in a surface layer of a nanocrystalline, low dielectric constant phase. Such surface layers, and reactions at the electrode/ceramic interface, can have a serious effect on the properties of sol-gel derived thin layers. Recent work from this laboratory has demonstrated the effectiveness of depositing a final layer of a lead oxide precursor before firing onto sol-gel derived PZT and PLZT thin layers to compensate for any loss of volatile components. It is clear from Fig. 4 that specimens prepared with the PbO coating display higher values of dielectric constant than those without. The effects of interfacial reactions which result in low K' phases can be modelled as series capacitors. In this case, there are

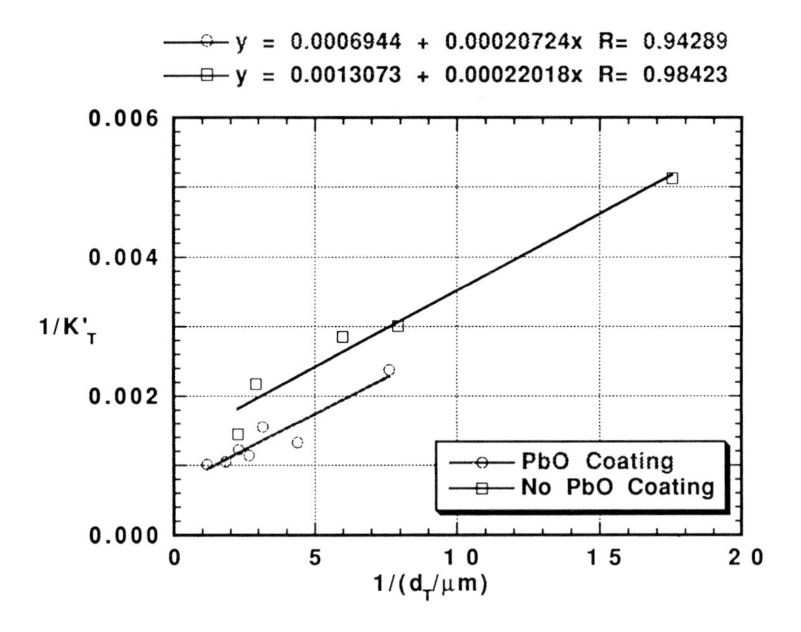

Figure 5 Series capacitor model applied to sol-gel derived PZT 53/47 thin layers.

two interfacial regions, namely, the substrate/ceramic interface, and the ceramic/air interface. However, a 2-layer capacitor model can be used to a first approximation giving:

$$\frac{1}{K'_T} = \frac{1}{K'_1} + \frac{d_2}{d_T}\left(\frac{K'_1 - K'_2}{K'_1 K'_2}\right) \tag{1}$$

where K'_T is the calculated dielectric constant value for the thin layer, K_i, and d_i are the dielectric constant and dielectric thickness, respectively, for each layer. Figure 5 shows this model applied to our data. The y axis intercepts give values for the dielectric constant for the bulk phase of ~1440, and ~760 for coated and uncoated specimens respectively. This implies that there is more than one interfacial process to be considered, since the intercept for uncoated specimens differs from that of bulk materials of the same composition (~1160). In addition, the value calculated for coated specimens is higher than expected, which may imply the existence of residual stresses in the layers.[16] This is currently under investigation.

Finally, examination of the ferroelectric properties has shown that the PbO coating improves the polarisation switching behaviour. The coercive field is reduced (in agreement with the series capacitor model), and the P–E loops tend to saturate more readily, and are less lossy.[8] The properties of coatings of varying thicknesses are summarised in Fig. 6. It appears from these data that the remanent polarisation (P_r) of coated and uncoated layers are similar and tend to increase with increasing thickness of the layers. Also, the coercive fields (E_c) of both coated

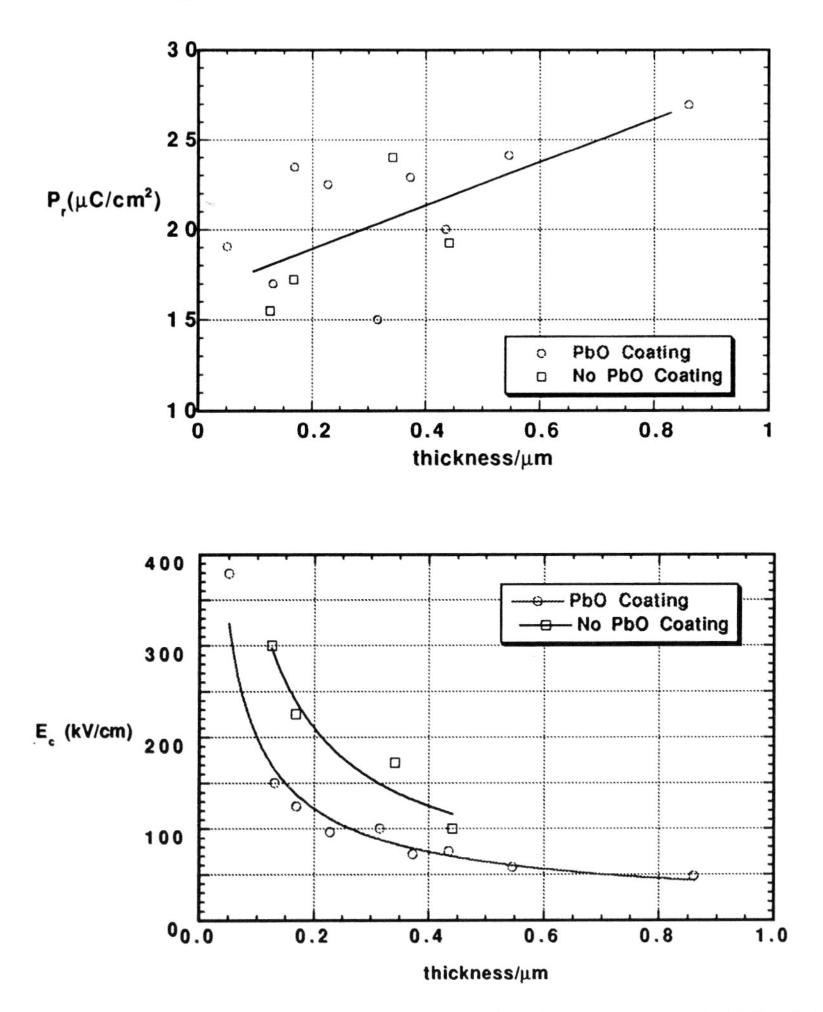

Figure 6 Apparent thickness dependence of ferroelectric properties of PZT 53/47 thin layers.

and uncoated layers decrease with increasing thickness. However, for PbO treated coatings, the values of E_c are lower than those for untreated specimens. The comparatively high values of E_c for these layers and lower values of P_r may also be indicative of residual stresses in the layers.

CONCLUSIONS

In summary, the sol-gel processing of PZT thin layers involves many variables which lead to a wide variation in property values reported in the literature. The heating rate and firing temperature, as well as interfacial reactions such as PbO volatility, and electrode/ceramic interactions may all contribute to the scatter due to variations in microstructure, and crystallinity. A coating of a PbO precursor prior to firing helps to minimise lead loss from the layers during firing and result in improved properties.

ACKNOWLEDGEMENTS

This work was supported by the US Department of Energy under contract number DEFG02-91ER45439 and made use of facilities in the Center for Microanalysis of Materials at the University of Illinois. We thank Mr Lee Kammerdiner of Ramtron Corporation for supplying the substrate materials, and Mr Andrew Crook and Professor Tom DeTemple for use of the RTP system.

REFERENCES

1. For example, see Proc. ISIF-91; Proc. ISAF, 1990, **VII** and 1992, **VIII**; Proc. Am. Ceram. Soc. Symp., 1991; Proc. Mater. Res. Soc., 1990, **200** and 1992, **243**.
2. C. D. E. Lakeman and D. A. Payne: *J. Am. Ceram. Soc.*, 1992, **75**, 3091; R. W. Schwartz, B. C. Bunker, D. B. Dimos, R. A. Asink, B. A. Tuttle, D. R. Tallant and I. A. Weinstock: Proc. ISIF-91, 1991, 535.
3. T. Tani and D. A. Payne: Mater. Res. Soc. Symp. Proc., 1993, to be published; K. F. Etzold, R. A. Roy and K. L. Saenger: Proc. Mat. Res. Soc. Symp., 1991, **243**, 489.
4. B. A. Tuttle, R. W. Schwartz, D. H. Doughty and J. A. Voigt: Proc. Mater. Res. Soc. Symp., 1990, **200**, 159.
5. D. R. Uhlmann, G. Teowee, J. M. Boulton and B. J. J. Zelinski: Proc. Mater. Res. Soc. Symp., 1990, **180**, 645.
6. B. Jaffe, W. R. Cook and H. Jaffe: *Piezoelectric Ceramics*, Academic Press, London, 1971.
7. For example, see L. F. Francis and D. A. Payne: *J. Am. Ceram. Soc.*, 1991, **74**, 3000.
8. T. Tani, C. D. E. Lakeman, J.-F. Li and D. A. Payne: Submitted to Ceramic Transactions, December, 1993.
9. J. Chen, K. R. Udayakumar, K. G. Brooks and L. E. Cross: *J. Appl. Phys.*, 1992, **71**, 4465.
10. C. J. Brinker and G. W. Scherer: *Sol-Gel Science: The Physics and Chemistry of Sol-Gel Processing*, Academic Press, San Diego, CA, 1990.
11. K. D. Budd, S. K. Dey and D. A. Payne: Proc. Br. Ceram., 1985, **36**, 107.
12. R. Pascual, M. Sayer, C. V. R. Vasant Kumar and L. Zou: *J. Appl. Phys.*, 1991, **70**, 2348.
13. B. J. J. Zelinski, B. D. Fabes and D. R. Uhlmann: *J. Non-Cryst. Solids*, 1986, **82**, 307.
14. P. C. Panda and R. Raj: *J. Am. Ceram. Soc.*, 1989, **72**, 1564.
15. K. Matsusita, T. Komatsu and R. Yokota: *J. Mater. Sci.*, 1984, **19**, 291.
16. W. R. Buessem, L. E. Cross and A. K. Goswami: *J. Am. Ceram. Soc.*, 1966, **49**, 33.

Ferroelectric Thin Films for Integrated Device Applications

A. PATEL, D. A. TOSSELL, N. M. SHORROCKS, R. W. WHATMORE and R. WATTON*

GEC–Marconi Materials Technology Ltd, Caswell, Towcester, Northants, NN12 8EQ, UK

*DRA, St Andrews Road, Malvern, Worcs, WR14 3PS, UK.

ABSTRACT
Lead-based thin ferroelectric films have been prepared using both sol-gel and dual ion beam sputtering (DIBS) processes. Material compositions within the $PbTiO_3$ and PLZT system have been deposited by both techniques onto metallised silicon. By using a standard sol-gel prepared solution, modified with acetylacetone and spin coating, 1 μm thick fully perovskite layers, were obtained at lower temperature (450°C) with some preferred orientation. The grain size was in the range 0.2–0.4 μm. A dielectric constant of 400 and a reversible pyroelectric coefficient of $1.2 \times 10^{-4}\ Cm^{-2}\ K^{-1}$ were obtained. In contrast, a range of capping layers (SiO_2, Al_2O_3, BPSG) on silicon have been investigated using the DIBS process. Highly crystalline (100) and (111) films were readily produced at temperatures in excess of 550°C, at a growth rate of 0.3 μm/hour. Control of stoichiometry has also been studied in detail, by sputtering of a composite metal–ceramic target with a high-energy Kr beam and by bombarding the growing film with a low-energy oxygen ion beam. Dielectric constants of 200–300, losses below 0.015 and resistivities above $10^{10}\ \Omega m$ have been achieved. A pyroelectric coefficient of the order of $2.5 \times 10^{-4}\ Cm^{-2}\ K^{-1}$, pre-poled for a La-doped film on BPSG capped Si, was obtained, which did not increase significantly on poling.

INTRODUCTION

The sensing of long wavelength infrared radiation is of growing interest for a wide range of applications, from the detection of flames for fire alarms to the detection through emitted heat of intruders, and thermal imaging applications.

The two wavebands of particular interest are from 3 to 5 μm and 8 to 14 μm. Both correspond to regions of low atmospheric absorption, and the latter is considered to be of particular interest as it corresponds to the peak in black body radiation spectrum for bodies at around 300 K. Compared to normal photo effect sensors, i.e., those based on GaAs or $Hg_xCd_{1-x}Te$, ferroelectric sensors can be operated at ambient temperatures, requiring low power and cheap detector technologies.[1] Pyroelectricity,[2] the release of charge due to a materials change in temperature, occurs in polar materials. As a group within the polar materials, ferroelectric crystals exhibit the largest effects and hence have been the subject of intensive research.[3] Because the pyroelectric signal voltage which is proportional to the temperature variation of the element, increases with a decrease in the element thickness and hence heat capacity, it therefore follows that a thin film up to 10 μm thick would be desirable as a pyroelectric sensor element. A wide array of thin film deposition techniques have been investigated such as MOCVD,[4] sol-gel,[5] RF[6] and ion beam sputtering.[7] In particular, full monolithic integration of such films with silicon or GaAs would yield significant advantages in terms of increased speeds, reduced voltages and improved response. In this paper, we report the use of the sol-gel and the emerging PVD technique of dual ion beam sputtering (DIBS) for the synthesis of undoped $PbTiO_3$ (PT) and La doped ($(Pb_{1-x}La_{2x/3})TiO_3$) (PLT) where $x = 0$–0.2. The $PbTiO_3$ family of materials are attractive as detector materials[8] because they show large pyroelectric coefficients P, small dielectric constant ε_r and small temperature coefficient of P. Poling of pure $PbTiO_3$, however, requires the application of high electric fields. The addition of La[9] has been shown to lower the T_c and the tetragonality, leading to a drop in the coercive field whilst maintaining good levels of polarisation.

EXPERIMENTAL PROCEDURE

The precursors used to prepare the sol-gel deposition solution consisted of lead acetate trihydrate, titanium *n*-butoxide and lanthanum acetate in 2-methoxyethanol as a solvent.

The preparation procedure was based on the technique originally proposed by Budd *et al.*,[10] however, the final stock solution, nominally 0.6 M, was further modified with acetylacetone (acac). The addition of acac decreased the tendency for premature hydrolysis and also improved the film quality and surface wetting characteristics. Details of processing conditions are given in Ref. 11 and therefore will not be discussed here.

The PVD deposition system consisted of a Nordiko 3450 DIBS system (Fig. 1). The system has been described fully previously,[7] except that the argon sputtering beam has been replaced by a heavier krypton beam. The DIBS process has a number of advantages, including low pressure deposition, low film contamination, good film adhesion, high film density and refractive index. For PLT deposition, an adjustable composite target has been developed and is shown in Fig. 2. Pieces of ceramic lead oxide and PLT are fixed to a titanium backing plate. By suitable variation of the configuration, films of the desired stoichiometry were obtained. The substrate temperature was up to 600°C, a growth rate of 0.33 μm/h was typically obtained and this was largely independent of target configuration and substrate temperature.

In both cases, films were deposited on platinum (1000 Å) coated silicon provided with a thermal oxide barrier layer or (borophosphosilicate glass) BPSG. The films were characterised by a number of techniques, including X-ray diffraction (XRD) using Cu_α radiation, electron probe microanalysis (EPMA) and scanning electron microscopy (SEM) and (STEM). Electrical properties were determined using a Wayne Kerr 6425 LCR meter. Hysteresis loop measurements were performed on a Radiant Technologies RT66A thin film tester. The pyroelectric coefficients were

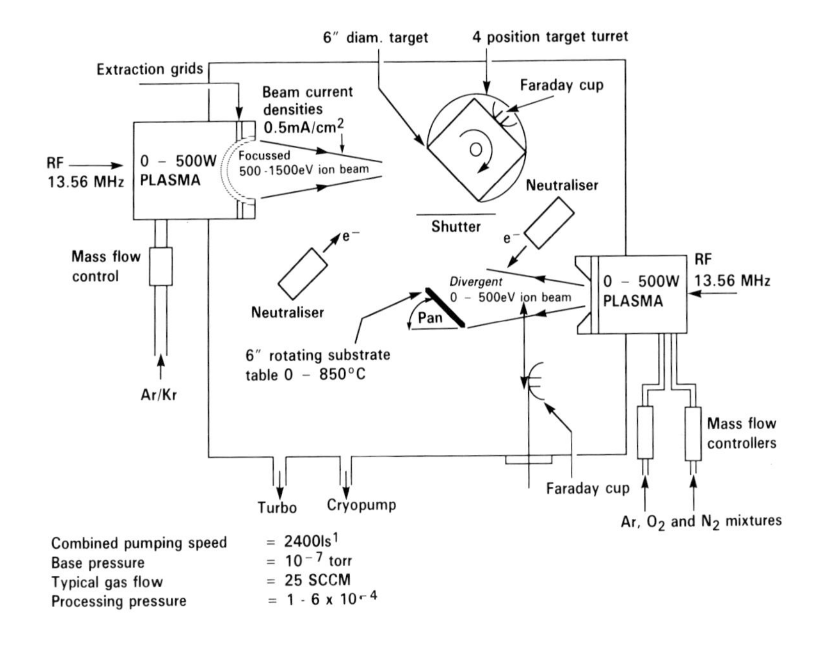

Figure 1 Schematic of DIBS system.

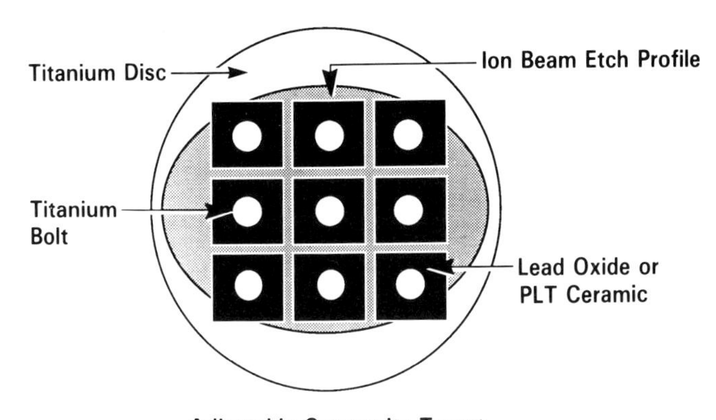

Figure 2 Adjustable composite sputtering target.

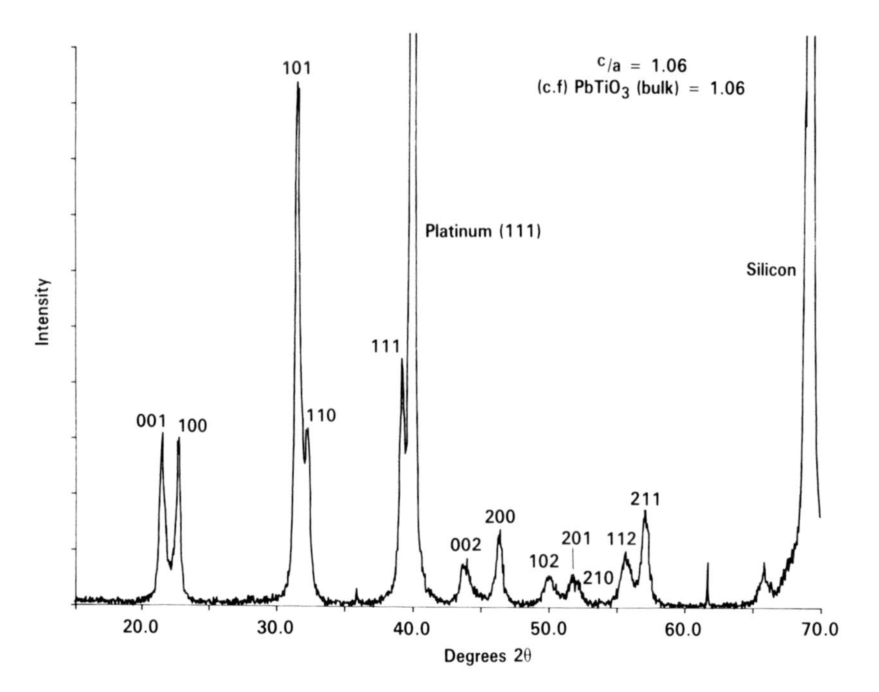

Figure 3 X-ray diffractogram for sol-derived PT film.

calculated from the ac current produced by modulating the substrate temperature at frequencies from 5 to 50 mHz. Poling was limited to a temperature maximum of 150°C and to fields up to 10^7 Vm^{-1}.

RESULTS AND DISCUSSION

An X-ray diffractogram trace for a PT film grown at 450°C from solution is shown in Fig. 3, indicating the expected tetragonal phase. The calculated c:a ratio was determined to be 1.056 which is close to the value of bulk PT (1.06). The intensity of the (001) and (101) lines decreases with increase in the thermal treatment up to 700°C. Similarly, a PLT film with $x = 0.1$, deposited at 450°C gave a very highly preferred (111) orientated film as shown in Fig. 4. Apart from the (111) the only other line present is the (100). This is not unexpected as the La would favour reduced tetragonality and thus the intrinsic favouring of a particular orientation for a given stress state. Also, the film was expected to lattice match the highly preferred (111) platinum bottom electrode.

La doping also favours the (111) orientation in films grown using the DIBS system. A typical trace is shown in Fig. 5 for a film grown at a temperature of 500–600°C. The deposition of DIBS PT films occurs above the Curie temperature of 490°C, and the degree of stress at T_c can greatly influence and fix the observed orientation after cooling. The majority of the PT films produced show a tendency

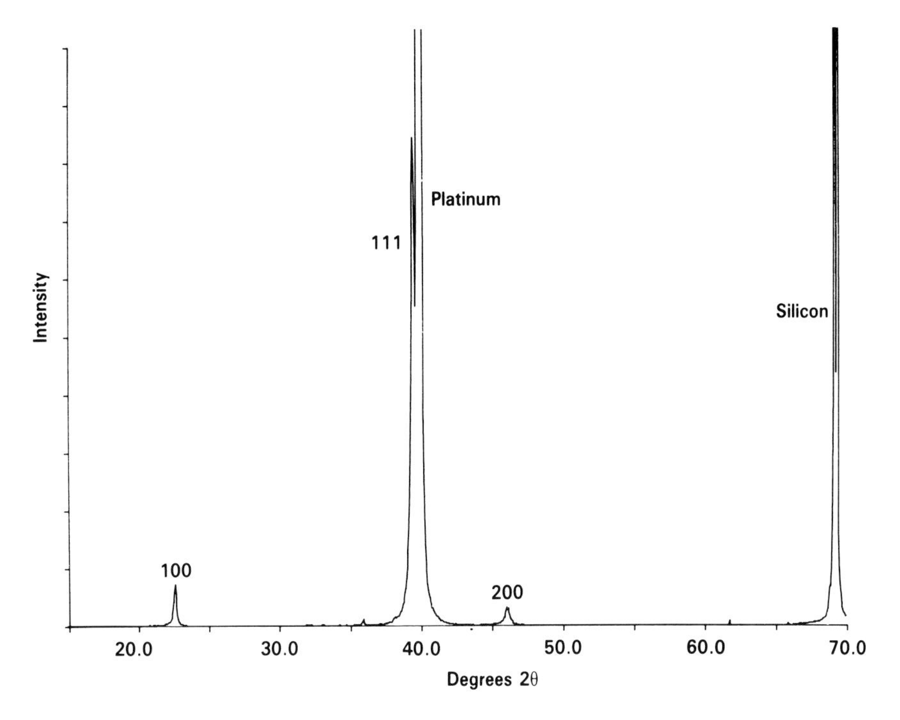

Figure 4 X-ray diffractogram for sol-derived PLT film.

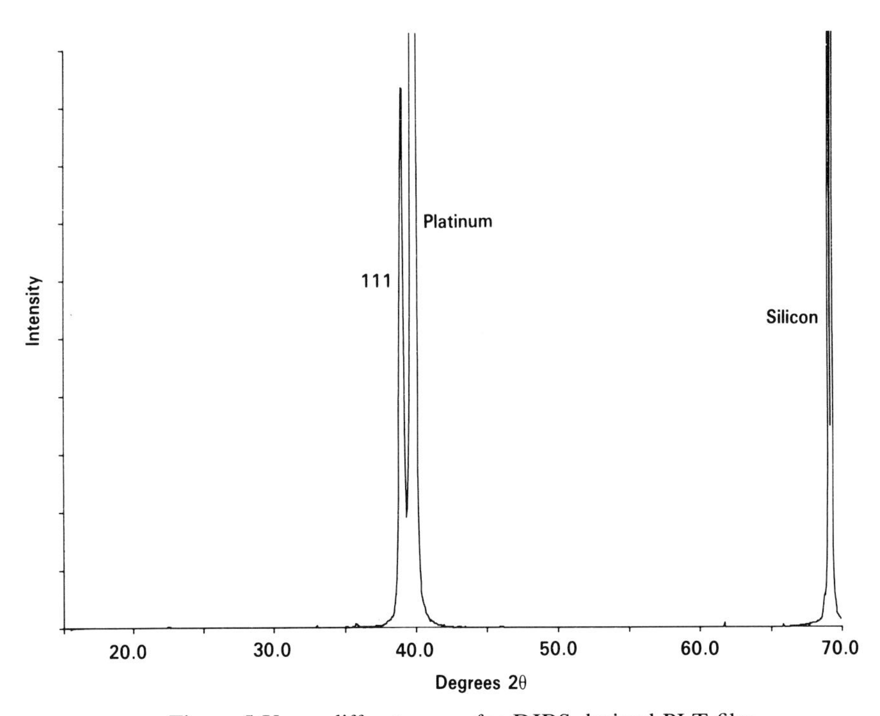

Figure 5 X-ray diffractogram for DIBS-derived PLT film.

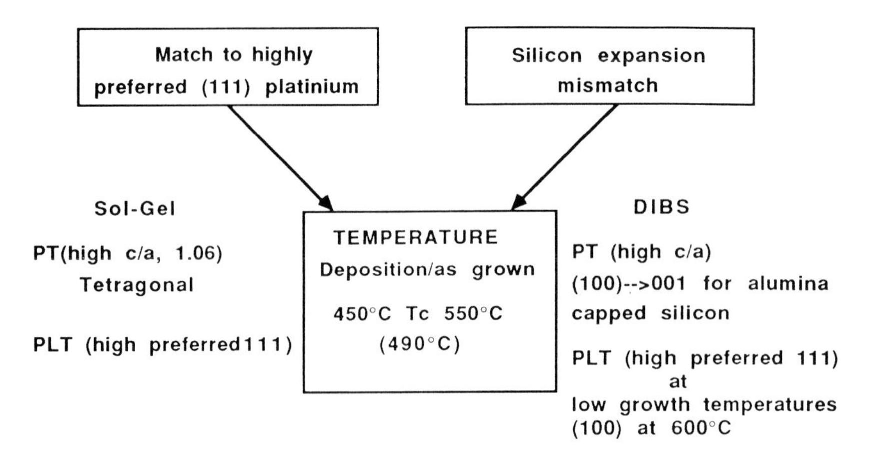

Figure 6 Orientation and temperature effects.

towards (100) orientation, however capping the silicon with an amorphous alumina layer does force the orientation towards (001) as compared with plain silicon. For La doped films, higher temperature (~600°C) favours (100) whereas lower temperature (~550°C) favours (111). It should be noted that there is no gradual changeover from (100) to (111); mixed (100)/(111) films do not occur. It seems that at higher temperature, a critical stress is exceeded that enables the film to overcome the restraint of being matched to the (111) Pt. This is shown schematically in Fig. 6.

The structure of the sol-gel films is illustrated in Figs 7 and 8 for PT and PLT, respectively. Individual strata corresponding to the intermediate baking at 450°C can clearly be seen. The grain size was determined to be $\leqslant 0.1\,\mu m$ across. The surface was also composed of smooth areas interspaced with small 'blisters', which were probably caused by stress relief. In contrast, the PLT film (Fig. 8) showed a smooth surface with a columnar through film structure confirming the high preferred (111) orientation seen by XRD. The grain structure was again small ($<0.2\,\mu m$).

Although the strata corresponding to individual bakes were not observed by SEM, these were clearly seen when the sample was examined by high-resolution TEM. It is thought that these 'strata' could be responsible for the weak pyroactivity seen in PLT films even after poling at high fields.

Similar SEM and TEM examination of DIBS films indicated smooth surfaces for both PT and PLT films, with La giving films with better optical properties. Cross-sectional SEM examination of a PLT film again showed a columnar grain morphology and no visible porosity. Also, the grain diameter (200 nm) seems fairly constant through the film thickness, through matching the Pt grain diameter, which is of the order of 100–150 nm, and then undergoing limited grain growth.

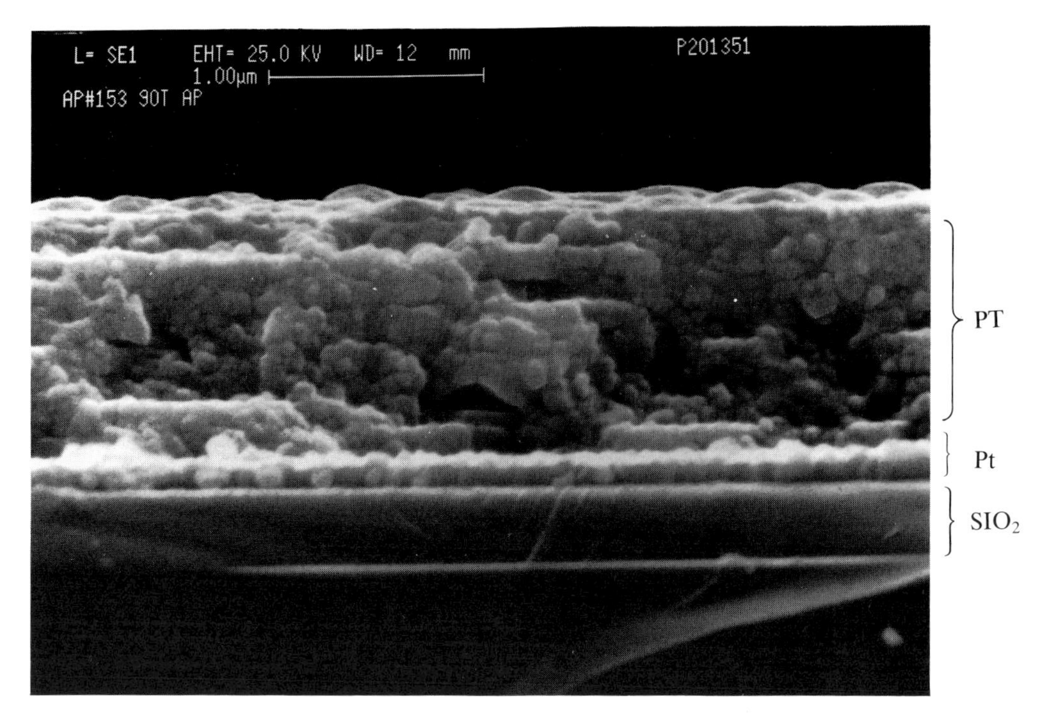

Figure 7 SEM cross-section of PT film on Pt/SiO_2ISi, showing typical 'strata'.

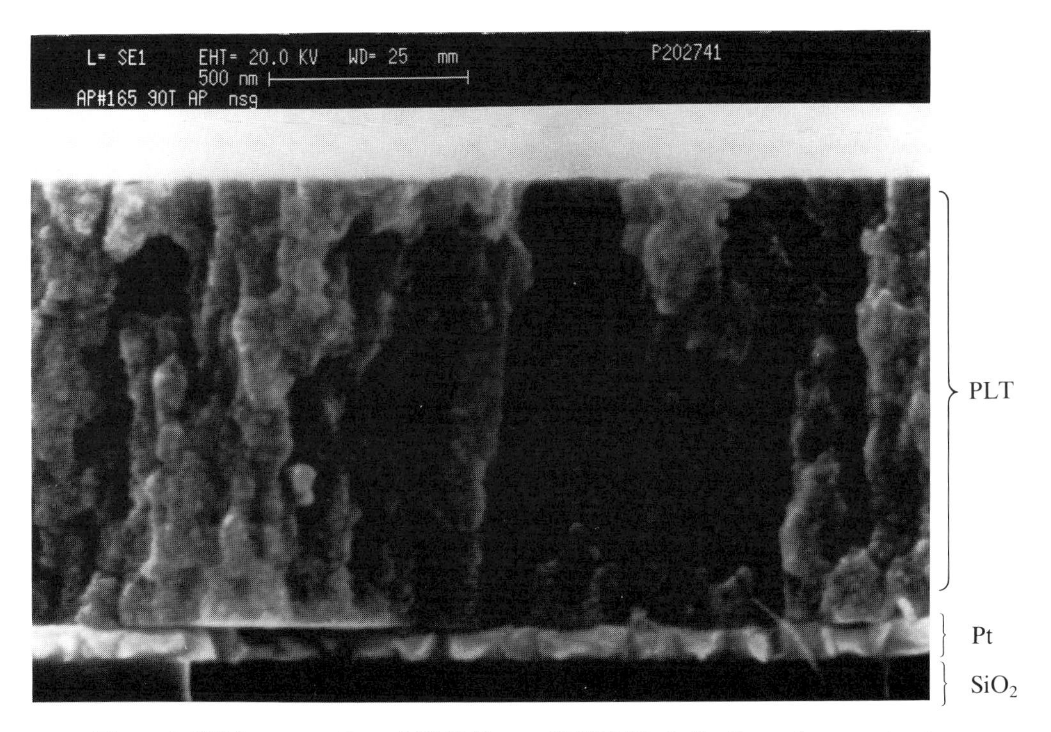

Figure 8 SEM cross-section of PLT film on Pt/SiO_2/Si, indicating columnar structure.

PYROELECTRIC MEASUREMENTS

Samples for characterisation were prepared by first evaporating Cr–Au or Pt dots (1–2 mm diameter) onto the surface of the film through a metal shadow mask. The bottom electrode in the case of sol-gel films was exposed by etching away the film at a corner. Contact to the bottom and top electrodes was made by silver epoxy wire bonds. The dielectric constant and loss (1 kHz) measured on the sol-gel PT films ranged from 100 to 200 and 0.8–2% respectively for film thicknesses of 0.3–1 μm. The PLT films, however, gave ε_r values of $\geqslant$400 with losses of ~5%. In both cases the resistivity was up to 10^{10} Ωm. The measured maximum pyroelectric coefficient obtained for PT was 1.45×10^{-4} C/m²/K after poling at up to 18 V/μm, which is in good agreement with that obtained by Kani *et al.*[5] The PLT film, however, gave a low value of 6×10^{-5} C/m²/K, this may be a consequence of the 'strata' in the films causing localised breakdown. None of the sol-gel films showed any pre-pole pyroelectric activity. The DIBS films gave similar dielectric constants but reversed losses for PT and PLT; values of 200–300 and $\leqslant$1.5% for the PLT. The resistivity value was $>10^{10}$ Ωm. The PLT films did, however, show pre-pole pyroelectric activity with values up to 2.5×10^{-4} C/m²/K which remained unchanged on poling. These films had greater breakdown fields and poling at 15×10^6 Vm^{-1} gave 85% reversal of the pyro response. This corresponds to a figure of merit F_D (1 kHz) $(p/C_v\sqrt{\varepsilon_0\varepsilon_r \tan\delta})$ of 2.2×10^{-5} Pa$^{-1/2}$. Films from both deposition systems yielded slim and non-saturating hysteresis loops (with an apparent Pr of 0.1–0.2 μC/cm² and E_c of 10 kV/cm). This is not surprising as previous studies[12] have shown that very high fields up to 500 kV/cm at 1 kHz were required to switch PT films.

CONCLUSIONS

The present work has demonstrated that good-quality highly crystalline lead titanate and La-doped lead titanate films can be readily deposited at low temperatures (450–600°C) using the sol-gel and by the DIBS processes. Both processes give highly (111) oriented films on metallised silicon, and showed good pyroelectric behaviour. With further development, this method is likely to become a preferred route for the preparation of integrated large-area 2-D arrays for IR detectors.

ACKNOWLEDGEMENTS

This work has been carried out with the support of the Defence Research Agency, Ministry of Defence.

REFERENCES

1. R. W. Whatmore: *Ferroelectrics*, 1991, **118**, 241.
2. S. T. Lui and D. Long: Proc. IEEE, 1978, **66** (1), 14.
3. R. Watton: *Ferroelectrics*, 1989, **91**, 87–108.
4. C. J. Brierley, C. Trundle, L. Considine, R. W. Whatmore and F. W. Ainger: *Ferroelectrics*, 1989, **91**, 181–192.
5. K. Kani, H. Murakami, K. Watari, A. Tsuzuki and Y. Torii: *J. Mat. Sci. Letts.*, 1992, **11**, 1605–1607.
6. R. Takayama, Y. Tomita, K. Iijima and I. Ueda: *J. Appl. Phys.*, 1987, **61** (1), 411.
7. D. A. Tossell, N. M. Shorrocks, J. S. Obhi and R. W. Whatmore: *Ferroelectrics*, 1992, **134**, 297–302.
8. C. Ye, T. Tamagawa and D. L. Polla, *J. Appl. Phys.*, 1991, **70** (10), 5538–43.
9. K. Keizer, G. J. Lansink and A. J. Burggraf: *J. Phys. Chem. Solids*, 1978, **39**, 59–63.
10. K. D. Budd, S. K. Dey and D. A. Payne: *Br. Ceram. Proc.*, 1985, **36**, 107–21.
11. A. Patel, D. A. Tossell, N. M. Shorrocks, R. W. Whatmore and R. Watton: Proc. MRS Symp., 1993, **310**, 53–58.
12. C. Ye, T. Tamagawa, Y. Lin and D. L. Polla: Proc. MRS Symp., 1992, **243**, 61–66.

Sol-Gel Derived PLZT Thin Layers Crystallised with Epitaxy on Surface-Modified Platinum Electrodes

T. TANI* and D. A. PAYNE†

*Toyota Central Research and Development Laboratories, Inc., Nagakute, Aichi, 480-11, Japan

†Department of Materials Science and Engineering, Materials Research Laboratory, and Beckman Institute, University of Illinois at Urbana-Champaign, Urbana, IL 61801, USA

ABSTRACT

Epitaxial (i.e., azimuthally oriented) (100) PLZT thin layers were formed on (100) Pt electrodes by sol-gel processing. The Pt layers were deposited with epitaxy by RF sputtering on heated (100) MgO single crystal substrates. Azimuthal orientation was confirmed by X-ray phi-scan analysis in a four-circle diffractometer. The extent of epitaxy was examined by X-ray diffraction rocking-curve measurements. Epitaxy became more dominant after the deposition of a thin Ti layer at the interface between Pt and PLZT. The improved epitaxy was attributed to the *in-situ* information of a Pt–Ti alloy at the interface. Prior heat treatment of the Ti–epi-Pt–MgO substrate before sol-gel deposition negated the effect of the thin Ti layer on epitaxy. This is the first report for epitaxially derived PLZT thin layers crystallised on metallised electrodes by sol-gel processing.

1. INTRODUCTION

Recent developments in sol-gel technology have enabled the chemical preparation of complex oxides with compositional and structural uniformity.[1] For example, lead zirconium titanate (PZT) and La-modified PZT (PLZT) have been integrated with various substrate materials, including semiconductors.[2] The properties of integrated ferroelectrics, in thin-layer form, are known to be dependent on texture and residual stress. Ferroelectric thin layers with preferred orientation or epitaxy will exhibit improved properties if the polar axis is oriented with respect to the substrate,[3] similar to the reports for vapour phase grown materials.[4] Iijima *et al.* determined that the pyroelectric coefficient for RF-sputtered $PbTiO_3$ was sensitive to the degree of crystalline orientation.[4] Tuttle *et al.* reported that the remanent polarisation (P_R) for sol-gel derived c axis oriented PZT (40/60) was 61 $\mu C/cm^2$, which was 50% greater than layers with random crystallite orientation (41 $\mu C/cm^2$).[3] By controlling texture it should be possible to design properties for thin-layer ferroelectrics.

Preferred orientation has been reported for sol-gel derived perovskite materials deposited on Pt–Ti–SiO_2–Si (a commonly used substrate), where the Pt layer has a preferred (111) orientation.[5–9] Okuwada *et al.* reported a strong dependence of preferred orientation on heating rate for lead magnesium niobate (PMN); (111) at slow heating rates and (100) for faster heating rates.[5] Francis observed a preferred

(100) orientation for PMN when the Pt layer was thick, and a (111) orientation when the Pt layer was thin on Ti–SiO_2–Si.[6] Chang and Desu investigated niobium-modified PZT (PNZT) and PLZT prepared by metallo-organic decomposition (MOD) methods, and reported a preferred (100) orientation.[7] Klee *et al.* reported that PZT layers could develop with a preferred (100) orientation on as prepared Pt–Ti–SiO_2–Si substrates, whereas (100) and (111) orientations developed on annealed substrates.[8] Previous work by the present authors indicated that different orientations could be attributed to different surface chemistries on Pt–Ti–SiO_2–Si substrates.[9] A preferred (111) orientation was always obtained for PLZT thin layers when the Pt layer had Ti on the surface. PLZT crystallised in the (111) direction by heteroepitaxial nucleation and growth off Pt_3Ti crystallites with lattice matching and B-site cation location matching (i.e., Ti).[9] In contrast, a preferred (100) orientation could be obtained off Ti-free surfaces for unannealed Pt–Ti–SiO_2–Si substrates. This was considered to be self-textured growth, in accordance with minimum surface energy conditions, for flat-faced surfaces. In both cases, –B–O– periodic bond chains in the perovskite structure appeared to play an important role in determining interfacial and surface energies, and thus, in controlling texture.

In epitaxial nucleation, guest crystallites show both textural and azimuthal orientation with respect to the substrate structure. Preferred orientation has been reported for sol-gel deposition on lattice matched single crystal substrates, by standard X-ray diffraction (XRD) powder methods.[10] However, reports confirming epitaxy by four-circle XRD measurements are somewhat limited.[11] Perovskite-type materials crystallise readily with preferred orientations on perovskite single crystal substrates, e.g., $SrTiO_3$.[10,12] MgO and sapphire single crystals have also been used as substrates with lattice matching. Wang *et al.* reported highly (001) oriented $PbTiO_3$ on (100) $SrTiO_3$ by MOD, while the use of another perovskite-type substrate, (100) $LaAlO_3$, resulted in $PbTiO_3$ thin layers with mixed orientations.[13] Okuwada *et al.* reported sol-gel derived PMN thin layers were obtained with preferred (100) and (110) orientations on (100) and (110) MgO, respectively, but randomly oriented $Pb(Fe_{1/2}Nb_{1/2})O_3$ (PFN) under the same conditions.[14] These results suggested that both lattice matching and B-site cation matching (e.g. Mg) are preferred for heterogeneous nucleation and epitaxial growth. Barlingay and Dey confirmed (110) epitaxial crystallisation of PNZT on (01.2) sapphire by high-resolution TEM and XRD pole-figure analysis.[15]

Deposition of ferroelectric thin layers, with epitaxy, on electrodes, has necessitated the development of epitaxial Pt on single crystal oxide surfaces. Pt layers can be epitaxially grown on heated (600°C) oxide single crystals by sputtering in an Ar–O_2 gas mixture.[16] Epitaxial $PbTiO_3$ and PZT layers were successfully grown on epitaxial Pt layers, deposited on single crystal oxides by physical vapour methods.[4,17] However, for sol-gel processing, epitaxial growth off noble metal electrodes is thought to be less likely to occur, due to the low mobility of nuclei in the solid matrix, and competitive homogeneous and heterogeneous nucleation and growth reactions. By use of the sol-gel route, Tuttle *et al.* prepared highly oriented

PZT layers on (100) Pt‖(100) MgO.[3] PZT layers with a Zr–Ti composition of 40:60 showed a pseudocubic (100) orientation (actually (001) in the tetragonal structure), with minor (111) and (110) X-ray diffractions. It was claimed that the crystal structure of the PZT layer mimicked the structure of the underlying Pt, but no evidence was given for epitaxy, i.e., azimuthal orientation.[18]

In this paper, we report the preparation of sol-gel derived PLZT thin layers on (100) Pt‖(100) MgO substrates (designated epi-Pt/MgO) and investigate the azimuthal orientation by four-circle XRD measurements. In addition, the effect of modified surface chemistry on the Pt layers was examined in an effort to reduce the interfacial energy between the deposited PLZT layer, and the electrode material, by seeding B-site cations (i.e., Ti) at the interface.

2. EXPERIMENTAL PROCEDURE

2.1 *Preparation of Substrates*

Four types of electroded substrates were prepared: (i) Pt–MgO, (ii) Pt–Ti–MgO, (iii) epi-Pt–MgO, and (iv) Ti–epi-Pt–MgO. (i) Pt was deposited on unheated (100) MgO single crystals by electron-beam evaporation. (ii) A thin Ti interlayer was also deposited for adhesion purposes. On the other hand, (iii) epi-Pt was RF sputter deposited on (100) MgO single crystals at 500°C in an equal mixture of oxygen in argon; and (iv) an additional Ti layer was deposited on epi-Pt by electron-beam evaporation. The layer thicknesses were (i) 300, (ii) 300–100, (iii) 300 and (iv) 5–300 nm. In addition, several epi-Pt–MgO and Ti–epi-Pt–MgO substrate were heat treated at 300°C for 20 min. in air.

The substrates were characterised by XRD (Rigaku D/Max IIIA with Cu K_α) before the deposition of PLZT. Standard X-ray powder diffraction techniques were used to determine the preferred orientation and lattice parameters for the Pt electrode layers. Rocking curves were taken on the epi-Pt electrodes.

2.2 *Preparation of Thin Layers*

Figure 1 illustrates the processing cycle for sol-gel derived PLZT thin layers. A 1 M precursor solution was prepared from lead acetate $Pb(CH_3COO)_2$, lanthanum isopropoxide $La(O\text{-}iC_3H_7)_3$, zirconium *n*-propoxide $Zr(O\text{-}C_3H_7)_4$, titanium isopropoxide $Ti(O\text{-}iC_3H_7)_4$, and 2-methoxyethanol $CH_3OC_2H_4OH$. All reactions, prior to hydrolysis of the alkoxides, were carried out in a dry nitrogen atmosphere using a Schlenk line apparatus. The process followed Budd's original method,[2] with the exception that lead acetate trihydrate was dehydrated in vacuum at 100°C for 20 h before preparation of the lead precursor solution. The elimination of water was confirmed by FTIR analysis. Lead acetate was dissolved in methoxyethanol, refluxed for one hour, and distilled under vacuum. Titanium isopropoxide, zirconium *n*-propoxide and lanthanum isopropoxide were dissolved in methoxy-

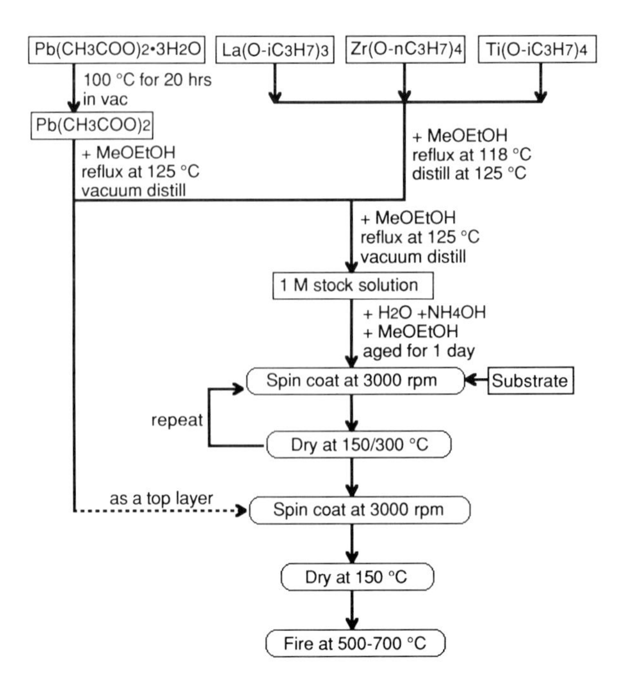

Figure 1 Sol-gel processing of PLZT thin layers.

ethanol and the solution was refluxed and thermally distilled. The two precursor solutions were combined, refluxed and distilled. Partially hydrolysed solutions were aged for one day prior to spin-casting onto substrates at 3000 rpm for 50 s. After each deposition, the coated substrates were placed on a hot plate at 300°C. A PLZT 8/65/35 composition is described in this work. An additional batch of a PbO containing 'cover-coating solution' was prepared separately.[19] Partially hydrolysed 0.25 M PLZT solutions containing ammonium hydroxide (NH_4OH) were deposited on Pt–Ti–SiO_2–Si substrates by spin casting at 3000 rpm. After each layer was deposited, the specimen was dried at 150°C for 15 s and pyrolysed at 300°C for 1 min. After 10 depositions, the layer was crystallised at 700°C for 5 min. in air in an electrically heated box furnace. The PbO cover coat was used to prevent the evaporation loss of PbO and the associated formation of additional phases on the surface. The final thickness of the fired layer was 0.3 μm.

Crystallisation behaviour was examined after heat treatment by XRD. Standard X-ray powder diffraction techniques were used to determine phase and texture development. Rocking curves were obtained for layers deposited on epi-Pt electrodes, and epitaxy was investigated by four-circle diffractometry (Elliot GX-21).

3. RESULTS

3.1 Characterisation of Substrates

Figure 2 gives XRD powder diffraction data for Pt electrodes deposited on single crystal MgO substrates, (a) by electron-beam evaporation on an unheated substrate, or (b) by RF sputtering in Ar–O_2 on a substrate heated at 500°C. The electron beam-deposited Pt had a preferred (111) orientation, which was, however, not azimuthally oriented. It was self-textured orientation with face-centred Pt cubic structure, since the close-packed layer had the lowest surface energy. By contrast, the RF sputter-deposited Pt had epitaxial (100) orientation with a sharp diffraction peak at d = 1.967 Å (2θ = 46.1°). There was an unknown diffraction peak at ~41° (d ~ 2.20 Å, shown in Fig. 2 (a and b) as unknown 1) which was sharp and had a peak height that correlated with the peak for (200) MgO. Another minor peak at ~44° (unknown 2 in Fig. 2(b)) was found only for the epi-Pt–MgO substrate and corresponded to a d spacing of 2.054 Å, which was between d = 1.967 Å for (200) Pt and d = 2.107 Å for (200) MgO. This may be diffraction from a buffer layer between MgO and Pt. The rocking curves for (200) MgO and epi-(200) Pt are shown in Fig. 3. The full width at half maximum (FWHM) is 0.04° for MgO (Fig. 3(a)) and 0.26° for epi-Pt (Fig. 3(b)). The latter value is comparable with the FWHM values (0.16–0.23°) for epi-Pt reported by Cui *et al.*[20] In addition, the electron-beam deposited Pt layer was found to peel off the MgO substrate when a Ti layer was not used. However, sufficient adhesion was obtained for the sputter-deposited epi-Pt layer on MgO.

3.2 Characterisation of PLZT Thin Layers

PLZT thin layers crystallised to a single-phase perovskite structure, without any additional phases, through use of the cover-coating technique. XRD data are given in Fig. 4, for PLZT thin layers deposited on (a) self-textured (111) Pt–Ti–(100)MgO, (b) epi-(100)Pt–(100)MgO, (c) Ti–epi-(100)Pt–(100)MgO, and (d) heat-treated Ti-epi-(100)Pt–(100)MgO substrates. PLZT on (a) self-textured (111)Pt–Ti–(100)MgO had random orientation, whereas PLZT on (b) epi-Pt–MgO had a strong (100) orientation (an unknown diffraction peak at d ~ 2.51 Å (2θ ~ 36°) may be attributed to residual PbO powder on the layer). A stronger preferred (100) orientation was measured for PLZT thin layers when crystallised on a (c) Ti–epi-Pt–MgO substrate. The strong alignment was lost when the layers were crystallised on a similar substrate (Ti–epi-Pt–MgO) which had been heat treated at 300°C in air for 20 min, prior to deposition (d). The heat treatment negated the effect of the Ti layer on texture development. Figure 5 compares XRD rocking curves for (100) PLZT on epi-Pt–MgO and on Ti–epi-Pt–MgO. The FWHM values were 2.15 and 1.55° for layers deposited on epi-Pt–MgO and Ti–epi-Pt–MgO, respectively. These values are similar to recent reports for PZT thin layers crystallised on unelectroded (100) MgO single crystals by MOD and

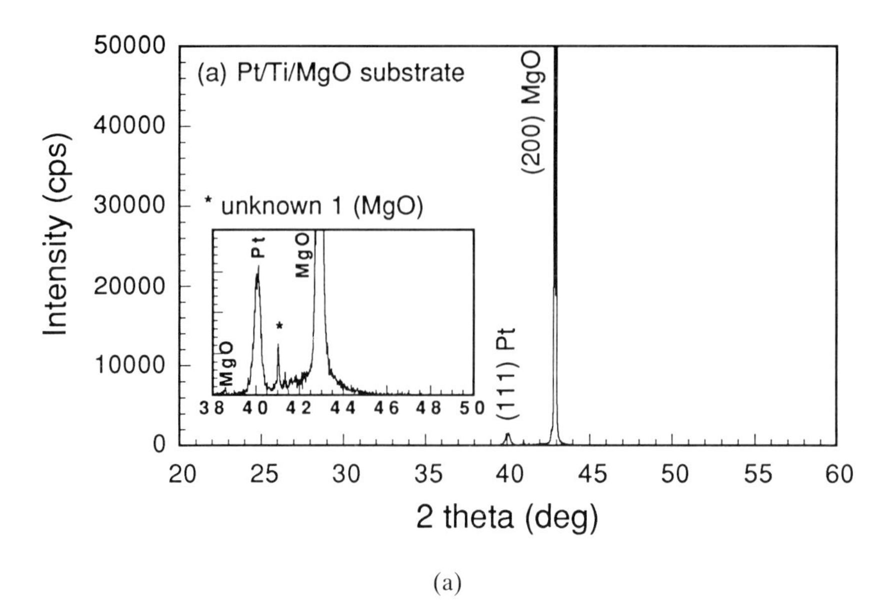

(a)

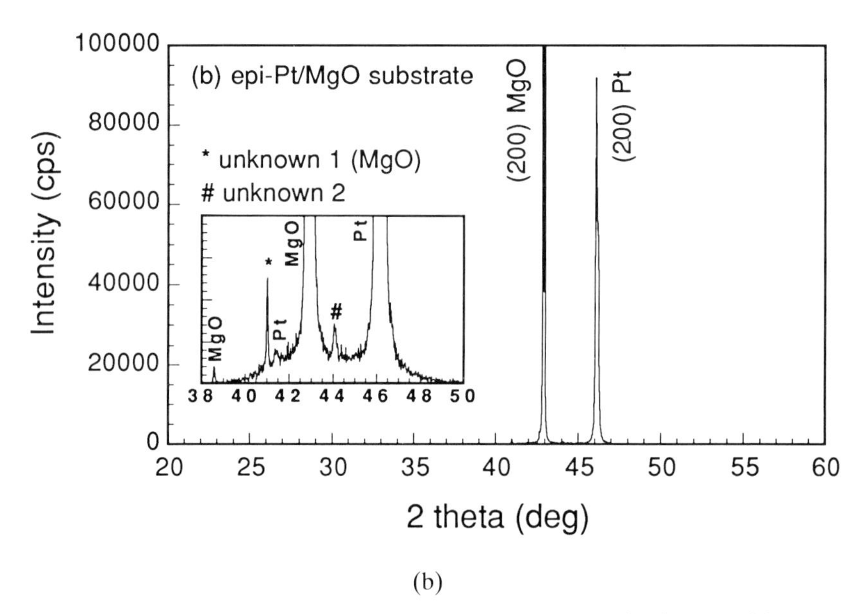

(b)

Figure 2 XRD data for Pt electrodes deposited on single crystal (100) MgO, (a) by electron beam evaporation on an unheated substrate, and (b) by RF sputtering in Ar–O_2 with the substrate heated at 500°C.

sol-gel methods (FWHM = 1.48–2.75).[21] Sharper rocking curves for layers crystallised on Ti–epi-Pt–MgO confirmed better epitaxy than for layers crystallised on epi-Pt–MgO substrates (i.e., without a Ti surface layer). The FWHM values for layers deposited on heat-treated substrates were 2.23 and 2.73° for epi-Pt–MgO

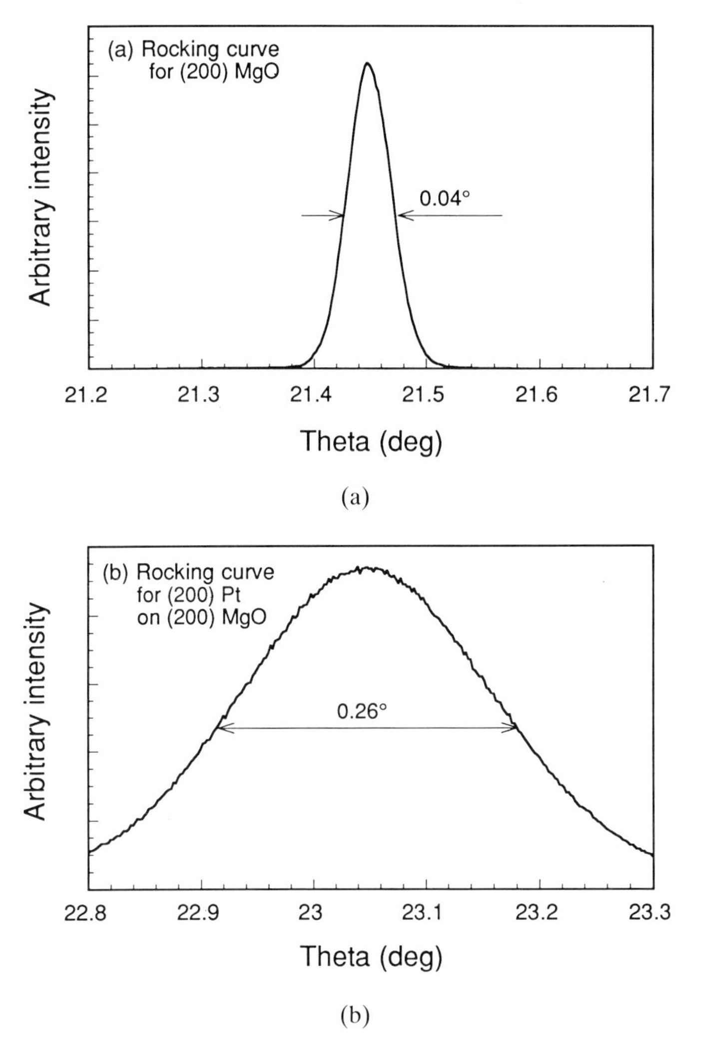

Figure 3 XRD Rocking curves for (a) single crystal (200) MgO and (b) (200) epi-deposited Pt on MgO.

and Ti–epi-Pt–MgO, respectively. Prior heat treatment of substrates degraded epitaxy for PLZT layers deposited on Ti–epi-Pt–MgO, whereas it did not affect layers crystallised on epi-Pt–MgO. XRD phi-scan analysis was carried out after the adjustment of the chi (i.e., tilting) angle at the (110) diffraction conditions for PLZT. Figure 6 gives phi-scan data for PLZT (110) diffraction off epi-Pt–MgO and Ti–epi-Pt–MgO substrates. The epitaxy, i.e., azimuthal orientation, is evident by the (110) peaks at every 90° in phi-scans for the (100) oriented layer. This is, to the author's knowledge, the first report for sol-gel derived perovskite thin layers deposited with epitaxy on metallised substrates, as confirmed by four-circle XRD measurements. The effect of the thin Ti layer on epitaxy is again indicated (Fig. 6) by the sharper diffraction peaks than for a layer crystallised on Pt alone.

(a)

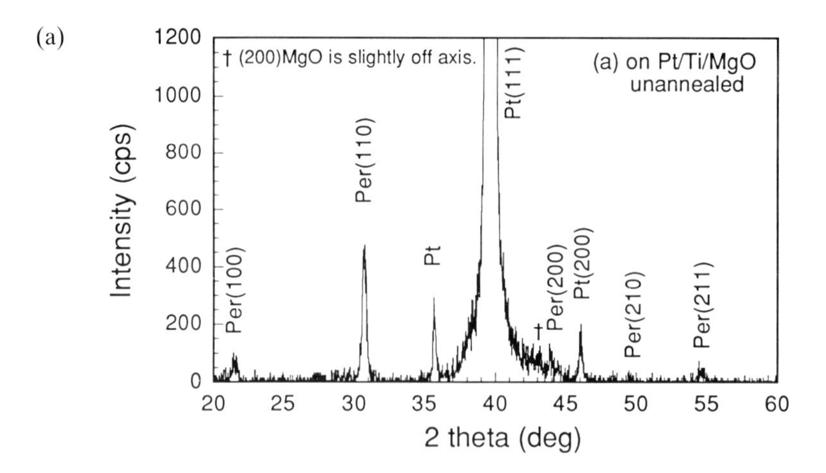

(b)

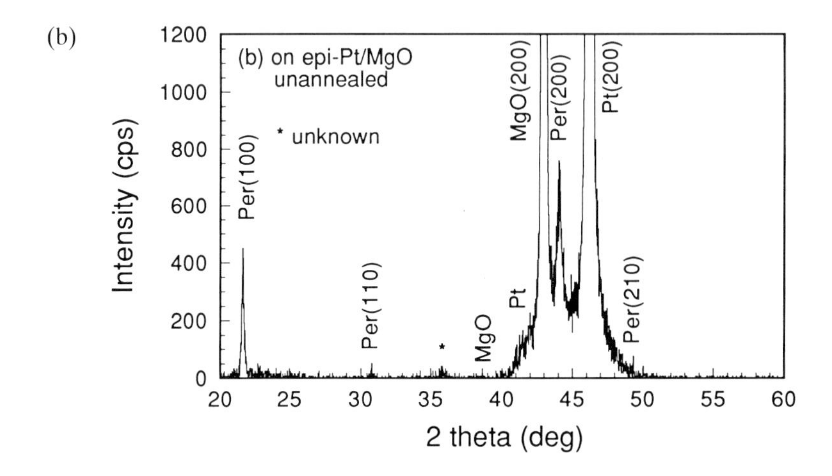

Figure 4 XRD data for PLZT 8/65/35 thin layers deposited on (a) Pt–Ti–MgO, (b) epi-Pt–MgO, (c) Ti–epi-Pt–MgO, and (d) heat treated Ti–epi-Pt–MgO substrates.

4. DISCUSSION

It was observed that PLZT thin layers could be grown epitaxially on Pt, without modification of the surface chemistry, if the Pt layer had azimuthal orientation (Fig. 6). On self-textured Ti-free (111) Pt, however, PLZT layers resulted in either random orientations (Fig. 4(a)) or self-textured (100) orientation.[9] These observations can be explained by heterogeneous nucleation.[22]

It is well known that the free energy change for the formation of a heterogeneously nucleated cluster, shaped like a spherical cap, is given by

$$\Delta G = \left\{ -\frac{4}{3}\pi r^3 \Delta G_V + 4\pi r^2 \gamma_{CM} \right\} S(\theta) \tag{1}$$

(c)

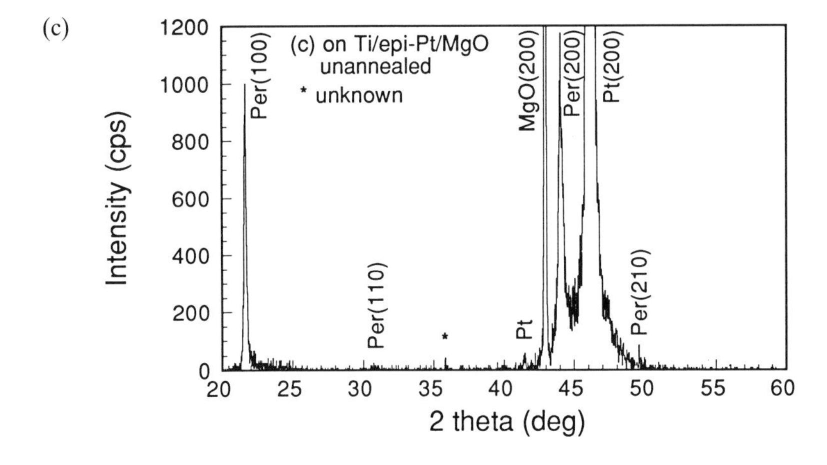

(d)

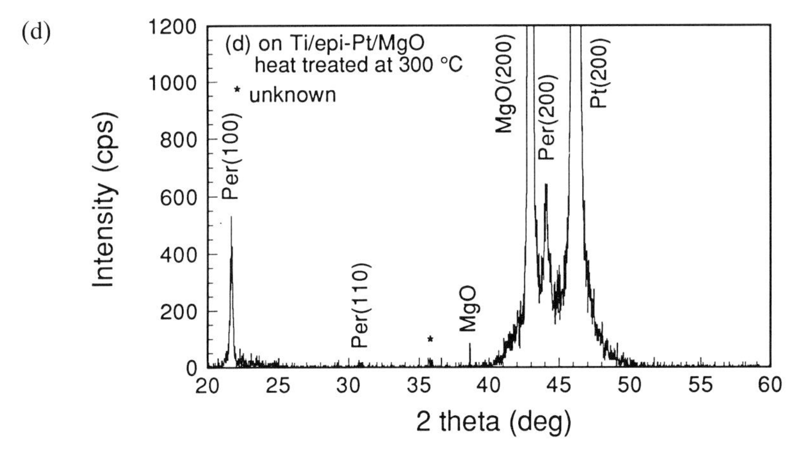

Figure 4 *contd.*

where ΔG_V is the volume free energy change for crystallisation, γ_{CM} is the interfacial energy between the crystallite and the matrix, and θ and r are the edge angle and radius for the spherical cap, respectively, with

$$S(\theta) = \frac{(2 + \cos\theta)(1 - \cos\theta)^2}{4} \tag{2}$$

and

$$\cos\theta = \frac{(\gamma_{SM} - \gamma_{CS})}{\gamma_{CM}} \tag{3}$$

where γ_{SM} and γ_{CS} are the substrate matrix and crystallite substrate interfacial

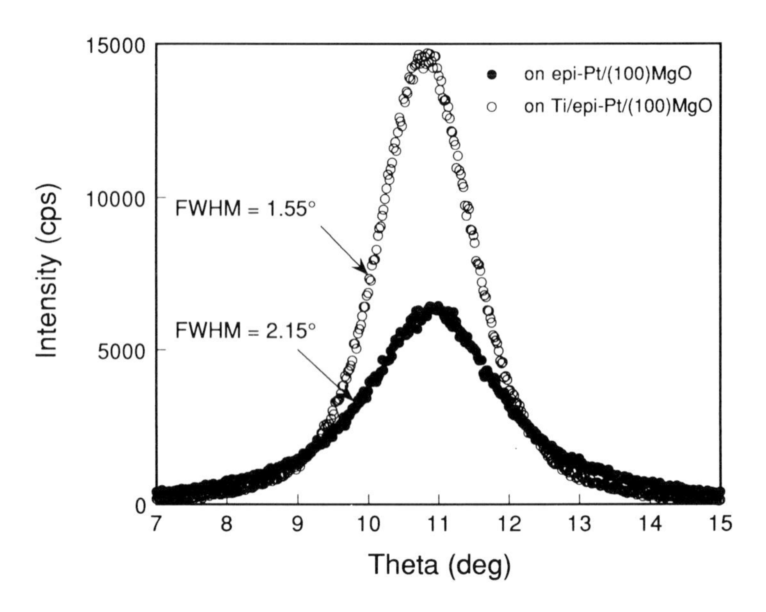

Figure 5 XRD rocking curves for PLZT 8/65/35 thin layers deposited on epi-Pt–MgO and Ti–epi-Pt–MgO substrates.

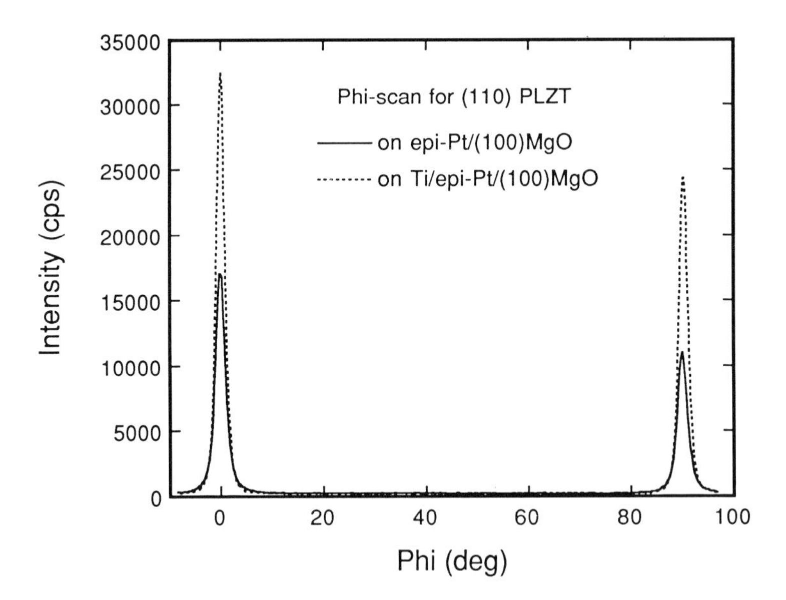

Figure 6 XRD phi-scan data for (110) diffraction from PLZT 8/65/35 thin layers deposited on epi-Pt–MgO and Ti–epi-Pt–MgO substrates.

energies, respectively. The critical radius r^* which gives the maximum free energy change ΔG_V is given by

$$r^* = \frac{2\gamma_{CM}}{\Delta G_V} \tag{4}$$

and the corresponding activation energy ΔG^* for stable growth is given by

$$\Delta G^* = \frac{16\pi\gamma_{CM}^3}{3\Delta G_V^2} \cdot S(\theta) \tag{5}$$

Since Pt is relatively inert to oxides, a large value of crystallite (guest)/substrate (host) interfacial energy γ_{CS} gives a large value of θ, and thus requires a relatively high activation energy for crystallisation to overcome ΔG^*. The fine crystallite size of self-textured Pt (20–60 nm) on the substrate restricts the cluster size for the epitaxially grown guest material (PLZT). Beyond this size, the interfacial energy γ_{CS} can be even higher due to the angle mismatch at the interface between the guest crystallite with a neighbouring crystallite of Pt, which increases the value of θ, and thus the value of ΔG^* for heterogeneous nucleation. On the other hand, epitaxially grown Pt surfaces provide a much larger area, with the same azimuthal orientation, so that the heterogeneous nucleation and growth of PLZT, with epitaxy, can be achieved more readily, than for self-textured (111) Pt with fine grains.

A thin Ti layer on epi-Pt further favours the epitaxial nucleation, presumably with a reduction in interfacial energy between (100) PLZT and (100) Pt. There are two possible explanations: (i) formation of a $PbTiO_3$ layer which has a smaller lattice mismatch with Pt than PLZT, or (ii) formation of a Pt–Ti alloy on the electrode surface. Prior heat treatment of the Ti–epi-Pt–MgO substrate in air did not lead to better epitaxy, presumably due to oxidation of Ti and a reduction in interfacial energy for epitaxy. This observation makes the latter explanation (i.e., (ii) the formation of a Pt–Ti alloy at the interface) more attributable for improved epitaxy. There is a report which mentions the role of Ti atoms in Pt electrodes to facilitate the crystallisation of PZT.[23] Abe *et al.* prepared single phase perovskite thin layers by RF sputtering on Pt–Ti alloys with Ti : Pt ratios between 0.02 and 0.17.[23] They suggested that Ti atoms on the surface of the electrodes could adsorb oxygen, and Pb atoms could remain on the surface for longer time than on pure Pt. That is, Ti atoms on the surface of Pt electrodes could reduce the interfacial energy between the electrode and the Pb-based perovskite thin layer. Furthermore, a proposed Pt_3Ti intermetallic compound has good lattice matching (5%) with perovskite and B-site cation location matching (i.e., Ti), as reported previously.[9]

Thus, the ability to crystallise an epitaxial layer is a function of the interfacial energy between the host and the guest materials, and the crystallite size of the host material. It is also proposed that the role of surface chemistry is important for the reduction of interfacial energy with noble metal electrodes, and the seeding of B-site atoms is an effective way of facilitating heteroepitaxial nucleation and growth of perovskite materials.

5. SUMMARY

Epitaxial (100) PLZT thin layers crystallized on epitaxial (100) Pt electrodes deposited on (100) MgO single crystal substrates. The azimuthal orientation was confirmed by X-ray phi-scan analysis in a four-circle diffractometer. XRD rocking-curve measurements gave FWHM values of 2.15 and 1.55° for (100) PLZT deposited on epi-Pt–MgO and Ti–epi-Pt–MgO, respectively. Prior heat treatment of the Ti–epi-Pt–MgO substrate negated the effect of epitaxy, which was attributed to oxidation of the Ti layer. The improved epitaxy with a Ti layer on Pt was attributed to the *in-situ* formation of a Pt–Ti alloy at the surface of the Pt electrode. Thus, the seeding of B-site cations (i.e., Ti) reduced the interfacial energy with Pt, and facilitated the heteroepitaxial nucleation and growth of PLZT. This is the first report for the crystallisation of Pb-based perovskite thin layers deposited on metallised electrodes by sol-gel processing.

ACKNOWLEDGEMENTS

The authors are grateful to Mr Masahiro Toyoda of Murata Manufacturing Company for the preparation of various metal layer combinations, and to Ms Joyce McMillan for four-circle X-ray diffraction analysis. The research was supported by DOE DMR-FG02-91ER45439. The use of facilities in the Center for Microanalysis of Materials at the University of Illinois at Urbana-Champaign is gratefully acknowledged.

REFERENCES

1. C. D. E. Lakeman and D. A. Payne: *Mater. Chem. Phys.*, accepted (1994).
2. K. D. Budd, S. K. Dey and D. A. Payne: *Brit. Ceram. Soc. Proc.*, 1985, **36**, 107–121.
3. B. Tuttle, J. A. Voigt, D. C. Goodnow, D. L. Lamppa, T. J. Headley, M. O. Eatough, G. Zender, R. D. Nasby and S. M. Rodgers: *J. Am. Ceram. Soc.*, 1993, **76**, 1537–1544.
4. K. Iijima, Y. Tomita, R. Takayama and I. Ueda: *J. Appl. Phys.*, 1986, **60**, 361–367.
5. K. Okuwada, M. Imai and K. Kakuno: *Jpn. J. Appl. Phys.*, 1989, **28**, L1271–1273.
6. L. F. Francis: Ph.D. Thesis, University of Illinois at Urbana-Champaign, 1991.
7. J. Chang and S. Desu: in *Ferroelectric Films*, A. S. Bhalla and K. M. Nair eds, 155–167, American Ceramic Society, Westerville, OH, 1992.
8. M. Klee, R. Eusemann, R. Waser, W. Brand and H. van Hal: *J. Appl. Phys.*, 1992, **72**, 1566–1576.
9. T. Tani, Z. Xu and D. A. Payne: in *Ferroelectric Thin Films III*, E. R. Myers, B. A. Tuttle, S. D. Desu and P. K. Larsen eds, 269–274, Materials Research Society, Pittsburgh, PA, 1993.
10. S. L. Swartz, P. J. Melling and C. S. Grant: in *Optical Materials: Processing and Science*, D. B. Poker and C. Ortiz eds., 227–232, Materials Research Society, Pittsburgh, PA, 1989.
11. D. S. Hagberg: M.S. Thesis, University of Illinois at Urbana-Champaign, 1991.
12. T. Suzuki, M. Matsuki, Y. Matsuda, K. Kobayashi and Y. Takahashi: *J. Ceram. Soc. Jpn.*, 1990, **98**, 754–758.

13. Y. Wang, P. Zhang, B. Qu and W. Zhong: *J. Appl. Phys.*, 1992, **71**, 6121–6124.
14. K. Okuwada, S. Nakamura, M. Imai and K. Kakuno: *Jpn. J. Appl. Phys.*, 1990, **29**, 1153–1156.
15. C. K. Barlingay and S. K. Dey: *Appl. Phys. Lett.*, 1992, **61**, 1278–1280.
16. K. Iijima, R. Takayama, Y. Tomita and I. Ueda: *J. Appl. Phys.*, 1986, **60**, 2914–2919.
17. M. Adachi, T. Matsuzaki, T. Yamada, T. Shiosaki and A. Kawabata: *Jpn. J. Appl. Phys.*, 1987, **26**, 550–553.
18. M. Gebhardt: in *Crystal Growth: An Introduction*, P. Hartman ed., 106, Elsevier, New York, 1973.
19. T. Tani and D. A. Payne: *J. Am. Ceram. Soc.*, 1994, in press.
20. G. Cui, P. C. V. Buskirk, J. Zhang, C. P. Beetz, J. Steinbeck, Z. L. Wang and J. Bentley: in *Ferroelectric Thin Films III*, E. R. Myers, B. A. Tuttle, S. D. Desu and P. K. Larsen eds, 345–350, Materials Research Society, Pittsburgh, PA, 1993.
21. K. Nashimoto: presented at PAC RIM Int. Symp. on Ferroelectric Thin Films, American Ceramic Society, Honolulu, 1994. *Ceram. Trans.*, in press.
22. D. Turnbull: in *Solid State Physics Vol. 3*, F. Seitz and D. Turnbull eds, 226–306, Academic Press, New York, 1956.
23. K. Abe, H. Tomita, H. Toyoda, M. Imai and Y. Yokote: *Jpn. J. Appl. Phys.*, 1991, **30**, 2152–2154.

Thin Films of PZT and Ca–Pt Prepared by a Sol-Gel Method

Y. L. TU, S. CHEWASATN, R. HOLT and S. J. MILNE
School of Materials, University of Leeds, Leeds, LS2 9JT, UK

ABSTRACT
A sol-gel route using propanediol as the starting solvent has been used to prepare films of $PbZr_{0.53}Ti_{0.47}O_3$ (PZT) and two calcium modified $PbTiO_3$ compositions, $Pb_{0.9}Ca_{0.1}TiO_3$ and $Pb_{0.85}Ca_{0.15}TiO_3$ (Ca–PT). All films were made from a single deposition and firing procedure. The microstructures of $\sim 0.5\ \mu m$ thick PZT films were composed of $\leqslant 1\ \mu m$ rosette-like grains and very fine $\ll 0.1\ \mu m$ grains. The Ca–PT microstructures exhibited clustering of $\sim 0.2\ \mu m$ grains into $\leqslant 0.5\ \mu m$ units that were similar in appearance to the structures that we have observed previously in unmodified PT films. Electrical data are reported for the three film compositions.

1. INTRODUCTION

Ferroelectric thin films of $PbTiO_3$ and $PbZr_xTi_{1-x}O_3$-based compositions are of interest for a range of electronic device applications including non-volatile memories, IR detectors and piezoelectric transducers.

Sol-gel methods offer a promising means of preparing thin films, as capital costs are low and it is possible to exert tight control over the chemical stoichiometry of the starting sols. A variety of methods have been reported but they are most commonly based on the procedure first reported by Gurkovich and Blum,[1] and Budd, Dey and Payne.[2] In these, lead acetate titanium tetrapropoxide and zirconium tetrapropoxide are mixed in a methoxyethanol solvent and react to form a polymeric gel precursor sol. Solvent exchange reactions play an important part in the chemistry of the process, and gelation is understood to proceed via hydrolysis and condensation reactions involving metal–alkoxide or acetate groups.

Various workers have added acetic acid to methoxyethanol-derived sols to inhibit precipitation, and a method has been demonstrated by Yi, Wu and Sayer[3] in which acetic acid is used in place of methoxyethanol to produce the gel precursor solutions, or sols.

We have adopted an approach in which diols are used as the starting solvent.[4–8] In the present paper we report on the properties of $PbZr_{0.53}Ti_{0.47}O_3$ (PZT) and Ca modified $PbTiO_3$ films (Ca–PT) prepared by modified diol sol-gel processes.

2. EXPERIMENTAL

The solution synthesis procedure for PZT sols is shown in Fig. 1. For Ca–PT, calcium nitrate was used as a starting material and small quantities of water and

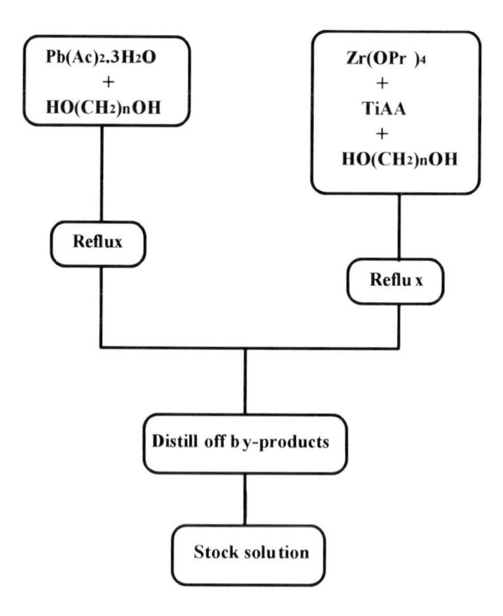

Figure 1 Flow diagram of solution synthesis process for PZT sol.

acetic acid were added to avoid solubility problems. Films were spin coated onto Si–SiO_2–Ti–Pt substrates in a class 100 clean room and were fired by placing them directly in a tube furnace set at the required temperature.

Phase analysis was performed at room temperature using X-ray diffraction (Philips APD 1700). Microstructures and film thickness were examined using scanning electron microscopy (Hitachi S700).

The top electrodes were applied by sputtering 0.6 mm diameter gold dots onto the surface of the films. Polarisation electric field (*P-E*) data were obtained using a Sawyer–Tower circuit; sinusoidal waveform voltages were applied at 60 Hz. Relative permittivity, ε_r, and dissipation factor, D, were measured using a HP4192A impedance analyser. Electrical resistivity was measured at 1 V after 1 min. using a Keithley 617 electrometer.

3. RESULTS AND DISCUSSION

3.1 PZT

X-ray diffraction experiments showed that single-phase perovskite-type PZT films could be produced by firing the coated substrates for >5 min at 700°C. The crystalline films were oriented preferentially as indicated by the enhanced intensity of the 111 reflection.

SEM examinations of transverse sections indicated the thickness of PZT single-layer films to be ~0.5 μm for a starting sol of metals concentration ~1 M, Fig. 2, and 0.25 μm for a sol concentration of ~0.5 M.

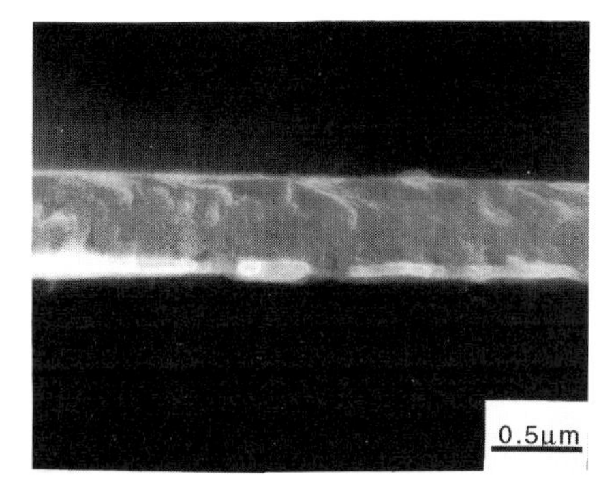

Figure 2 SEM micrograph of cross-section of the PZT film.

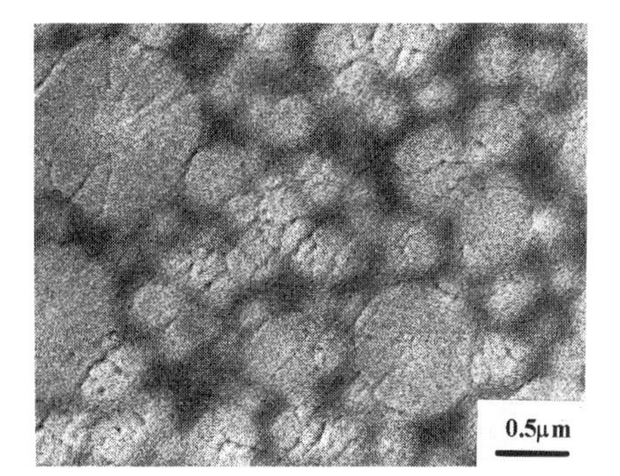

Figure 3 Microstructure of 0.5 μm PZT (53/47) film prepared from sols containing 10 mol.% excess Pb and fired at 700°C for 15 min.

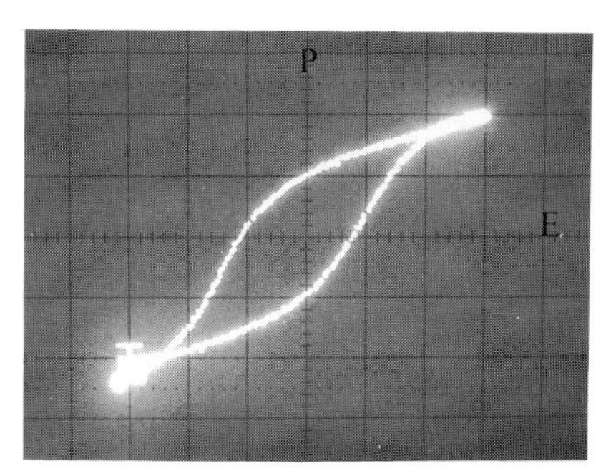

Figure 4 *P-E* hysteresis loop of 0.5 μm PZT (53/47) film prepared from sols containing 10 mol.% excess Pb and fired at 700°C for 1 h. ($P = 18\,\mu C\,cm^{-2}$/div, $E = 100\,V\,cm^{-1}$/div).

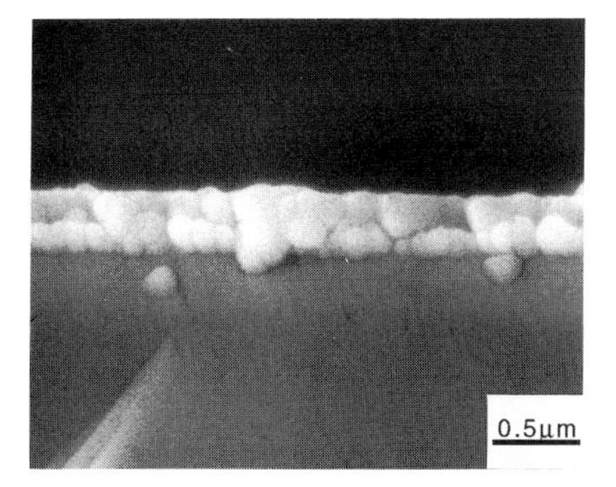

Figure 5 SEM micrograph of cross-section of Ca–PT film.

Figure 6 Microstructure of $Pb_{0.85}Ca_{0.15}TiO_3$ film fired at 650°C for 1 h.

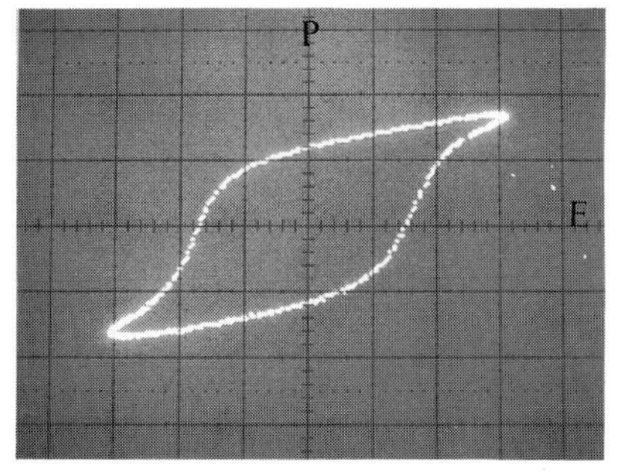

Figure 7 Typical *P-E* hysteresis loop of $Pb_{0.85}Ca_{0.15}TiO_3$ film fired at 650°C for 1 h. ($P = 18\,\mu C\,cm^{-2}$/div, $E = 55\,V\,cm^{-1}$/div).

Table 1
Electrical properties of $PbZr_{0.53}Ti_{0.47}O_3$ films prepared from starting sols containing 10 mol.% excess lead and fired at 700°C for 15 min.

Sample	0.5 μm films	0.25 μm films	
Ageing period	1 day	3 days	
Applied field (KV cm^{-1})	300	200	300
P_r (μC cm^{-2})	+15/−16	+21/−19	+26/−24
E_c (kV cm^{-1})	+85/−90	+90/−75	+120/−90
ε_r	850	800	800
D	0.13	0.17	0.17
P (Ω cm)	7×10^{11}	2×10^{12}	2×10^{12}

Table 2
Electrical properties of Ca-modified $PbTiO_3$ films prepared from starting sols containing <1 mol.% excess lead and fired at 650°C for 1 h.

Sample	$Pb_{0.9}Ca_{0.1}TiO_3$	$Pb_{0.85}Ca_{0.15}TiO_3$
Ageing period	7 days	7 days
Applied field (kV cm^{-1})	330	125
P_r (μC cm^{-2})	26	19
E_c (kV cm^{-1})	175	80

The microstructural features of the surface of the films are shown in Fig. 3; films were composed of rosette-type grains <1 μm in diameter existing in a fine grain matrix. The grain size of the latter was too small to be resolved by SEM.

The electrical properties of 0.25 μm and 0.5 μm PZT films made from sols containing 10 mol.% excess lead, and fired at 700°C for 15 min., are summarised in Table 1. The *P-E* response was measured 1 day after the gel to ceramic conversion for the 0.5 μm film and 3 days after the firing of the 0.25 μm film. We state these times because we have very recently noted an ageing effect which is under further investigation. An example of a typical slightly asymmetric hysteresis loop is shown in Fig. 4. Notwithstanding possible ageing effects, for starting sols containing 10 mol.% excess lead, at an applied field of 300 kV cm^{-1}, the thinner PZT film exhibited a higher remanent polarisation, $+P_r = 26\,\mu C\,cm^{-2}$, compared to 15 μC cm^{-2} for the 0.5 μm thick sample; the respective coercive fields $+E_c$ were ~120 kV cm^{-1} and ~85 kV cm^{-1}. These values represent the average for 4 dot electrodes; values varied over a range of <5%. We estimate the error in measuring film thickness and hence E_c to be ±10%. The measured relative permittivities were 850 for the 0.6 μm sample and 800 for the 0.25 μm sample; corresponding dissipation factors were 0.13 and 0.17. The ferroelectric properties of a 0.25 μm PZT film at an applied voltage of 5 V (200 kV cm^{-1}) are also included in Table 1.

The P_r and E_c values of the films were within the range quoted in a number of publications in the literature (see, for example, Refs 9, 10, 11). However the values of dissipation factor D for the films are much higher than expected, since values <0.05 are quoted by others.[9–13]

An examination of the effect on the value of D of changing key process variables, especially firing conditions, is currently under investigation.

3.2 Ca–PT

For both Ca–PT compositions tetragonal single-phase films were obtained after firing at 650°C for 60 min. The Ca–PT films were prepared from 0.4 M sols, and were ~0.3 μm thick as shown in Fig. 5.

The films exhibited microstructures, Fig. 6, similar to those which we have previously observed for unmodified PT films made from longer chain diols, e.g. butanediol.[8] A typical *P-E* hysteresis loop for a $Pb_{0.85}Ca_{0.15}TiO_3$ film is shown in Fig. 7. The P_r and E_c values, averaged for 6 dot electrodes, were 19 μC cm^{-2} and 80 kV cm^{-1}, respectively, whilst for a $Pb_{0.9}Ca_{0.1}TiO_3$ composition, P_r = 26 μC cm^{-2} and $E_c \sim$ 175 kV cm^{-1}. The P_r values are higher and the E_c values lower than those reported by Tsuzuki *et al.*[14] for ~1 μm Ca–PT films made by a methoxyethanol route and deposited on platinum metal substrates. For example they quote a value of P_r = 17 μC cm^{-2} and E_c = 230 kV cm^{-1} for a $Pb_{0.9}Ca_{0.1}TiO_3$ composition. However other workers[15] using rf magnetron sputtering reported values of P_r = 55 μC cm^{-2} and E_c = 120 kV cm^{-1} in highly *c*-axis oriented $Pb_{0.9}Ca_{0.1}TiO_3$ films.

4. CONCLUSIONS

Films of PZT and Ca–PT can be prepared using a relatively simple sol-gel processing route in which propanediol is used as the solvent medium. In this paper ~0.5 μm thick PZT and ~0.3 μm thick Ca–PT samples were produced from a single coating on platinised substrates.

Further refinements of the diol processing route are in progress in order to optimise electrical parameters.

REFERENCES

1. J. B. Blum and S. R. Gurkovich: *J. Mater. Sci.*, 1985, **20**, 4479–4483.
2. K. P. Budd, S. K. Dey and D. A. Payne: *Brit. Ceram. Proc.*, 1985, **36**, 107–121.
3. G. Yi, Z. Wu and M. Sayer: *J. Appl. Phys.*, 1988, **64**, 2717–2724.
4. Brit. Pat. Appl. 9114476.6, 1991.
5. N. J. Phillips and S. J. Milne: *J. Mater. Chem. Lett.*, 1991, **1**, 893–894.

6. N. J. Phillips, M. L. Calzada and S. J. Milne: *J. Non-Cryst. Solids*, 1992, **147 & 148**, 285–290.
7. M. L. Calzada and S. J. Milne: *J. Mater. Sci. Lett.*, 1993, **12**, 1221–1223.
8. M. L. Calzada, N. J. Phillips, A. Beitollahi, I. P. Wadsworth and S. J. Milne: *J. Am. Ceram. Soc.* (submitted).
9. C. D. E. Lakeman and D. A. Payne: *J. Am. Ceram. Soc.*, 1992, **75**, 3091–3096.
10. G. Yi, Z. Wu, M. Sayer, C. K. Jen and J. F. Bussiere: in *Ceramic Thin and Thick Films*, B. V. Hiremath ed., 363–374, Am. Ceram. Soc., 1990.
11. N. Tohge, S. Takahashi and T. Miami: *J. Am. Ceram. Soc.*, 1991, **74**, 67–71.
12. H. Hu, L. Shi, V. Kumar and S. B. Krupanidhi: in *Ferroelectric Films*, A. S. Bhalla and K. M. Nair eds, 113–120, Am. Ceram. Soc., 1992.
13. U. Selvaraj, K. Brooks, A. V. Prasadarao, S. Komarneni, R. Roy and L. E. Cross: *J. Am. Ceram. Soc.*, 1993, **76**, 1441–1444.
14. A. Tsuzuki, H. Murakami, H. Kani, K. Watari and Y. Torii: *J. Mater. Sci. Lett.*, 1991, **10**, 125–128.
15. E. Yamaka, H. Watanabe, H. Kimura, H. Kanaya and H. Ohkuma: *J. Vac. Sci. Technol.*, 1988, **A6**, 2921–2928.

Aqueous and Sol-Gel Synthesis of Submicron PZT Materials and Development of Tape Casting Systems for Multilayer Actuator Fabrication

D. HIND* and P. R. KNOTT†

*School of Materials, The University of Leeds, Leeds, LS2 9JT, UK
†Morgan Materials Technology Ltd, Stourport-on-Severn, DY13 8QR, UK

ABSTRACT
Novel routes for the chemical synthesis of lead zirconium titanate (PZT) materials have been investigated. The compositions selected were lanthanum-doped PZT, $Pb_{0.91}La_{0.09}(Zr_{0.65}Ti_{0.35})_{0.98}O_3$ (electrostrictive: PLZT 9/65/35), and neodymium-doped PZT, $Pb_{1.00}Nd_{0.02}(Zr_{0.55}Ti_{0.45})O_3$ (piezoelectric: PNZT 2/55/45).
Two routes were developed, namely, an aqueous synthesis using nitrate–alkoxide precursors and an organic synthesis using acetate–alkoxide materials. Multiple batch production of powder for each composition was achieved by each route, in particular materials of well-controlled and characterised crystallite sizes in the region 0.1–0.5 μm.
The development of laboratory doctor-blade tape casters and organic-based slip systems is described.
Characterisation of slips, according to viscosity and wetting properties, and of the resultant tapes by thickness, handleability and ease of release is discussed.
The fabrication of multilayer actuators produced from thin ($\leqslant$20 μm) tapes is outlined.

1. INTRODUCTION

Piezoelectric and electrostrictive actuators are of great interest as microdisplacement transducers for producing small movements ($\approx$0–100 μm) with high resolution (<50 nm), rapid response ($\approx$10 μs) and large forces ($\approx$1000–5000 N). They have the advantages of modest drive power, low noise and heat dissipation, compact size, low creep and backlash, and adaptability to severe environments, such as vacuum, cryogenic conditions and radiation fields. Consequently these devices are finding novel application in diverse areas such as automative, printing, optical communications, machine tools and high-speed machinery.

By fabricating devices using largely well-established multilayer (ML) technologies, drive voltages have been brought down to $\approx$75–150 V and below, for which high-stability power sources with good frequency response and dynamic range are available. This approach obviates the arcing and discharge problems characteristic of the high-voltage bulk piezoelectric devices. Most importantly, it enables unit costs to be cut dramatically by the use of mass-production techniques.

Requirements for ML actuators operating at such voltages (active ceramic layers of the order of 100 μm thick) are fairly readily satisfied by conventional mixed oxide processing. There are now however economic pressures to reduce layer thicknesses to around 20 μm and even as low as 5 μm, in order that direct drive can be achieved using low-cost microelectronic circuitry. Indeed this goal must be attained if the very large markets predicted for these devices are to be realised. Such thin layers imply the use of submicron size powders which can hardly be achieved by normal ceramic technology.

In this work two doped lead zirconate–titanate (PZT) materials were synthesised by alternative aqueous and organic chemical routes. The aim was to achieve homogeneous and sinter-active powders, with crystallite sizes in the range 0.1–0.5 μm, by methods which could be scaled up, and using precursors commercially available in bulk. Subsequently organic-based slips were developed and tapes cast to a target green thickness of 20 μm. After electroding by screen printing, tapes were cut, stacked in registration and laminated. Burnout and cofiring of prototype actuators were successfully demonstrated.

2. EXPERIMENTAL METHOD

2.1 Powder Synthesis

The materials chosen were lanthanum-doped PZT, composition $Pb_{0.91}La_{0.09}(Zr_{0.65}Ti_{0.35})_{0.98}O_3$ (PLZT 9/65/35), and neodymium-doped PZT of composition $Pb_{1.0}Nd_{0.02}(Zr_{0.55}Ti_{0.45})$ (PNZT 2/55/45). As is well known, PLZT 9/65/35 is electrostrictive in character,[1,2] while the PNZT is piezoelectric.[3,4]

Both compositions are lead-rich, to promote sintering. The PLZT, which is stoichiometric, has been formulated assuming there are B site vacancies; whether or not this is actually the case, the Pb-excess arises in a natural way. However the PNZT is non-stoichiometric: formulation is based on the conventional (and more likely) assumption that there are present A site vacancies, and the inherent Pb deficit is eliminated by adding an arbitrary amount of PbO (about 3 at.% Pb per formula unit). This question is discussed in Ref. 5.

A number of batches between 100 and 750 g in scale (as finished PLZT, PNZT powders) were prepared according to the two routes described below.

Inorganic chemical synthesis routes reported to yield powders of controlled particle size (0.1–0.5 μm) and of precise chemical composition were reviewed.[6–10] The selected route, referred to here as the 'nitrate route', is based on that outlined by Thompson[6] with modifications by one of the present authors. In this process a clear aqueous solution of PLZT or PNZT precursor is prepared using lead nitrate and lanthanum nitrate (or neodymium nitrate) salts plus zirconium nitrate and titanium isopropoxide solutions. Initial experimental evaluation demonstrated that, contrary to Ref. 6, the zirconium nitrate could not be dissolved to give an aqueous solution of the required concentration, namely 1 M (PLZT). The modification involved replacement of this by a commercially available high-purity aqueous solution of zirconium nitrate; this was selected in particular so that one of the prime objectives of the present work — scale up to commercial exploitation — could be met. Spray drying of the clear aqueous solution of PLZT/PNZT was carried out.

An organic synthesis route was also sought, proceeding via an intermediate sol or solution of all required elements, with hydrolysis to form a monolithic gel, thereby fixing chemical composition, and subsequent heat treatment to generate a submicron powder. The basic 'alkoxide–acetate' method for PZT and related compounds, described notably by Payne *et al.*[11–13] and Cross *et al.*[14,15] was chosen. In outline, lead acetate and, optionally, lanthanum or neodymium acetates are dissolved in 2-methoxyethanol (MOE) with refluxing and subsequent distillation to eliminate the water from the hydrated salts, which would otherwise bring about premature alkoxide hydrolysis. After cooling the required amounts of titanium and zirconium alkoxides (usually propoxides) and MOE are added. The liquid is refluxed again to promote final complexation and the excess solvent, byproduct alkyl acetates etc. are distilled off. Hydrolysis of the clear sol formed is often vigorous prior to gelation. Reaction rate is increased by base catalysis.

The scheme was simplified to some extent. The main modification was to use anhydrous lead subacetate in place of the trihydrated lead acetate. By this means the preliminary multiple distillations with additions of MOE were unnecessary, and stable sols could be made by refluxing alone. As reported previously,[11] lanthanum acetate was soluble only with difficulty in the presence of the lead salt. Gelation was conveniently achieved over a period of 1–2 h. Gels were then dried; at this stage they were quite friable, and easily reduced to fine powder by dry ball milling for a short period. Powders were calcined in open dishes.

In order to meet the primary particle size target — 0.1–0.5 μm — it was thought necessary to incorporate a milling operation for the nitrate route powders. The calcined materials were processed by air impact pulverisation (jet milling), using equipment fitted with a polyurethane liner which would not introduce any significant contamination. The alkoxide route powders after calcination were not milled but taken on to the next processing step (die pressing or slip preparation).

Scanning electron microscopy (SEM) was used to examine particle morphology before and after calcining. Thermogravimetric analysis (TGA) and differential scanning calorimetry (DSC) in conjunction with X-ray diffraction analysis (XRD) were used in order to follow decomposition of the dried intermediates and development of the perovskite phase. Particle sizing and specific surface area measurements (BET) provided physical characterisation of powders after calcining. Chemical analysis for the principal cations was done primarily by inductively coupled plasma (ICP) emission spectroscopy, supplemented by gravimetric (Pb, Zr) and ultraviolet absorption spectroscopy (Ti).

Sintering studies were carried out by die pressing 15 mm ϕ pellets of the milled PLZT/PNZT powders at up to 100 MPa and firing at temperatures between 1050 and 1200°C for between 1 and 6 h. In order to minimise loss of lead from the samples during sintering the atmosphere control technique illustrated in Fig. 1 was used.

Materials were characterised by density measurements, SEM of polished and etched sections, and dielectric and piezoelectric measurements.

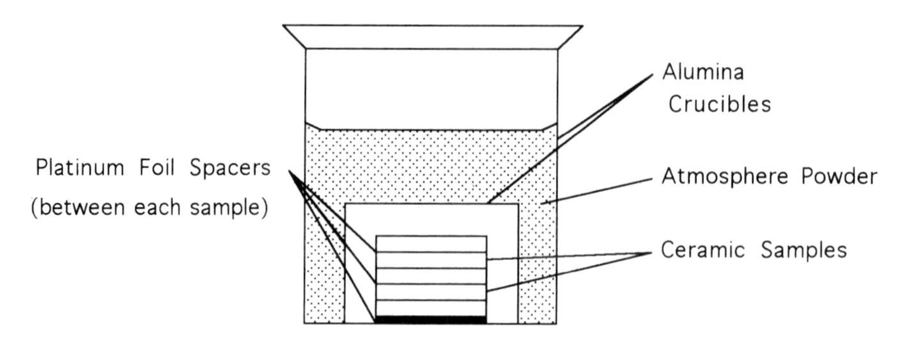

Figure 1 Schematic diagram of controlled atmosphere sintering technique.

2.2 Slip Development and Tape Casting

From a review of the literature[16–19] two organic binder/solvent-based tape-casting systems were selected: (a) polyvinyl butyral (PVB) base and (b) acrylic base. The slip systems were comprised of ceramic powder, organic solvents, dispersant, binder and plasticiser.

The first system was developed using a proprietary copolymer closely related to PVB, and a fatty acid ester as dispersant. In parallel, a slip system was developed using an acrylate–methacrylate copolymer as binder. This has good dispersing properties, making addition of a separate dispersant unnecessary. The required amounts of PLZT/PNZT powder, binder, plasticiser and solvents were blended and the slip prepared by ball milling (ZrO_2 media) for periods up to 72 h. A three-solvent mixture was used, simply to provide a range of drying rates.

Slip characterisation was done by computer-controlled rheometry. Specifically, measurements were made of shear stress/viscosity for shear rates up to 500 s^{-1}, and of viscosity at fixed shear rate as a function of dispersant level. Slip formulations were subsequently adjusted to achieve near-Newtonian behaviour.

Tape-casting trials were carried out using 6 mm-thick float glass plates, with slip dispensed from a hand-held paint film applicator. Difficulty was immediately found in releasing cast tape from the float glass. Subsequently, rubbing the glass surface with a dispersion of vegetable oil extract before casting proved a very satisfactory releasant for acrylic-based tapes. However for the PVB base, several different polymer films were tried as alternative casting surfaces to obtain the best results (wetting and release).

Tapes were assessed subjectively for general handling qualities — relative strength, plasticity, flexibility or brittleness — and further small adjustments were made to slip formulations.

2.3 Production of Thin Ceramic Tapes

Two identical fixed-bed laboratory tape-casters were used for parallel work on PVB- and acrylic-based slips. Each caster was built on an extremely stiff aluminium alloy frame. A lead screw was driven by a velocity servomotor; this provided the lead screw with a range of linear velocities, 0–45 mm s^{-1}. The screw in turn drove a carriage on which, as originally designed, was mounted a doctor blade head (single blade). Slips were to be cast on the 6 mm float glass plates. The blade gap was set using feeler gauges, with the glass surface as reference.

In the first casting trials, difficulty was experienced in producing tapes of repeatable or uniform thickness. A dispenser consisting of a PTFE box was made to meter slip to the doctor blade. The box was pushed along by the doctor blade head; an adjustable blade at the trailing edge presented a pool of slip of constant depth to the doctor blade.

This still did not result in adequate thickness control. The doctor blade was abandoned and replaced by a paint film applicator with gap heights of 25–200 μm in 25 μm steps. Though a precision instrument, this applicator is very simple in concept and design. A major advantage is that, because of the finite gap length, shearing action is imparted to the slip (Couette flow), resulting in thinner tapes compared with a sharp edge blade.

Acrylic-based slips were then cast very satisfactorily using the glass prepared as above. For the PVB base slips the other caster was modified to accommodate the preferred polymer film as the casting surface. This enabled the film, dispensed from 100 mm-wide rolls, to be supported by and tensioned along the float glass plate.

Tapes were cast to a target dry thickness of ~20 μm. They were assessed for thickness variation and green density. Thermal analysis of tape samples was carried out as an aid to determining burnout conditions in subsequent fabrication work.

2.4 Screen Printing and Laminating

The model actuator design (square type) shown in Fig. 2 incorporates electrodes 14 mm square with a 3 mm square re-entrant at one corner.

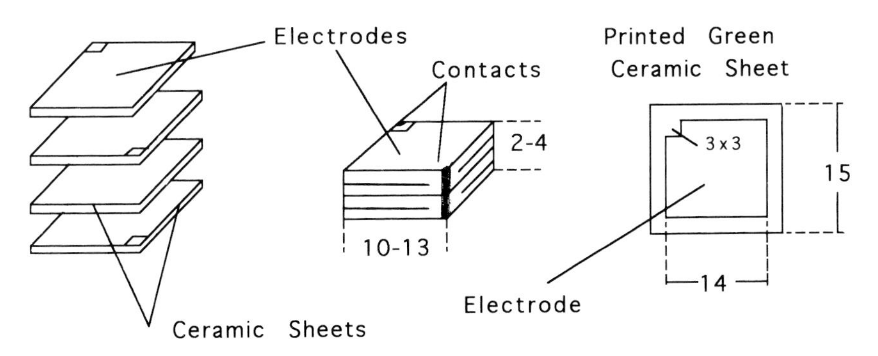

Figure 2 Schematic diagram of model actuator design (square type).

Screen printing of electrode inks on the green tapes was carried out using bench top manual and semi-automatic machines. They were equipped with 495 mesh polyester 45° screens, emulsion thickness 10 μm. These were supplied with exposed electrode patterns, generated in sets of four by standard photoresist techniques.

Pieces of tape were held in place for printing by means of a vacuum bed. Due to the sheer vulnerability of such thin pieces to any handling operation, electroding posed considerable problems. One particular difficulty arose because of tape deformation due to suction into the vacuum apertures; where these coincided with the exposed screen pattern, small holes resulted in the print layer. These problems were satisfactorily eliminated for the PVB base tapes by screening the pattern onto the tape before it was stripped from the polymer carrier.

Printed tapes were subsequently cut as required to fit 15 or 50 mm square laminating dies. Up to 50 layers were stacked in 180° registration and warm laminated.

2.5 Burnout and Cofiring

Experiments were carried out using PLZT, PNZT acrylic-base laminates only.

Binder burnout is a difficult procedure for ML stacks and especially for the thicker parts. Organic components have to be eliminated slowly and smoothly, which entails very slow ramp up in a non-oxidising atmosphere.

According to the thermal analysis of PLZT acrylic tape indicated above, most organic material is removed between 150 and 350°C; however some carbonised/part-pyrolysed material persists to ≈500°C. Above ~550°C PZT ceramics become sensitive to reducing atmospheres. Laminates 20–30 layers thick were accordingly burned out by heating, in a variety of atmospheres.

It is well known that Ag/Pd electrodes react at ~1130°C and above with PZT ceramics, due to alloying of Pd and Pb.[4] Reactions between this electrode system and various lead-base electroceramics have recently been investigated in detail.[20]

In sintering experiments therefore, burned-out PLZT, PNZT–Ag/Pd laminates were cofired at trial temperatures below 1135°C, using a similar atmosphere control technique to that employed for bulk ceramic test pieces (see Fig. 1).

Fired pieces were sectioned and examined for signs of delamination or any electrode–ceramic reaction, using optical microscopy, SEM and energy-dispersive X-ray analysis (EDX).

3. RESULTS

3.1 Powders

After drying, nitrate route powders were pure white and hygroscopic, while the alkoxide powders were typically a pale fawn colour.

TGA/DSC analysis of PLZT nitrate and alkoxide precursors to 600°C revealed quite different behaviour. The nitrate powder lost over 40% by weight in total: loss

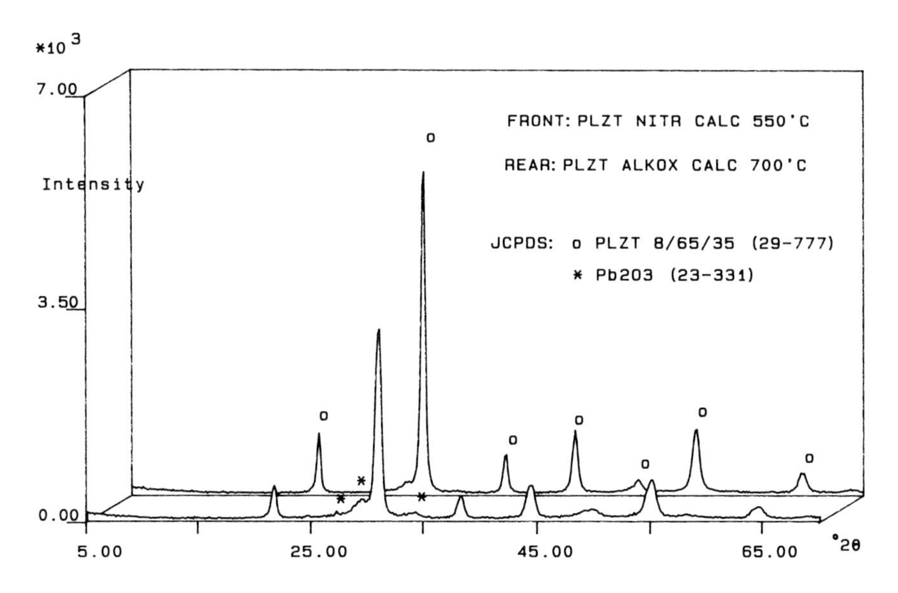

Figure 3 XRD spectra of calcined PLZT nitrate and alkoxide powders.

was continuous with temperature, and was essentially complete by 450°C. Endotherms at about 75, 130 and 230°C were presumably due to loss of absorbed/hydrated water. The large endotherm at 420°C is believed to represent one-stage nitrate decomposition. The alkoxide powder weight loss was ≈18–20% overall, and was complete or nearly so by 600°C. It occurred in two distinct stages — continuous to about 300°C, with a further small step at about 500°C. A large exotherm at ≈285°C was due to wholesale pyrolysis of the alkoxide complex, while two small ones around 500°C represented final decomposition of the organic residues.

XRD spectra were examined for samples calcined at different temperatures. Traces for the PLZT powders matched well with published data (JCPDS 29–777: PLZT 8/65/35 and 23–331: lead (sesqui) oxide Pb_2O_3); see Fig. 3. Some samples showed a reddish hue, presumably due to small amounts of the lead oxide, which were progressively eliminated with increase in calcination temperature. Traces for corresponding PNZT materials indicated even less residual lead oxide; in that sense they were perhaps better reacted. For instance, of alkoxide powders calcined at 700°C; the PLZT showed a little of the minor phase (see Fig. 3) while the PNZT showed none at all (i.e. perovskite phase alone). The reason for this slight but systematic difference is unclear.

SEM pictures showed no significant change in morphology on calcining, at least up to 700°C. Figure 4 shows micrographs of PNZT nitrate and alkoxide powders calcined at respectively 550° and 650°C. The nitrate-derived powder displayed spherical agglomerates, many apparently hollow and ranging from ≈0.2–5 μm in size, characteristic of spray drying, while the alkoxide showed agglomerates rather

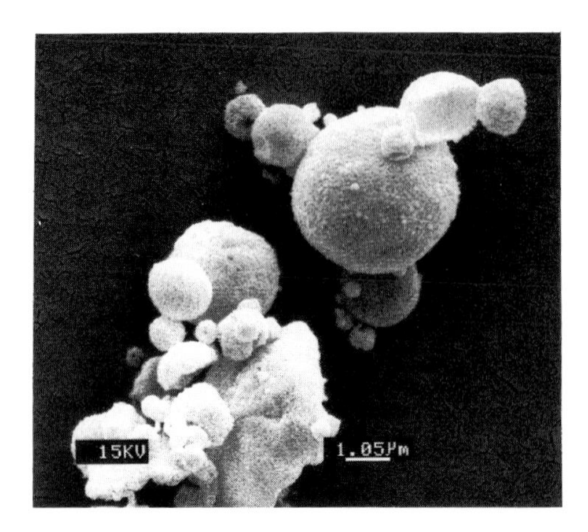

Figure 4 SEM micrographs of calcined PNZT nitrate (left) and alkoxide powders (right).

ill-defined in shape or size. Both samples revealed a ≈0.1–0.2 µm crystallite texture.

The specific surface area (BET) of the nitrate sample powder was 5.9 m^2/g, consistent with a soft-aggregate, dispersible powder, but that for the alkoxide was only 1.6 m^2/g, showing that some hard aggregation had occurred. Characterisation of the jet milled powder showed that, typically, material with a mean (d_{50}) size of about 0.5 µm and a mean equivalent primary particle size, calculated from BET data, of 0.073 µm could be obtained.

Chemical analysis of reacted PLZT samples for Pb, La, Zr and Ti demonstrated that, within a 5% error band, the formulation of the 9/65/35 composition was achieved.

3.2 Bulk Ceramics

Ceramic densities varied rather little with firing temperatures. Density values for the alkoxide route materials were however generally slightly lower and more variable than for the nitrate. For instance for pieces fired under the same conditions, PLZT nitrate route densities were 98.7–98.8% theoretical, but PLZT alkoxide route densities were 97.3–98.8% (3 results in each case). Again, for the same sintering conditions, PNZT nitrate density was 98.9% (1 result) and PNZT alkoxide densities were 96.8 and 97.3% (2 results).

Figures 5 and 6 show polished and etched sections for nitrate and alkoxide route materials, respectively. Mean grain sizes for the alkoxide pieces were 1.28 µm for PLZT and 0.69 µm for PNZT. It is interesting to note the similar PLZT/PNZT grain size relationship for both synthesis routes.

Some preliminary electrical measurements have been made. These include relative permittivity (εr) determinations at 1 kHz and room temperature. Typical values were as follows: for nitrate route, εr (PLZT) ≈7300 and εr (PNZT) ≈1550;

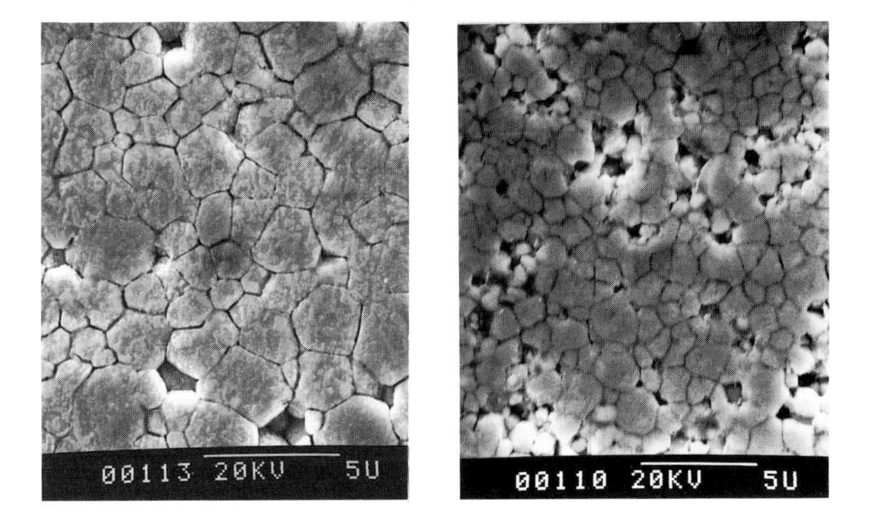

Figure 5 Microstructures of sintered nitrate route PLZT (left), PNZT (right).

for alkoxide route, εr (PLZT) ≈4900 and εr (PNZT) ≈1350. The discrepancies are particularly large for the PLZT and cannot be explained by the small density differences. Finally, piezoelectric measurements were carried out on alkoxide route PNZT samples fired at different temperatures. Typical low field values were: planar coupling factor $k_p \approx 0.62$, mechanical quality factor $Q_m \approx 65$ and charge constant $d_{33}^{B} \approx 400\,\mathrm{pCN^{-1}}$. One sample (SN9/5) yielded a high field d_{33} coefficient of $650\,\mathrm{pmV^{-1}}$ at $1\,\mathrm{kV\,mm^{-1}}$ (poling field $3\,\mathrm{kV\,mm^{-1}}$). These figures seem to compare very favourably with good commercial piezoelectrics (mixed oxide route materials).

3.3 Slips and Tapes

Typical formulations for PVB- and acrylic-based slips are given in Fig. 7. These were developed in order to achieve a working viscosity of ≈1–2 Pa.s and near-Newtonian behaviour up to shear rates of $500\,\mathrm{s^{-1}}$, as required. It is interesting to note that much higher ceramic solids could be accommodated in the acrylic-based system.

Thickness uniformity was excellent, being about ±1 μm variation in ≈20 μm over a cast area of 430 mm × 76 mm. Fig. 8 compares green tape properties for PVB and acrylic systems.

Thermal analysis in N_2 of PLZT tape samples revealed slight differences between PVB-based and acrylic-based systems. Weight loss from the PVB tape sample was complete by about 550°C with a large and diffuse exotherm around 400°C. The acrylic sample was burned out at a lower temperature (450°C) with a large but rather sharp exotherm, again at about 400°C. Smaller endotherms (230

(a)

(b)

Figure 6 Microstructures of sintered alkoxide route PLZT (a), (b) PNZT.

and 290°C, PVB-based; 230 and 330°C, acrylic-based) are probably due to evaporation of individual constituents.

3.4 Screen Printing and Lamination

The specific Pt and 70Ag/30Pd electrode inks used were procured on the advice of the respective manufacturers. No compatibility problems were experienced with the Pt ink and PVB-based tapes. Some reaction (wrinkling of tape under the wet ink) occurred between the Ag/Pd ink and acrylic-based tapes. The effect was

SLIP COMPONENTS	PVB SLIP VOLUME %	ACRYLIC SLIP VOLUME %
Ceramic PLZT / PNZT	15.0	30.0
Solvent	69.0	54.0
Plasticiser	4.0	4.5
Dispersant	3.0	-
Binder	8.6	11.5

Figure 7 Slip formulation — PVB and acrylic systems.

PROPERTY	PVB BASE VOLUME %	ACRYLIC BASE VOLUME %
Powder fraction V_p%	~39	~54
Residual organic fraction V_b%	40	28
Residual gas fraction V_g%	21	18
Drying rate	Slow	Fast
Uniformity of texture	Excellent	Very good
Relative mechanical strength	Fairly strong plastic	Strong, slightly brittle

Figure 8 Green PLZT/PNZT tape properties.

actually quite negligible for thin tapes ($\geqslant 60\ \mu m$), but for the thick layers being developed in this work there remained a slight puckering of the unelectroded areas after the ink had dried. This naturally made accurate registration very difficult.

Lamination was satisfactorily achieved with both tape systems. The somewhat higher pressures and temperatures needed to consolidate the acrylic-based tapes arose at least in part because of the rather higher-volume proportion of ceramic solids.

3.5 Burnout and Cofiring (Acrylic-Based Laminates)

Good results on burnout were finally achieved with all three non-oxidising atmospheres, although particular conditions sometimes resulted in partially carbonised or discoloured pieces.

Some early PLZT-, PNZT–Ag/Pd stacks were cofired. In all cases some delamination occurred, usually associated with the formation of metallic globules

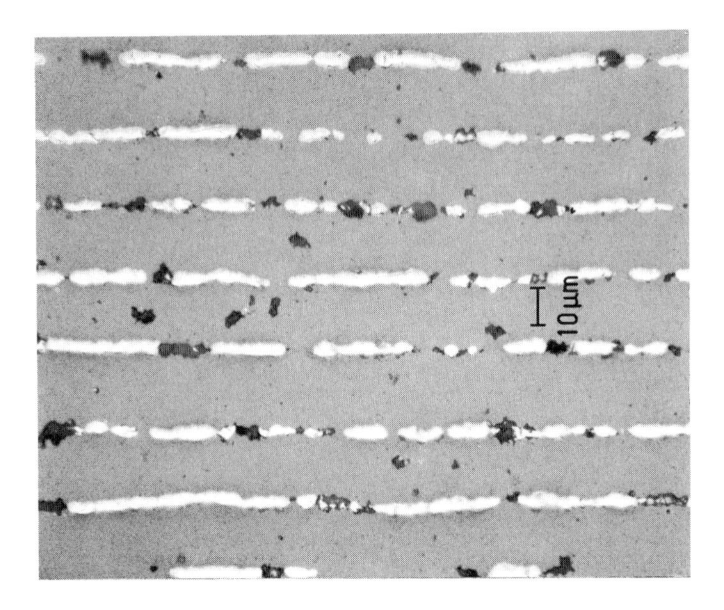

Figure 9 Polished section of PNZT-Ag/Pd stack (prototype actuator).

just discernible to the naked eye. EDX analysis showed these were likely to be Ag–Pd–Pb alloy.

Finally some stacks were fired at slightly lower temperatures without visible degradation. Figure 9 shows a polished section through one of these (PNZT–Ag/Pd stack). The electrode layers are relatively thin and incomplete but the ceramic–electrode bonding appears to be good.

The ceramic itself contains a number of fairly large (a few μm) and irregular pores due to the alkoxide powder being partly aggregated, as mentioned above.

4. CONCLUSION

The nitrate and alkoxide–acetate routes have been shown to yield homogeneous La- and Nd-doped PZT powders, of the required compositions, with primary particle sizes well down in the range 0.1–0.5 μm, and sintering to high densities ($\approx$97–99% theoretical). Both processes are amenable to scale up.

Nitrate route powders were evidently dispersible in this work, and it is probably unnecessary to have a milling operation separate from that required for slip preparation. As noted above, the alkoxide route powders showed some persistent hard aggregation, leading to $\approx\mu$m voidage in sintered parts and not quite such good consistency as the nitrate powders. Possibly this might be overcome with a better understanding of the hydrolysis process, and especially the aspect of catalysis.

The preparation of bulk materials by both routes with good electrical properties has been demonstrated, as has the fabrication of mechanically good multilayer stacks/prototype actuators with active ceramic layers ≈15–20 μm thick. It is hoped to carry out further device fabrication coupled with electrical testing in the future.

ACKNOWLEDGEMENTS

This work formed part of a BRITE/EURAM project (Contract No. BREU-0079), support for which from the Commission of the European Communities is appreciated. The authors wish to record their thanks to colleagues, especially Mr A. Searle for preparatory work on the alkoxide–acetate synthesis route, Mr R. Fries for assistance with formulation and rheological characterisation of slips, and Mr M. Rowlands for much general technical assistance. Finally PRK would like to thank the directors of Morgan Materials Technology Limited for permission to publish.

REFERENCES

1. G. H. Haertling and C. E. Land: *J. Am. Ceram. Soc.*, 1971, **54** (1), 1–11.
2. K. Furuta and K. Uchino: *Adv. Ceram. Mater.*, 1986, **1** (1), 61–63.
3. W. Rossner, K. Lubitz and G. Tomandl: *Silicates Industr.*, 1985, **3–4**, 31–34.
4. W. Wersing, H. Wahl and M. Schnöller: *Ferroelectrics*, 1988, **87**, 271–294.
5. G. H. Haertling: in *Ceramic Materials for Electronics*, R. C. Buchanan ed., Marcel Dekker Inc, New York, 1986.
6. J. Thompson: *Am. Ceram. Soc. Bull.*, 1974, **53** (5), 421–433.
7. J. A. Little and P. C. Yao: *Communication Am. Ceram. Soc.*, 1984, C29–31.
8. Y. Yoshikawa *et al.*: *J. Mat. Sci.*, 1988, **23**, 2729–2734.
9. R. Lal *et al.*: *Ferroelectrics*, 1989, **100**, 43–55.
10. Y. Yoshikawa and K. Tsuzuki: in *Euro Ceramics 2*, 2229–2303, Proc. 1st European Ceram. Soc. Conf., Maastricht, June, 1989.
11. K. D. Budd, S. K. Dey and D. A. Payne: *Brit. Ceram. Proc.*, 1985, **36**, 107–121.
12. K. D. Budd, S. K. Dey and D. A. Payne: in *Better Ceramics Through Chemistry II*, Proc. MRS Symp., vol. **73**, C. J. Brinker, D. E. Clark and D. H. Ulrich eds, 711–716, Materials Research Society, Pittsburgh, PA, 1986.
13. S. D. Ramamurthi and D. A. Payne: *J. Am. Ceram. Soc.*, 1990, **73** (8), 2547–2551.
14. Z. Q. Zhuang, M. J. Haun, S.-J. Jang and L. E. Cross: *Adv. Ceram. Mater.*, 1988, **3** (5), 485–490.
15. Z. Q. Zhuang, M. J. Haun, S.-J. Jang and L. E. Cross: *IEEE Trans. Ultrason. Ferroelectr. and Freq. Contr.*, 1989, **36** (4), 413–416.
16. *Advances in Ceramics, Vol. 26, Ceramic Substrates and Packages for Electronic Applications*, M. F. Yan, *et al.* eds, 525–541, American Ceramic Society, 1989.
17. R. R. Landham *et al.*: *Am. Ceram. Soc. Bull.*, 1987, **66** (10), 1513–1516.
18. J. C. Williams: in *Treatise on Materials Science and Technology, Vol. 9, Ceramic Fabrication Processes*, F. F. Y. Wang ed., 173–198, Academic Press, New York, 1976.
19. J. S. Reed: in *Introduction to the Principles of Ceramic Processing*, 395–399, Wiley Interscience, 1988.
20. S.-F. Wang and W. Huebner: *J. Am. Ceram. Soc.*, 1993, **76** (2), 474–480.

Dielectric Properties and Ageing of Fe-doped PZT Ceramics Prepared by the EDTA-Gel Method

H.-W. WANG, D. A. HALL and F. R. SALE
Materials Science Centre, University of Manchester, Grosvenor Street, Manchester, M1 7HS, UK

ABSTRACT
Fine-grained Fe-doped lead zirconate titanate (PZT) ceramics were prepared by the EDTA-gel method. Uniform microstructures with grain sizes in the range 0.6–3.7 μm were obtained at sintering temperatures of 1150–1280°C. Ferroelectric hysteresis measurements showed that the fine-grained specimens exhibited a much smaller loop area than the larger-grained specimens, giving much reduced values for the remanent polarisation (4 $\mu C/cm^2$ for the 0.6 μm grain size specimen compared with 18 $\mu C/cm^2$ for the 2.4 μm grain size specimen). The ageing of dielectric properties for both unpoled and poled materials was measured as a function of temperature, from 30 to 120°C. It was found that the ageing effects became more pronounced with an increase in grain size and decreased with an increase in temperature for unpoled samples. The observed ageing behaviour of poled samples was more complex. The results are discussed in terms of the development of an effective internal bias field associated with the stabilisation of the ferroelectric domain walls during ageing.

1. INTRODUCTION

The term 'ageing' is used to indicate the gradual changes which occur in the properties of ferroelectric materials as a function of time in the absence of either external mechanical or electrical influences, or temperature changes. Ageing is a reversible process in that an aged sample can be de-aged by the application of an a.c. field higher than the coercive field E_c for an extended time (known as 'hysteresis relaxation') or by heat treatment at temperatures higher than the Curie point (thermal de-ageing). New ageing processes initiate after the de-ageing process has finished. It has been found that ageing effects become more pronounced as the concentration of oxygen vacancies increases.[1] In lead-based perovskite ceramics, such as PZT, the excess charge introduced by aliovalent dopants is usually compensated by the formation of either lead vacancies (in the case of donor-doped materials) or oxygen vacancies (for acceptor-doped materials). Therefore, the ageing effects in PZT are reduced by donor dopants and increased by acceptors.

Various models have been proposed to describe the ageing behaviour in $BaTiO_3$ and PZT ceramics, as reviewed by Schulze and Ogino.[2] Of these, it has been suggested[3–9] that the dominant effect in $BaTiO_3$ and PZT is a 'volume' stabilisation of the ferroelectric domain structure by the ordering of acceptor ion–oxygen vacancy ($A''_{Ti} - V_O^{\cdot\cdot}$) defect associates. This model was described by Lambeck and

Jonker[4] and further developed by Arlt *et al.*[5–9] to give a quantitative treatment of the ageing and de-ageing processes.

The principle underlying the model is that the defect associate constitutes an electric and elastic dipole which can be reoriented with respect to the spontaneous polarisation in a given domain in order to reduce the electric and elastic interaction energies. The dipole reorientation can occur by the oxygen vacancy 'jumping' to one of the adjacent sites in the octahedron surrounding the acceptor ion. The observed ageing effects are then attributed to a reduction in the 90° domain wall mobility which leads to a reduction in the domain wall contribution to the dielectric and piezoelectric coefficients.[5–9]

The ageing of the dielectric constant can be expressed for ageing time t as:

$$\varepsilon(t) = \varepsilon(0) - A \log t \tag{1}$$

where $\varepsilon(0)$ and $\varepsilon(t)$ are the dielectric permittivity measured immediately after de-ageing and at time t after ageing, respectively. A is the absolute decrease in one decade, called the absolute ageing rate. The relative ageing rate is expressed as:

$$R = \frac{A}{\varepsilon(0)} \times 100 \tag{2}$$

The unit of relative ageing rate is % per decade. Both the absolute ageing rate and the relative ageing rate are used to characterise the ageing behaviour.[11] It has been found that ageing is a thermally activated process, the activation energies for the ageing process of acceptor-doped $BaTiO_3$ ceramics being of the order of 1.1 eV, which is said to be related to the jumping of the oxygen vacancies between adjacent sites having different energies.[6]

There is a distinct relation between the ageing of the low-field dielectric and piezoelectric coefficients and the presence of an effective internal bias field E_i, which is evident in high field *P-E* hysteresis measurements. As the dielectric properties decrease during ageing, so the internal bias field builds up at the same time.[2–9] The ageing behaviour may be attributed to the domain wall clamping effect which is associated with the internal bias field. The value of E_i depends on the type of dopants used and is generally found to increase with an increasing dopant concentration.[3,4] This is attributed to a dependence on the concentration of oxygen vacancies and thus the concentration of defect associates present. As the dipoles re-orient to become parallel to the spontaneous polarisation, the electric and elastic interaction energies are lowered and the internal bias field becomes established.

For materials which have been aged in the unpoled state, with a random domain structure, the internal bias field causes the *P-E* hysteresis loop to become constricted in the centre, as shown in Fig. 1. Such a loop can be viewed as comprising two sub-loops, which are shifted by $\pm E_i$ with respect to the origin. Here, the presence of an internal bias field opposes changes in polarisation in either sense. Materials which have been aged in a state of remanent polarisation exhibit a deformation and shift of the hysteresis loop along the field axis (Fig. 2). In thie case, the internal bias field acts to stabilise the domain configuration associated with the polarised state and opposes changes in polarisation in the opposite sense.

The two definitions of E_i are not equivalent, since they refer to different states of polarisation. However, for a given material it is usually assumed that they arise from the same physical phenomenon and therefore that they should show similar dependencies on time, temperature, and dopant concentration.

Empirical rules for the time dependence of E_i during ageing and de-ageing have been given as:[3,6]

$$E_i(t) = A \log t + B \text{ (ageing)} \quad (3)$$

and

$$E_i(t) = E_i(0) \cdot \exp(-t/\tau) \text{ (de-ageing)} \quad (4)$$

Lohkamper *et al.*[6] noted that the build-up of E_i in Ni-doped $BaTiO_3$ ceramic was substantially lower at 90°C relative to that at 50°C. On the other hand, Carl and Hardtl[3] showed that the build-up of E_i in Mn-doped PZT was greater at 90°C than at 50°C. Both results[3,6] were obtained for ceramics aged in the poled state. The apparent contradiction between these two results has not been fully explained. It may be expected that the ageing processes occur more rapidly at higher temperatures as found by Carl and Hardtl,[3] while Lohkamper *et al.*[6] interpreted their results in terms of a reduction in the ultimate, or 'saturation', value of E_i due to a reduction in the energy difference between the sites occupied by the oxygen vacancies (i.e. the splitting energy ΔW in Ref. 6).

Although the ageing characteristics of $BaTiO_3$ and PZT are now well established, there are still uncertainties regarding the effects of grain size and temperature on ageing behaviour. The present study was carried out to investigate these effects in Fe-doped PZT ceramics based on fine powder derived from the EDTA-gel method.

2. EXPERIMENTAL PROCEDURE

The detailed preparation procedure for the PZT–EDTA precursors has been described previously.[12] In the present case, the composition chosen for investigation was $(Pb_{1.00}Sr_{0.05})[(Zr_{0.53}Ti_{0.47})_{0.97}Fe_{0.03}]O_3$ (i.e. containing 5 mol.% excess PbO), this being representative of 'hard' PZT ceramics.[1] The appropriate metal nitrate solutions were mixed in stoichiometric proportions and then gradually added to the EDTA solution while the pH was maintained at a value of 5 by the addition of ammonia solution. The liquid was dried in a vacuum oven at 80°C in order to produce the desired precursor. The as prepared precursor was calcined at 800°C for 4 h in air in order to remove the organic component and form the required oxide powder. The calcined powder was then ball milled for 4 h in a polypropylene bottle using MgO-stabilised ZrO_2 balls and 99.8% ethanol as the milling liquid. Green compacts of the PZT powder were prepared by uniaxial die-pressing at a pressure of 250 MPa. Sintering was carried out at 1150–1280°C for

2–8 h in a closed Al_2O_3 crucible in order to produce specimens with various grain sizes. A PbO-rich atmosphere was produced by enclosing a $PbZrO_3$ compact within the crucible. Grain sizes were determined using the line intercept technique from polished, chemically etched sections, observed by SEM. The crystal structure were determined by X-ray diffraction using a Philips PW1710 powder diffractometer.

Electrodes were applied to disk-shaped specimens (1 mm in thickness and 12 mm in diameter) using fired-on silver paint (DuPont 7095). The dielectric properties were measured as a function of ageing time at a series of fixed temperatures using a HP 4284A precision LCR meter. The specimens were heated to a temperature over 450°C for more than 2 h prior to the measurement to ensure complete thermal de-ageing (T_c = 330°C for the investigated composition). The measurements of dielectric permittivity and loss were carried out immediately after cooling the specimen from 450°C (≈2 min) to the ageing temperature (30, 60, 90 and 120°C) for a period up to 1000 min. Similar measurements were carried out on polarised specimens after first de-ageing under a high a.c. field at the ageing temperature. High-temperature dielectric measurements (up to 450°C) were carried out in an alumina tube furnace using a Wayne–Kerr B6425 LCR meter.

Hysteresis measurements were carried out using a 1 Hz triangular waveform, produced by a Thurlby Thandar TG 1304 Function Generator, which was amplified by a factor of 1000 by a HVA1B high-voltage amplifier (Chevin Research, Otley, UK). The subsequent waveform (±5 kV) was then applied to the test specimens held in a heated silicone oil bath. The induced current was converted to a measurable voltage (±10 V) using an I–V converter, both voltage and current being monitored by an Opus PC via an Amplicon PC30B interface card. Integration of the measured current yielded the electric charge and hence the polarisation of the specimen at a given time. After first de-ageing by the application of a continuous high a.c. field, two complete cycles of the waveform were applied at each chosen time during subsequent ageing. The generator output was switched off between these measurements in order to minimise any reduction in E_i caused by hysteresis relaxation. This procedure allowed the determination of E_i for polarised specimens as a function of time at a number of different ageing temperatures (30, 60, 90 and 120°C). For the unpoled (thermally de-aged) specimens, E_i was determined by applying a single measurement cycle after ageing for a period of 20 h. In each case, the values of E_i were determined from the positions of the peaks in the *I–E* curves, as indicated in Figs 1 and 2.

3. RESULTS

3.1 Microstructure and Initial Hysteresis Measurements

Table 1 lists the grain sizes and densities of the specimens chosen for the ageing measurements. The materials were found to have a tetragonal structure with a $c:a$ ratio of 1.016. The microstructures of these materials, as observed by SEM, are shown in Fig. 3. It is clear that there was a second phase (liquid phase) at the grain

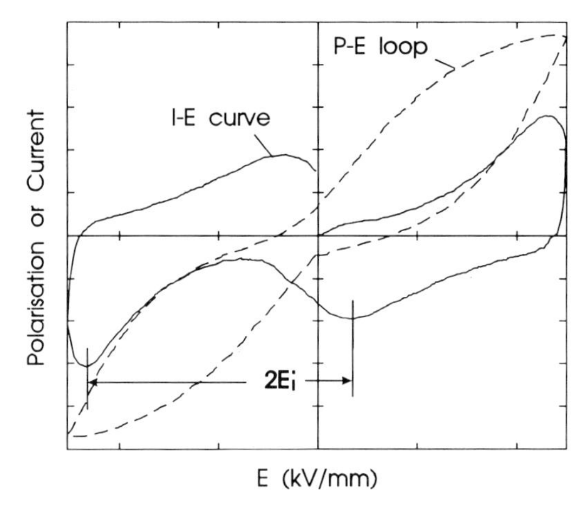

Figure 1 Typical high-field hysteresis characteristics for unpoled, aged PZT ceramic showing definition of effective internal bias field E_i.

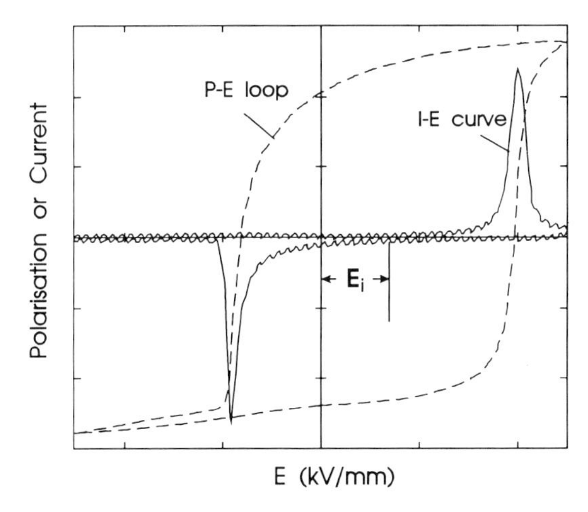

Figure 2 Typical high-field hysteresis characteristics for poled, aged PZT ceramic showing definition of effective internal bias field E_i.

Table 1
Summary of physical and high field hysteresis characteristics of Fe-doped PZT ceramics prepared under various sintering conditions

Sintering temperature	Sintered density	Grain size (μm)	E_c (kV/mm)	P_r (μC/cm^2)
1150°C/2 h	93.5%	0.6	1.0	4
1200°C/2 h	97.5%	0.8	1.0	6
1250°C/8 h	96.2%	2.4	1.2	18
1280°C/8 h	93.0%	3.7	1.5	16

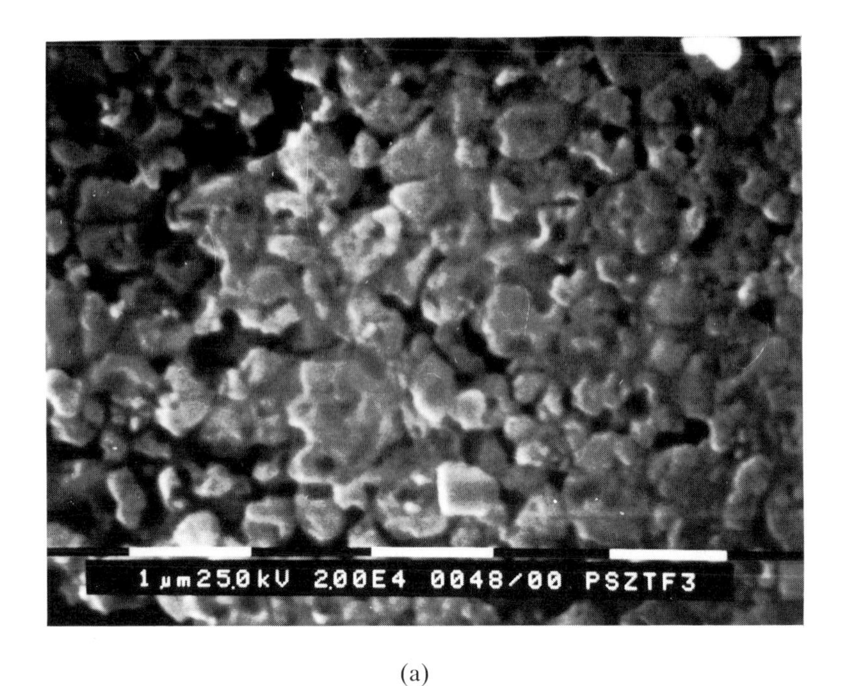

(a)

(b)

Figure 3 SEM micrographs of specimens sintered at (a) 1150°C/2 h (b) 1200°C/2 h (c) 1250°C/8 h (d) 1280°C/8 h.

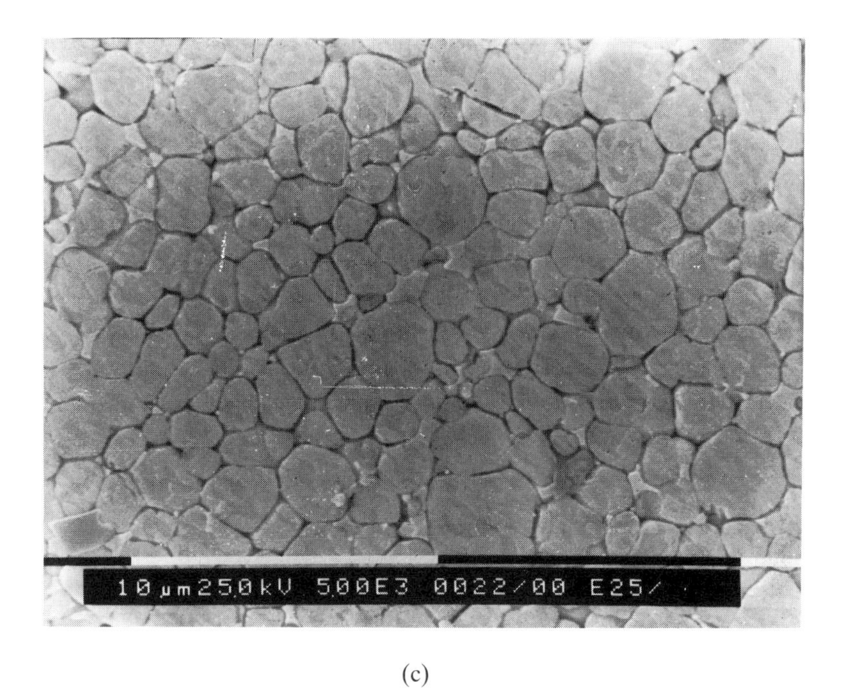

(c)

(d)

Figure 3—*contd.*

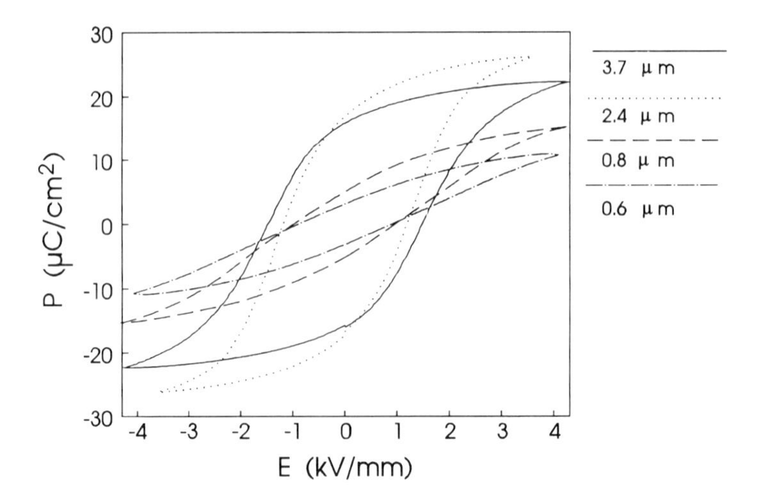

Figure 4 *P–E* hysteresis loops for Fe-doped PZT ceramics with various grain sizes, measured at 20°C.

boundaries which became more clearly evident as the sintering temperature increased. This is particularly pronounced for the sample sintered at 1280°C. EDX analysis showed that this second phase was mainly enriched in Fe, with some Mg and Al also being present. The minor impurities, Mg and Al, are thought to be due to contamination from the ball-milling process or from the raw materials. These elements are known to act as acceptors for PZT ceramics and are assumed to exert little influence on the resulting properties of Fe-doped matcrials.

Figure 4 shows the hysteresis loops obtained for these four specimens at 20°C after complete de-ageing under the high a.c. field. The larger-grained specimens yielded larger and more open loops and thus higher P_r values than those obtained for the fine-grained materials. Both fine-grained materials gave smaller loops with much reduced values for the remanent and saturation polarisation, as shown in Table 1. Slightly lower P_r and higher E_c values were obtained for the specimens sintered at 1280°C (3.7 μm) than those sintered at 1250°C (2.4 μm), where it is apparent that the presence of substantial amounts of the continuous liquid phase resulted in an increase in E_c and a reduction in P_r. The lower value of P_r might arise from the reduction in density. However, the associated increase in E_c can only be explained in terms of a low permittivity grain boundary phase, which effectively reduces the electric field applied to the ferroelectric grains. It is clear that switching of the polarisation occurred more freely in the large-grained materials. These results are similar in many respects to those reported for $BaTiO_3$ ceramics[10] where it has been found that internal stress gives rise to much less mobile domain structures. In the present work, examination by TEM revealed clear domain structures in the larger-grained materials (Fig. 5(a)) but less evidence of domains within the grains in the fine-grained materials (Fig. 5(b)). Instead, irregular bands were observed within the grains which were identified as strain contours. Therefore, the slip-loop behaviour of the fine-grained materials may be

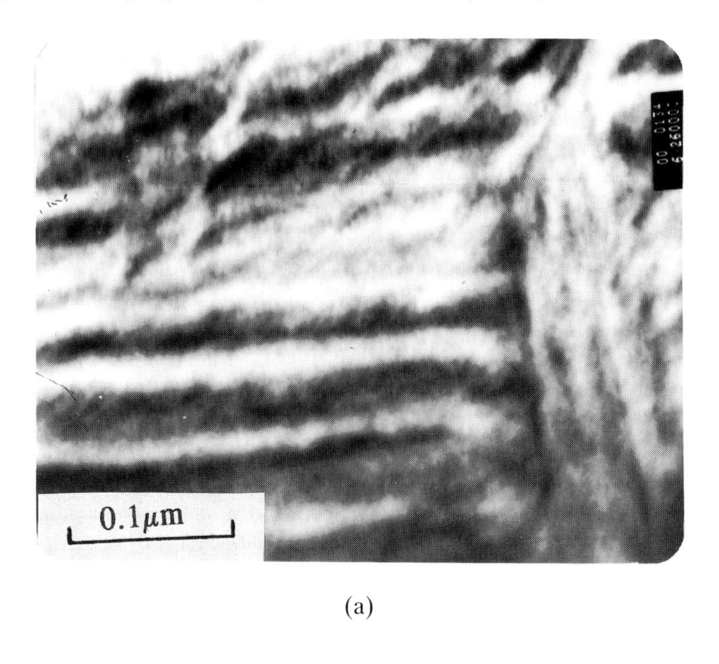

(a)

(b)

Figure 5 TEM micrographs of Fe-doped PZT ceramics with grain sizes of (a) 2.4 μm and (b) 0.8 μm.

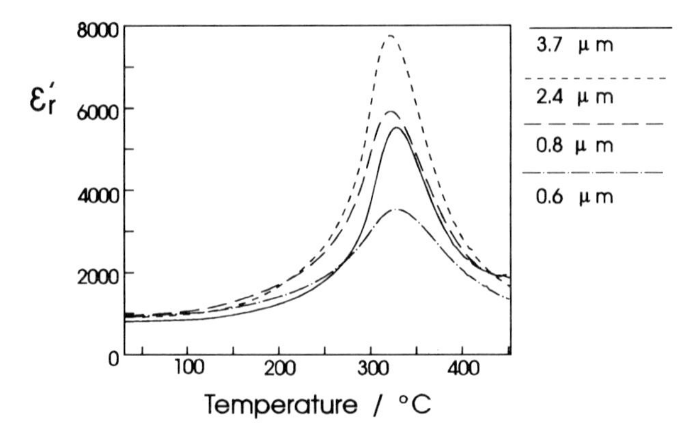

Figure 6 Temperature dependence of permittivity for Fe-doped PZT ceramics with various grain sizes.

attributed to a low incidence of ferroelectric domains and/or a reduction in the domain wall mobility.

3.2 *Dielectric Properties*

Figure 6 shows the permittivity vs temperature characteristics obtained for the four chosen specimens. It is evident that the fine-grained specimens (0.6 and 0.8 μm) gave broad peaks at 330°C corresponding to the tetragonal (ferroelectric) to cubic (paraelectric) phase transition. The larger-grained materials gave more well-defined peaks at approximately the same temperature. These results are similar to those reported for barium titanate ceramics in which fine-grained materials exhibit a broadened phase transition which is thought to be due to internal stress.[13] The specimen sintered at 1280°C shows a substantially lower permittivity in the ferroelectric region, which again can be related to the microstructure of this specimen which shows a continuous second phase at the grain boundaries.

The effects of temperature on the ageing of these specimens in the unpoled state were investigated. Typical results for the 0.8 and 2.4 μm grain size specimens are shown in Table 2, together with calculations of the average absolute and relative ageing rates. It is evident that the ageing process became less pronounced at the higher temperatures for all the materials investigated, if the measurements were carried out on the specimens in the unpoled state. In addition, larger-grained materials gave higher ageing rates than those of the fine-grained materials.

It is interesting to note that the above results did not hold for poled specimens. For a well-de-aged (by high a.c. field) 2.4 μm grain size specimen, the ageing of permittivity was slightly faster at 60°C and 90°C than at 30°C, as shown by the results presented in Table 3. It is evident that these results did not exhibit as clear a trend as was observed for the unpoled specimens. This may be due in part to the experimental procedure, where it was necessary to remove a specimen from the

Table 2
Average ageing rates of unpoled Fe-doped PZT ceramics measured at 100 kHz

T/°C	0.8 μm, A	0.8 μm, R	2.4 μm, A	2.4 μm, R
30	35.57	3.56%	39.57	3.76%
60	33.88	3.23%	41.26	3.75%
90	26.29	2.41%	34.82	3.03%
120	17.07	1.45%	30.81	2.43%
150	14.20	1.05%	22.27	1.54%

Table 3
Average ageing rates of poled Fe-doped PZT ceramics measured at 100 kHz

T/°C	0.8 μm, A	0.8 μm, R	2.4 μm, A	2.4 μm, R
30	43.3	3.96%	38.6	4.25%
60	40.3	3.61%	44.1	5.34%
90	24.3	2.10%	42.5	4.26%
120	8.4	0.68%	27.5	2.92%

heated oil bath, which was used for the high-field de-ageing process, and transfer it to the dielectric testing apparatus. Clearly, it would have been preferable to carry out the dielectric ageing experiments within the oil bath, thus avoiding any intermediate thermal excursions. Nevertheless, the ageing characteristics of the poled specimens do appear to be significantly different from those of the unpoled specimens.

3.3 Internal Bias Field

Figure 7 shows the high-field hysteresis results obtained for the unpoled 2.4 μm grain size samples after ageing for 20 h at 30, 60, 90 and 120°C. The values of E_i, as found from the separation of the two peaks in the *I–E* curves, were approximately 1.50, 1.35, 1.30 and 1.25 kV/mm for ageing temperatures of 30, 60, 90 and 120°C, respectively. It is clear that the effective internal bias field at a given time was lower at higher ageing temperatures for unpoled specimens.

For poled specimens, de-aged by hysteresis relaxation, it was possible to measure the hysteresis characteristics as a function of ageing time, as noted above. Typical results obtained for the 2.4 μm grain size specimens are shown in Figs 8 and 9. It is clear that the specimen aged at 120°C exhibited a more pronounced ageing effect than that obtained at 30°C. Therefore, the differences obtained in the ageing of dielectric properties of specimens in the poled or unpoled state are also reflected in the high-field ageing characteristics.

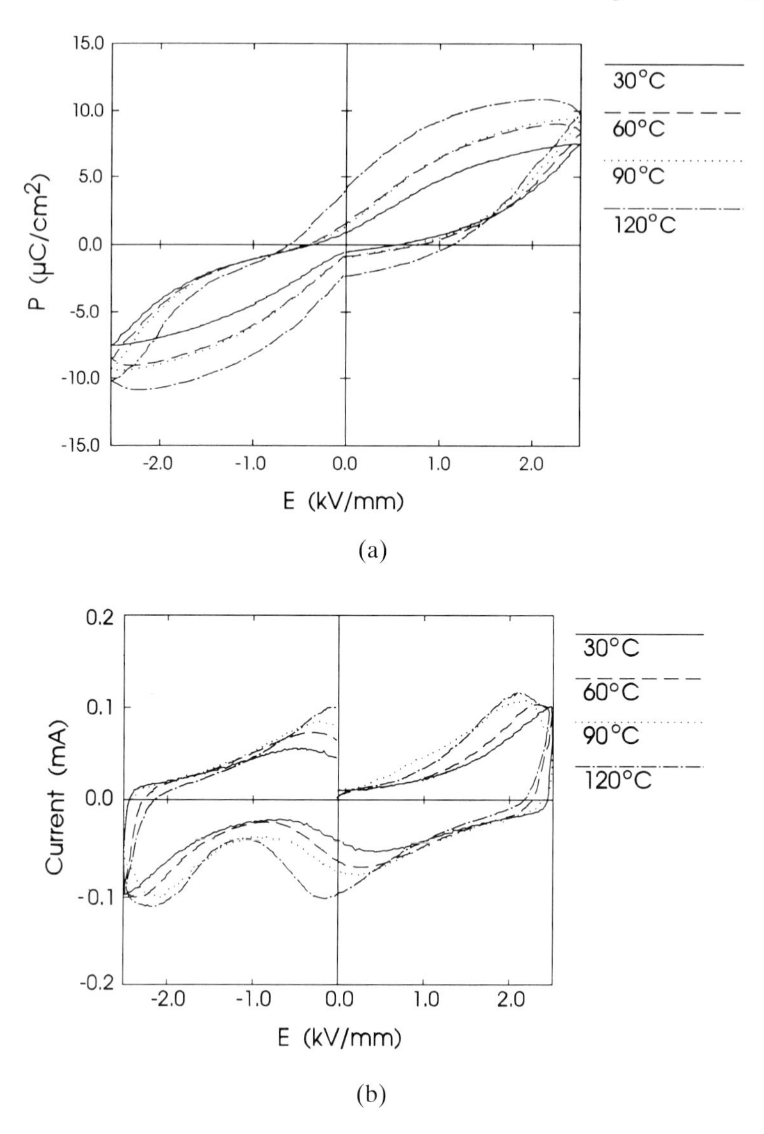

Figure 7 High-field hysteresis results for thermally de-aged 2.4 μm grain size specimen after ageing for 20 h at various temperatures. (a) *P–E* and (b) *I–E* curves.

The build-up of E_i as a function of time for the poled 2.4 μm grain size specimen, at various ageing temperatures, is presented in Fig. 10. It is apparent that the ultimate value of E_i (after ageing for 1000 min) was greater at higher ageing temperatures. However, the rate of increase of E_i appears to show little dependence on temperature, the most pronounced effect being the apparently discontinuous jump which occurred within the first minute of ageing. This effect

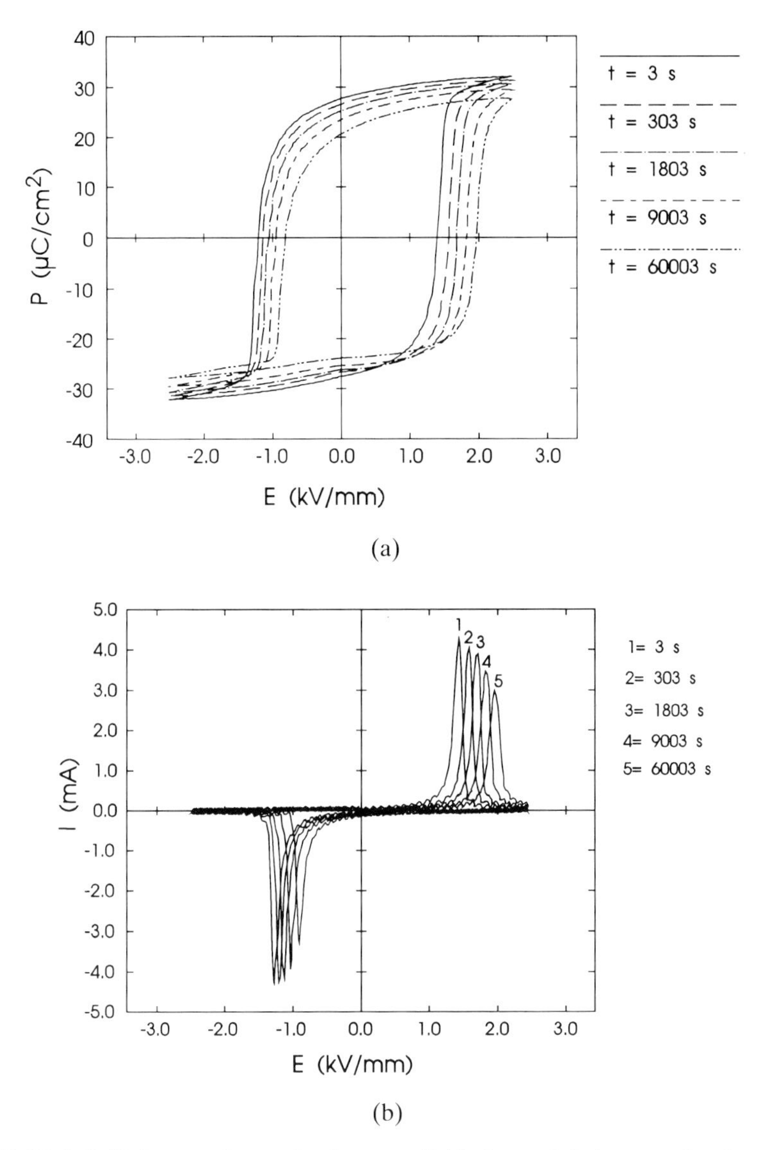

Figure 8 High-field hysteresis results for a.c. field de-aged 2.4 μm grain size specimen during ageing at 30°C. (a) *P–E* and (b) *I–E* curves.

may be related to a different ageing mechanism from that responsible for the gradual ageing process, since the ageing rate is so much faster.

The measurement of E_i for poled specimens was not so straightforward for the fine-grained materials. Firstly, it was found to be difficult to de-age the materials by hysteresis relaxation, this process often taking several hours as compared with just a few minutes for the larger-grained materials. Secondly, the ageing measurements

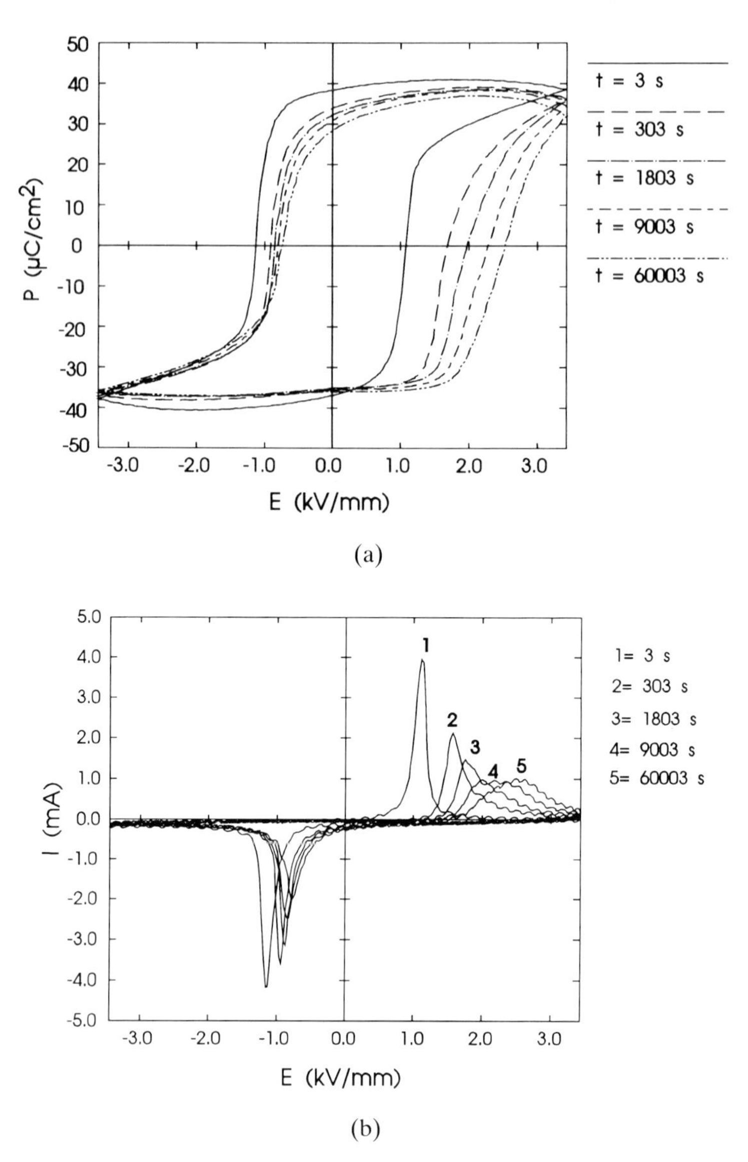

Figure 9 High-field hysteresis results for a.c. field de-aged 2.4 μm grain size specimen during ageing at 120°C. (a) *P–E* and (b) *I–E* curves.

showed the development of a constricted loop, characteristic of an unpoled specimen, which made it difficult to define a single value of E_i as shown in Fig. 11. This behaviour may be explained most simply in terms of the low value of P_r for the fine-grained materials, which reflects the almost random nature of the ferroelectric domains. It is also interesting to note that the constriction in the loop developed very rapidly (as shown in Fig. 11), which again seems to indicate two different ageing mechanisms.

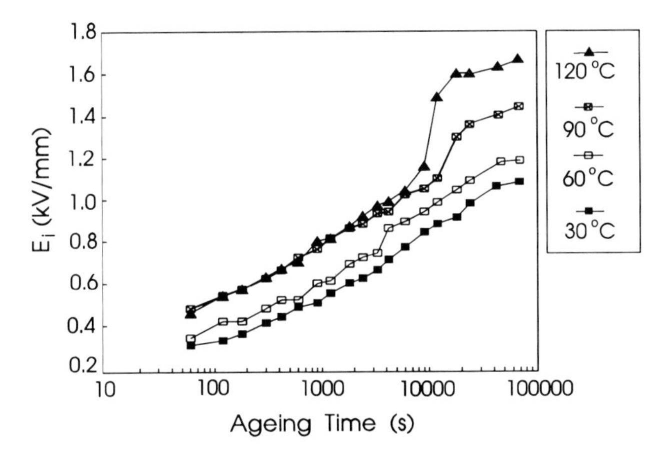

Figure 10 Build-up of internal bias field for a.c. field de-aged 2.4 μm grain size specimen during ageing at various temperatures.

4. DISCUSSION

It is well known that domain wall motion contributes to both the low-field dielectric permittivity and the high-field polarisation switching behaviour in ferroelectrics.[2–11] Two effects which will affect the contribution of domain wall motion need to be clarified, i.e. the grain size effect and the temperature effect.

4.1 The Effect of Grain Size on Ageing Behaviour

The domain structures observed by TEM for the fine-grained and larger-grained specimens were presented in Fig. 5. There is less evidence of well-defined domains in the fine-grained specimens, which may be attributed to the fine grain size (0.8 μm) and small tetragonality ($c{:}a$ ratio = 1.016). For the larger-grained specimens, the hysteresis loops were much larger, giving clear evidence for a larger contribution from domain wall motion to the 'switchable' polarisation for such materials, at least for high a.c. field strengths. During ageing, the larger-grained specimens showed a greater ageing rate for the dielectric properties due to the more pronounced reduction in the domain wall contribution to ε, caused by the stabilising effect of defect dipole re-orientation.

The larger-grained materials were rapidly de-aged by the application of a high a.c. field, this process taking up to 10 min at 30°C and only 30 s at 120°C. They also exhibited well-defined shifts of the hysteresis loop along the field axis, allowing the effective internal bias field E_i to be determined as a function of ageing time. This shift was accompanied by a slight deformation of the loop, resulting in reduced values for the saturation and remanent polarisation.

The increase in E_i during ageing was characterised by a discontinuous jump during the first few seconds, followed by a more gradual increase up to 1000 min,

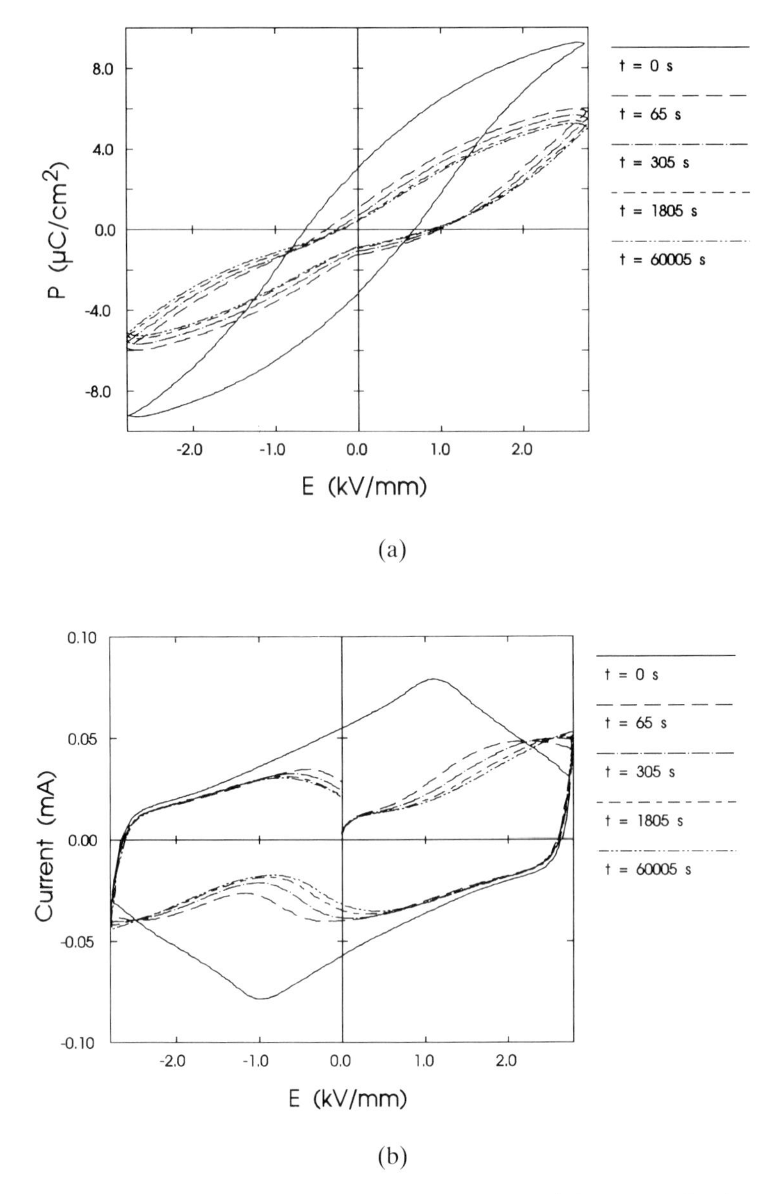

Figure 11 High-field hysteresis results for a.c. field de-aged 0.8 μm grain size specimen during ageing at 120°C. (a) *P–E* and (b) *I–E* curves.

as is evident from Fig. 10. This gradual ageing process may be attributed to domain stabilisation by the reorientation of $Fe'_{Ti} - V_O^{\cdot\cdot}$ defect pairs, as reported elsewhere.[4] It is likely that the early ageing behaviour may originate from a different effect, since the associated time constant appears to be so much shorter.

The hysteresis results for the fine-grained materials (Fig. 11) also showed a very pronounced change during the first minute of ageing, the most obvious effect here being a constriction of the hysteresis loop similar to that observed for aged, unpoled materials. It is clear from the observed slim loop behaviour that the polarisation in the fine-grained materials was not easily re-orientated. This, together with the TEM observations, indicates that one of the major factors opposing the domain switching is internal stress, which is present in the unpoled materials and which must be increased under the influence of the applied electric field. For these materials, there appears to be a very strong tendency to revert to an almost random domain structure on removal of the applied field, as indicated by the relatively low values of P_r. Therefore, it may be suggested that the cause of the early ageing behaviour in both large- and fine-grained materials is a relief of the internal stress by 90° domain switching.

4.2 The Effect of Temperature on Ageing Behaviour

For the unpoled specimens, the ageing of low-field dielectric properties and the internal bias field present after ageing were less pronounced for ageing at higher temperatures. Two effects must be considered in order to explain this behaviour.

The ageing processes (for example defect dipole re-orientation) are thermally activated and therefore are expected to occur more rapidly at higher temperatures. However, the ultimate value of E_i should be reduced at higher temperatures, due to thermal randomisation of the defect dipole directions, which will cause the internal bias field to reach a lower 'saturation' value at short ageing times. For the unpoled materials, it was observed (Fig. 7) that E_i was reduced at higher temperatures and therefore it may be concluded that the thermal randomisation effect is dominant in this case.

For the poled materials, the continuous increase in E_i as a function of ageing time, and the more evident ageing of the low-field dielectric properties at higher temperatures (60 and 90°C for the 2.4 μm grain size specimen) must mean that the thermal randomisation of the defect dipole orientation is relatively weak within the range of temperatures investigated. It may be concluded that for specimens aged in the poled state, electric and elastic interactions occur between defect dipoles in separated domains through the bulk of the material. In other words, the internal bias field is self-reinforcing in poled specimens, whereas in unpoled specimens the development of E_i within a given domain occurs in isolation from its neighbours. Therefore, the thermal randomisation effect is more pronounced in unpoled specimens, giving rise to the observed differences in ageing characteristics.

The ageing behaviour of the fine-grained materials can also be understood in these terms, since the low value of P_r means that the domain structure is almost random, giving rise to ageing characteristics similar to those of the unpoled materials. The reported differences in the temperature dependence of E_i for poled $BaTiO_3$ and PZT ceramics may thus be explained on the basis of the lower remanent polarisation in $BaTiO_3$ ceramics, which means that the thermal randomisation of the defect dipole orientation is also dominant in $BaTiO_3$.

5. CONCLUSIONS

The effects of grain size and temperature on the ageing characteristics of Fe-doped PZT ceramics were investigated. Evidence was found to suggest the existence of two different ageing mechanisms, with different associated time constants. The early ageing behaviour was thought to be a result of stress relief by 90° domain switching, while the more gradual longer-term ageing process was attributed to domain stabilisation by defect dipole re-orientation. It was observed that for specimens aged in the unpoled state, the ageing effects were less pronounced at higher temperatures, while the ageing of poled specimens exhibited a much less marked dependence on temperature. The effect of temperature on the ageing characteristics of the unpoled specimens may be understood in terms of thermal randomisation of the defect dipole orientations, which results in a reduction in the ultimate, or 'saturation', value of the effective internal bias field E_i. For poled specimens, the internal bias field within each domain is oriented in the same general direction. This appears to result in a cooperative effect which acts to stabilise E_i against thermal randomisation, at least within the temperature range investigated in the present study.

ACKNOWLEDGEMENTS

The authors would like to thank Professor G. Arlt for valuable discussions and encouragement during the present work. The financial assistance of The Royal Society and The Nuffield Foundation are also gratefully acknowledged.

REFERENCES

1. B. Jaffe, W. R. Cook and H. Jaffe: *Piezoelectric Ceramics*, Academic Press, London, 1971.
2. W. A. Schulze and K. Ogino: *Ferroelectrics*, 1988, **87**, 361–377.
3. K. Carl and K. H. Hardtl: *Ferroelectrics*, 1978, **17**, 473–486.
4. P. V. Lambeck and G. H. Jonker: *Ferroelectrics*, 1978, **22**, 729–731.
5. G. Arlt and H. Neumann: *Ferroelectrics*, 1988, **87**, 109–120.
6. R. Lohkamper, H. Neumann and G. Arlt: *J. Appl. Phys.*, 1990, **68**, 4220–4224.
7. H. Dederichs and G. Arlt: *Ferroelectrics*, 1986, **68**, 281–292.
8. H. Neumann and G. Arlt: *Ferroelectrics*, 1987, **76**, 303–310.
9. U. Robels and G. Arlt: *J. Appl. Phys.*, 1993, submitted.
10. G. Arlt: *Ferroelectrics*, 1987, **76**, 451–458.
11. K. Wu and W. A. Schulze: *J. Amer. Ceram. Soc.*, 1992, **75**, 3390–3395.
12. H. W. Wang, D. A. Hall and F. R. Sale: *J. Amer. Ceram. Soc.*, 1992, **75**, 124–130.
13. D. Hennings: *Int. J. High Tech. Ceram.*, 1987, **3**, 91–111.

Preparation of PLZT Powder by a Citrate Gel Technique

M. A. AKBAS and W. E. LEE
University of Sheffield, Dept. of Eng. Materials, Sheffield, UK

ABSTRACT
Homogeneous PLZT ceramic powder can be prepared from citrate source solution using a gelation process which avoids alcohol dehydration involving large volumes of expensive ethanol. Cl^- is a critical impurity in the gel evaporating during calcination as $PbCl_2$, rendering the powder PbO deficient and halting formation of PLZT. However, removal of the Cl^- enables single-phase PLZT to be formed at temperatures as low as 375°C, crystallisation being caused by the oxidation of carbon arising from the citrate organics.

1. INTRODUCTION

To obtain lanthanum doped lead zirconate titanate (PLZT) ceramics with high transparency and uniform electro-optic coefficients, chemically homogeneous starting powders having small particle size and high purity are needed. Various chemical routes have been used to achieve these goals including coprecipitation of alkoxides in the presence of PbO, coprecipitation of alkoxides with H_2O, coprecipitation of clear aqueous nitrate and chloride solution, preparation of PLZT powder by an oxalate or carbonate method, and alcohol dehydration of aqueous citrate solutions.[1–6]

Techniques for the preparation of multication aqueous citrate solutions and a coprecipitation process based on alcohol dehydration are known[7] and were first applied to production of chemically homogeneous and fine $BaTiO_3$ powders. Later, the same techniques were used to produce PLZT and PZT powders.[6,8] In the alcohol dehydration process, droplets of a PLZT source solution are atom sprayed into an alcohol bath to obtain PLZT citrate salt. The relative volumes of alcohol and source solution are critical; there must be at least 10 times greater volume of alcohol than solution to maintain the amount of water in the bath at a low level. The resulting powder can be filtered out from the bath, and after drying, calcined to obtain PLZT.

Unfortunately, the alcohol dehydration technique is impractical because of the large volumes of expensive alcohol required which cannot be reused without purification, and also because of the susceptibility of the resulting salt to moisture absorption.

The present study is concerned with the preparation of fine PLZT powder by a citrate gel technique to avoid the necessity of alcohol dehydration. The gelation of an aqueous citrate solution of PLZT, and its decomposition under various conditions, is described.

2. EXPERIMENTAL PROCEDURE

2.1 Preparation of PLZT Citrate Solution and its Gelation

Lead oxide (PbO, >99.9%), lanthanum oxide (La_2O_3, >99.99%), zirconium oxychloride octahydrate ($ZrOCl_2.8H_2O$, >99%) and titanium butoxide ($Ti[O(CH_2)_3CH_3]_4$, >99%) were the starting materials used in the chemical processing. Citrate solutions of each of the cations in PLZT i.e., La, Pb, Zr and Ti were prepared by mixing their respective oxides and hydroxides with citric acid solution and using ammonia gas to regulate the pH so as to obtain a clear solution.[7] The individual citrate solutions were then combined to obtain source solutions of PLZT 8, 9, 10/65/35 with excess Pb^{+2} to provide 3, 5, 7 and 10 wt% excess PbO to the composition after calcination. The PLZT source solutions contained Cl^- arising from the $ZrOCl_2.8H_2O$ used in the preparation of zirconium citrate solution. The chlorine in the zirconium citrate solution was removed by repeated precipitation of ammonium zirconyl citrate with large volumes of ethyl alcohol, washing the precipitate and redissolving in distilled water until there was no reaction with photographic paper. This Cl^- free zirconium solution was then used to produce Cl^- free PLZT source solution.

The resulting PLZT solutions were gelled by holding at 60°C for 16 to 20 h while magnetically stirring. The starting pH of the solutions was around 7. However, the pH gradually decreased below 6 due to ammonia loss at the gelation temperature, and a white precipitate formed in the solution at these low pH values. The precipitate was redissolved by adjusting the solution pH to 8 to 8.5 with ammonia solution. This process yielded translucent to transparent citrate gel of PLZT.

2.2 Decomposition of the Resulting Gel and Characterisation of the Product

Decomposition of the resulting gel was studied by DTA and TG (Du Pont thermal analyst 2000) at a heating rate of 10°C/min. To identify the intermediate phases formed, powders quenched after holding for 4.5 h at various temperatures, were analysed by XRD (Philips diffractometer equipped with 1710 diffractometer control unit). Consequently, direct comparisons between the thermal analysis temperatures and those for XRD are not possible. The morphology of the resulting powder was examined by scanning electron microscopy (SEM, Camscan series 2).

Early calcination experiments were conducted by holding the chlorine-containing gel for 3 h at 550°C in air. The resulting highly agglomerated powder was further calcined for 3 h at 650°C in an oxygen atmosphere to increase the homogeneity and to remove any residual carbon (arising from citric acid) in the powder (Method I).

This route yielded a very PbO-deficient powder and a technique was therefore developed to control the carbon combustion rate, and hence local excess heating of the powder during calcination. The PLZT citrate gel was first heated to 300°C for 1 h yielding a homogeneous black ash which was ground and reheated to 600°C at 10°C/min in nitrogen and held for another hour. The nitrogen atmosphere was then diluted by passing oxygen through the furnace and the powder further calcined for 4 h (Method II).

3. RESULTS AND DISCUSSION

3.1 Preparation and Decomposition of PLZT Citrate Gels

The citrate route yielded a clear and colourless solution of each individual component of PLZT. No precipitation was observed when individual solutions were mixed together to obtain PLZT source solution of the desired composition. After gelation, the colour of the gel varied between purple, milky brown and cloudy white for solutions prepared at different times. Although the solution chemistry is beyond the scope of this paper, this colour change may be attributed to slight changes in cation to citric acid ratio. No colour difference was observed between the Cl^--containing and Cl^--free gel, prepared from the same source solutions. All the solutions and gels were stable even after 7 months' storage in air-tight plastic containers.

TG (Fig. 1a and b) indicates that PLZT citrate gels decompose in two separate stages. Previous work by Courty *et al.*[9] recognised the existence of two types of pyrolysis for the citrate–nitrate precursors. Type I was characterised by a continuous and vigorous reaction which occurred with precursors containing metals with a strong catalytic activity in the oxidation process. Type I decomposition is characterised by a single-stage decomposition in TG. Type II is typified by a two-stage process in which an intermediate decomposition step occurs as a result of the formation of a metastable semidecomposed precursor which is thought to be a mixed citrate salt.

3.1.1 Decomposition of chloride-containing citrate gel

TG and DTA (Fig. 1a) indicate that the first stage of decomposition after the removal of water starts at around 120°C (endotherm *a* on Fig. 1a), and continues until 180°C. This part of the decomposition was associated with a 20% weight loss. Samples quenched from 160°C indicated a yellow ochre intermediate phase (Table 1) which was amorphous (Fig. 2a). Removal of ammonia and break-up of some citrate chains is believed to occur at this stage.[10] The second stage of decomposition is indicated by an endothermic peak at 180°C (endotherm *b* on Fig. 1a) and associated loss of 25 wt% in a relatively small temperature range of 180–225°C. Samples quenched from 200°C were dark brown to coffee coloured and still amorphous (Table 1 and Fig. 2a). Most decomposition of the citrate chains occurs at this stage. From 225 to 500°C, gradual weight loss is detected due to break-up of any remaining citrate chains. Samples quenched from 375°C were black and still amorphous (Table 1 and Fig. 2a) indicating complete break-up of citrate chains and the presence of carbon as a decomposition product. DTA indicated a strong exothermic reaction at 525°C (exotherm *d* on Fig. 1a) corresponding to vigorous oxidation of the remaining carbon. Samples quenched from 505, 600, 700 and

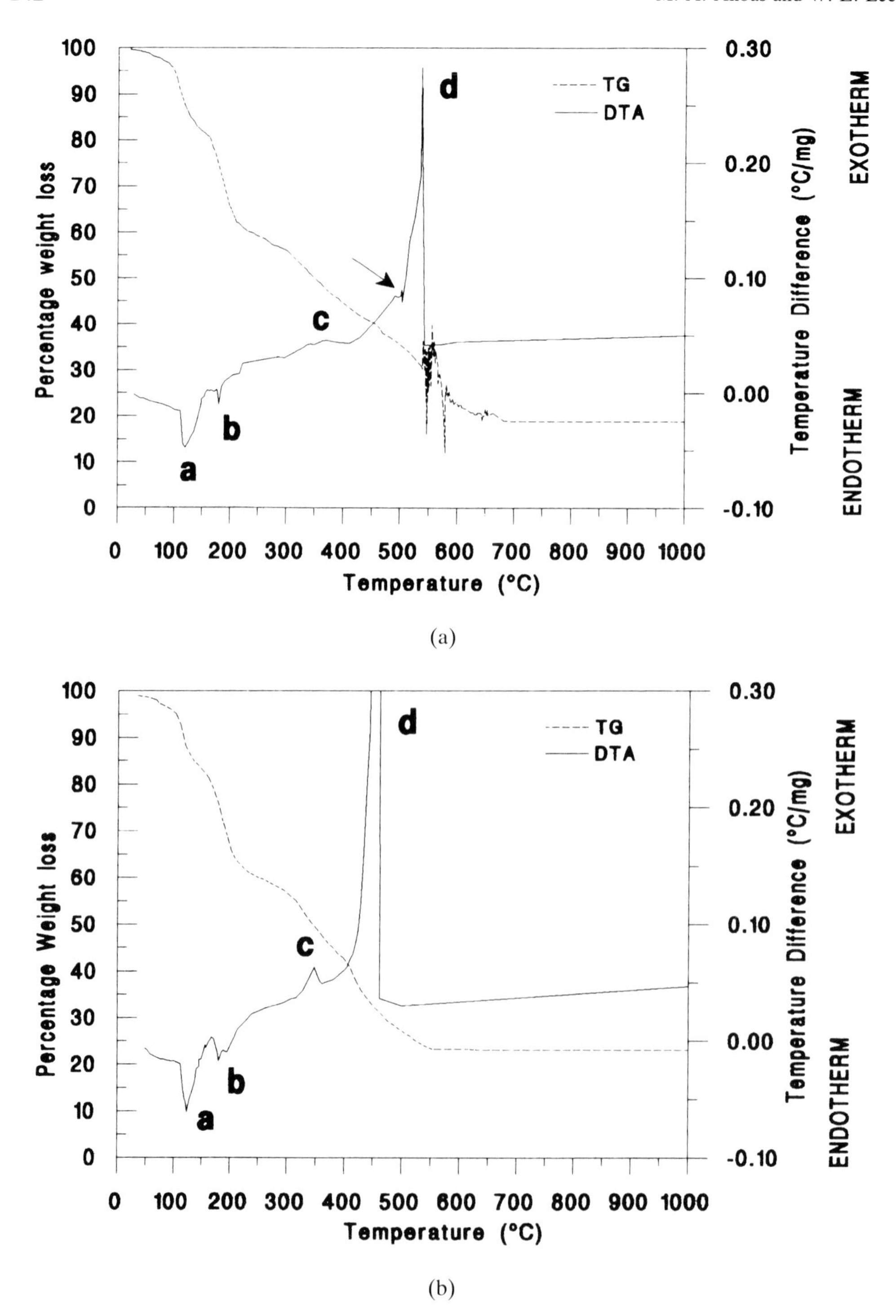

Figure 1 DTA/TG curves of citrate gels (a) Cl^--containing (b) Cl^--free.

Table 1
Appearance and phase analysis of samples quenched from various temperatures. Samples were held isothermally for 4.5 h at temperature.

Sample type and quenching temperature °C		Appearance	Phases present	Comments
160	Cl^--containing	yellow ochre	amorphous	120°C–180°C
	Cl^--free	yellow ochre	amorphous	Loss of H_2O and ammonia (some citrate break-up)
200	Cl^--containing	dark brown	amorphous	180°C–225°C
	Cl^--free	dark brown	amorphous	Decomposition of citrate chains
375	Cl^--containing	black	amorphous	Burning of carbon and
	Cl^--free	yellow	PLZT	crystallisation
505–800	Cl^--containing	white	$PbTiO_3$, $PbZrO_3$	Burning of carbon and
	Cl^--free	yellow	PLZT	crystallisation

800°C were white and crystalline (see section 3.2.1). This indicates that crystallisation of the powder is favoured by the heat produced by oxidation and occurred simultaneously. Although the onset temperature for this reaction was determined as 525°C by DTA, samples quenched from 505°C indicated that the reaction could start at lower temperatures if they were held isothermally for long times. No further DTA peaks were observed after 525°C and the weight of the sample remained constant from 600 to 1000°C.

3.1.2 Decomposition of chloride-free citrate gel

A typical two-stage decomposition was also observed in Cl^--free citrate gel (Fig. 1b). However, vigorous oxidation initiated at 420°C, as shown by the strong exotherm (exotherm *d* on Fig. 1b), a temperature almost 100°C lower than in the chloride citrate gel. Moreover, as indicated by Table 1, samples quenched from 375°C were yellow and crystalline. This indicates that the ignition temperature of carbon occurs at even lower temperature if held isothermally. The carbon was completely burned out during this oxidation, since little weight loss was observed for temperatures higher than 500°C.

The decomposition of chloride-containing citrate gel was not completed until a much higher temperature than was necessary for Cl^--free citrate gel. The reasons for this behaviour are not obvious. However, it is possible that chlorine bonds to one of the elements in the gel which would otherwise catalyse the ignition. As will be discussed further in the following sections, formation of $PbCl_2$ or $PbOCl_2$ may hinder Pb oxidation which itself ignites the carbon oxidation at low temperatures. Although the broadening and shift of the Pb oxidation exotherm (370°C, exotherm *c* on Fig. 1)[6] to higher temperatures for chloride-containing citrate gel supports this theory, no such intermediate crystalline phases ($PbCl_2$ or $PbOCl_2$) have been detected by XRD.

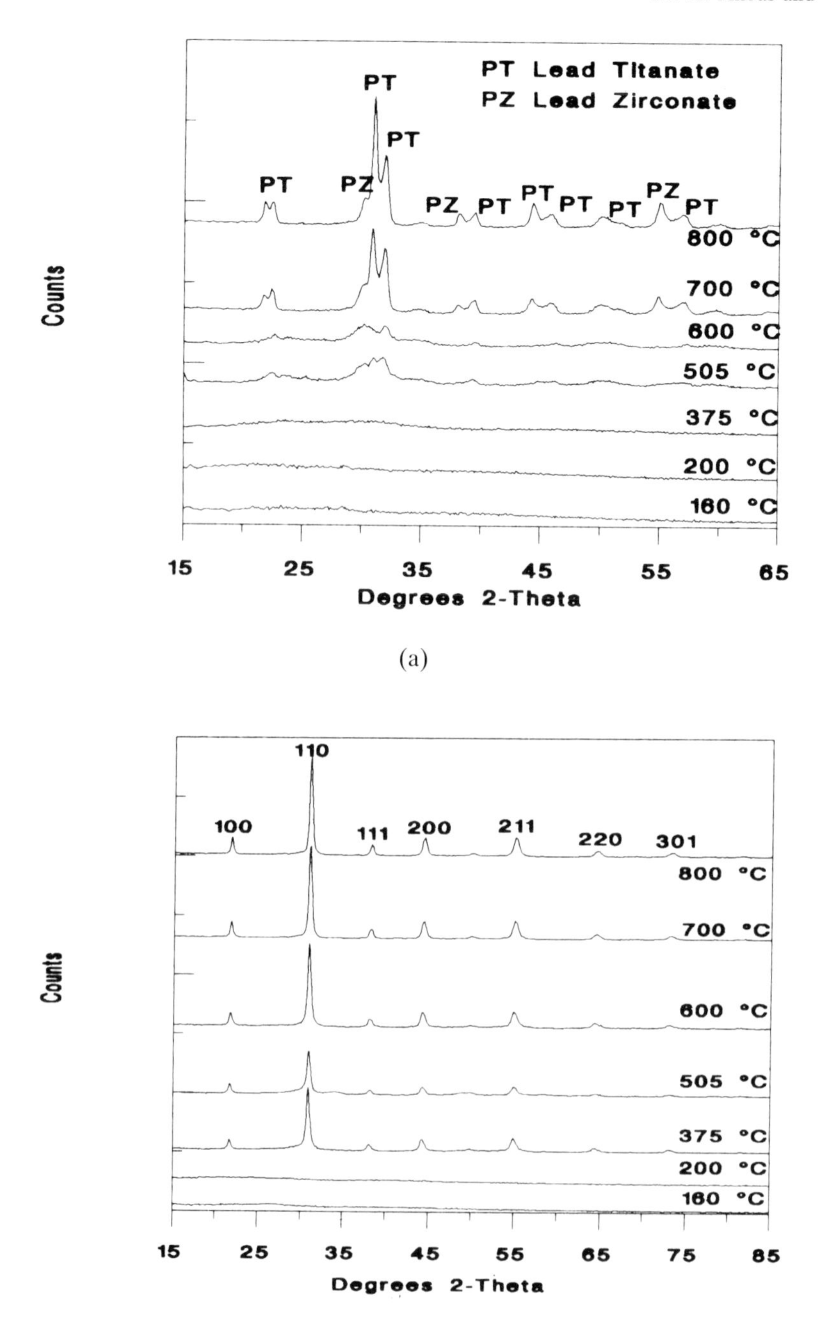

Figure 2 XRD patterns of (a) Cl^--containing (b) Cl^--free, samples quenched from various temperatures. Samples were held isothermally for 4.5 h at temperatures.

Table 2
Gravimetric analysis of powder calcined in air or oxygen-diluted nitrogen.

Method I		Calcination in air	
Composition	8/65/35	9/65/35	10/65/35
Excess PbO (wt%)	10	10	10
Theoretical weight (g)	17.8346	17.7867	17.7386
Actual weight obtained (g)	12.2500	12.4865	11.9484
Weight deficiency (wt%)	31.3	29.8	32.6
PbO loss (wt%)	53	55	50.3
Method II		Calcination in O_2 diluted N_2	
Compositions	9/65/35	9/65/35	9/65/35
Excess PbO (wt%)	3	5	8
Theoretical weight (g)	3.09	3.15	3.24
Actual weight obtained (g)	2.6139	2.6010	2.6476
Weight deficiency (wt%)	15.4	17.4	18.18
PbO loss (wt%)	24.1	27	27.8

3.2 Characterisation of PLZT Powder

3.2.1 Chloride-containing citrate gel

Early PLZT citrate gels were produced from the Cl^- containing source solution which was prepared by a technique described by Li *et al.*[6] PLZT citrate gel of compositions 8, 9 and 10/65/35 with 10 wt% excess PbO was prepared to yield 0.05 mole of powder which theoretically weighs roughly 17.7 g (Table 2). However, far lower quantities were obtained after calcination of the gels in air (Method I). As indicated in Table 2, all the compositions weigh approximately 30 wt% less than their theoretical weights. Using the known compositions of the starting materials, the mean total weight loss after calcination is 77.6 wt% due to water/organic loss. In practice a value of 85 wt% is obtained. The reasons for this difference will be discussed in the following sections.

In order to prevent this greater-than-expected weight loss, another calcination method was designed to control the combustion rate of carbon. In this method, citrate gels of PLZT 9/65/35 with 3, 5 and 8 wt% excess PbO were calcined first in nitrogen then in an oxygen-diluted nitrogen atmosphere (Method II). As indicated in Table 2, the percentage weight deficiency was reduced to approximately 17 wt%. The mean total weight loss of the citrate gel was around 81 wt% which was still higher than the theoretical value. Almost the same quantities were obtained for powders with different excess PbO contents and the percentage weight deficiency increased with increasing excess PbO. This suggests that the added excess PbO, which is not chemically bonded, is more susceptible to loss than the chemically bonded PbO and the addition of large quantities of excess PbO is not a viable method of compensating for PbO loss during calcination.

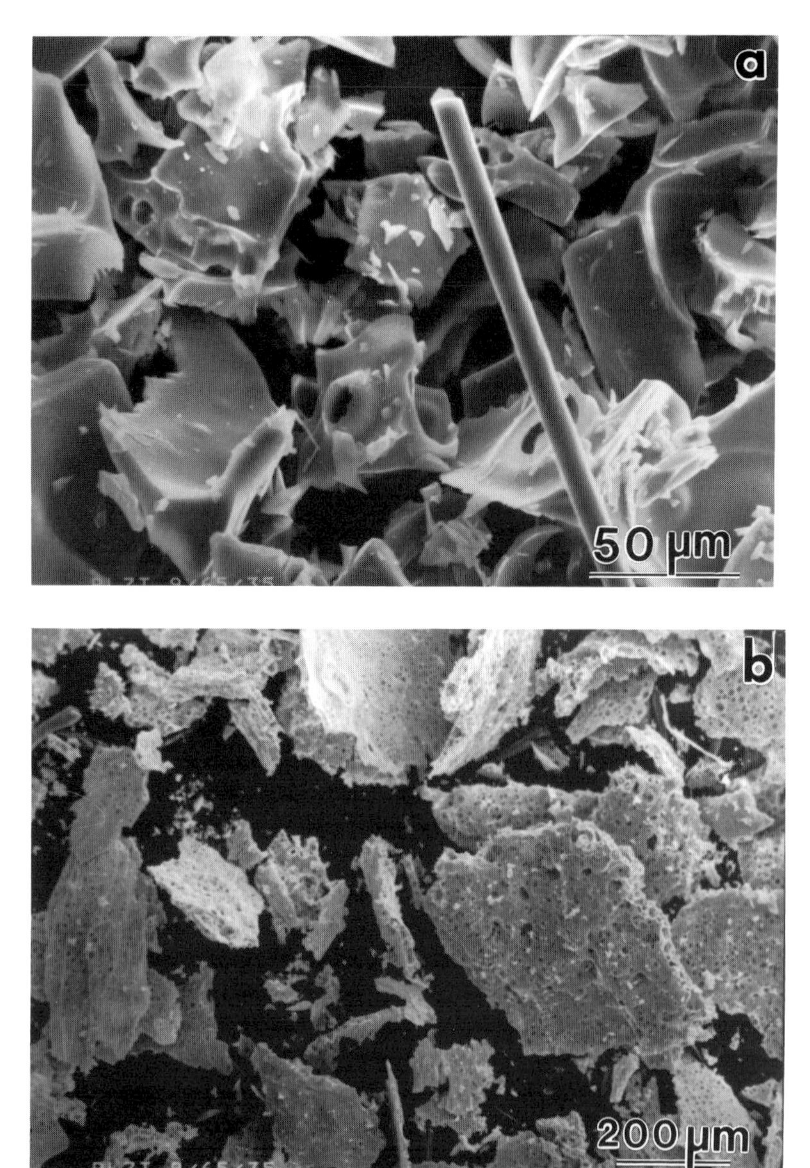

Figure 3 SEM electron micrographs of powder quenched from isothermal hold temperatures (a) 160°C (b) 375°C (c) 375°C (d) 505°C.

Such an unexpectedly high weight loss was also recorded during calcination of a chloride-EDTA precursor of PZT.[12] In that work, crystallisation of $PbCl_2$ was observed as a first intermediate phase and the weight loss was explained by the melting and subsequent evaporation of $PbCl_2$ in a temperature range of 500–600°C.

Figure 2a shows the XRD patterns of chloride-containing citrate gel powders

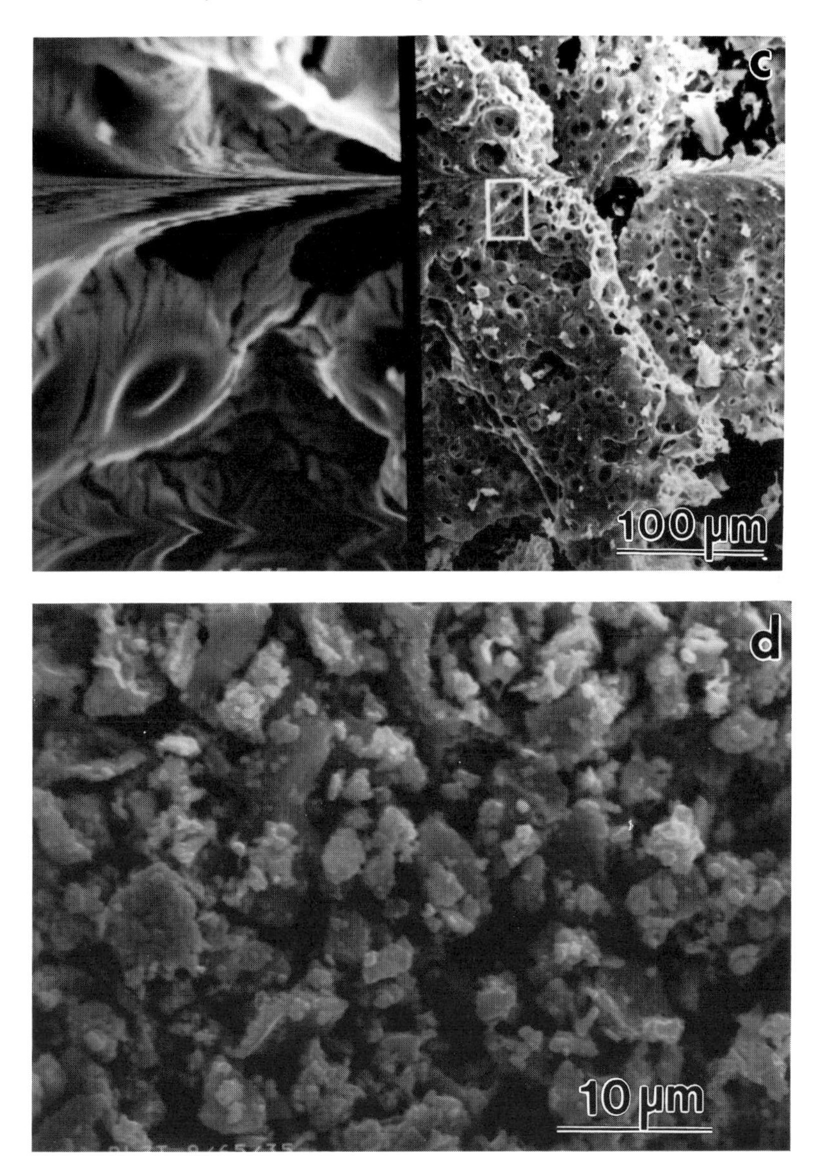

Figure 3–*contd.*

quenched from different temperatures. Although some trace amounts of ammonia salts were detected at 160°C, they decomposed at 200°C. The powder remained amorphous until the vigorous oxidation/decomposition reaction. No $PbCl_2$ was observed up to 375°C and after the oxidation reactions. However, DTA of the chloride-containing citrate gel showed an endothermic peak of 505°C (indicated by an arrow in Fig. 1a). No such peak was observed for Cl^--free gel. This peak was also detected by Wang *et al.*[12] and attributed to the melting of $PbCl_2$. Moreover, TG of the chloride-containing citrate gel showed fluctuations between 525°C and 600°C, perhaps indicating evaporation, and the total weight loss was 82 wt%. Such fluctuations, and excessive weight loss, were absent in the TG thermogram of Cl^--free gel. Therefore, it is possible that $PbCl_2$ forms either at temperatures higher than 375°C or simultaneously with the strong oxidation reaction and was subsequently melted and evaporated by the excess heat of carbon burning. This conclusion is further supported by the white colour of the resulting powder indicating the absence of PbO in the system.

The chloride-containing powder crystallised after the oxidation reaction at 505°C. However, the XRD peaks were very broad due to the small crystallite size. These peaks could be indexed to $PbTiO_3$ and $PbZrO_3$. At higher temperatures, $PbTiO_3$ peaks became clearer and sharper indicating increasing crystallite size and $PbTiO_3$ content (Fig. 2a). Even at 800°C, PLZT was not formed and the powder remained as a mixture of mainly $PbTiO_3$ and $PbZrO_3$. In the absence of enough PbO, the slow diffusing quadrivalent Zr and Ti ions could not form a homogeneous microstructure. Small shifts in the d spacings of the resulting $PbTiO_3$ were detected indicating the presence of dissolved Zr and La in the crystal structure.

3.2.2 Chloride-free PLZT citrate gel

Figure 2b shows a series of XRD patterns obtained from the Cl^--free product calcined at increasing temperatures. The powder remained amorphous up to 375°C and then crystallised into single-phase PLZT composition by a vigorous oxidation/decomposition reaction. All the diffraction peak positions and intensities match with those reported for PLZT.[11] With increasing temperature, the peaks get sharper and stronger indicating an increase in crystallite size and amount.

An SEM micrograph of the powder quenched from 160°C shows large (30 μm) platy, angular particles with smooth surfaces (Fig. 3a). Their smooth appearance indicates that not much citrate decomposition has occurred at this temperature. After the strong oxidation of carbon at 375°C, the particles contain many microchannels and pores indicating the decomposition of citrate chains (Fig. 3b and c). The large porous plates decompose, leaving fine (1 μm) agglomerates after subsequent calcination at higher temperatures (Fig. 3d).

4. CONCLUSIONS

Fine homogeneous PLZT ceramic powder has been prepared by a citrate gel technique. This method avoids the expensive alcohol dehydration process which may allow possible stoichiometry loss either in precipitation or during filtration.[7] The calcination temperatures were substantially lower than those required in conventional routes. The powder was crystallised into single-phase PLZT by the heat evolved during carbon oxidation which starts at temperatures as low as 375°C. The resulting powder is fine and highly agglomerated.

The unexpectedly high weight loss obtained in Cl^--containing citrate gel was explained by formation and subsequent evaporation of $PbCl_2$. Consequently, the resulting powder was PbO deficient and did not form into PLZT.

ACKNOWLEDGEMENTS

The authors would like to thank Dr Hong-Wen Wang and Dr David A. Hall for useful discussions and Mr Ayhan Mergen for his assistance in experimental work. This work is sponsored by the Turkish Ministry of Education in the form of a scholarship (MAA).

REFERENCES

1. G. H. Heartling and C. E. Land: *Ferroelectrics*, 1972, **3**, 269.
2. L. M. Brown and K. S. Mazdiyasni: *J. Am. Ceram. Soc.*, 1972, **55**, 541.
3. J. R. Thomson: *Bull. Am. Ceram. Soc.*, 1974, **53**, 421.
4. H. Yamamura, M. Tanada, H. Haneda, S. Shiraska and Y. Mariyashi: *Ceramic Int.*, 1985, **11**, 23.
5. W. K. Lin, M. S. Jov and Y. H. Chang: *Ceramic Int.*, 1988, **14**, 223.
6. C. E. Li, H. Y. Ni and Z. W. Yin: in *Ceramic Powders*, P. Vincenzini ed., 593, Elsevier Scientific Publishing, Amsterdam, 1983.
7. B. J. Mulder: *Am. Ceram. Soc. Bull.*, 1970, **49** (11), 990.
8. C. M. R. Bastos, J. R. M. Jafelicci, J. R. Varela and M. A. Zaghette: in *Ceramics Today – Tomorrow's Ceramics*, P. Vincenzini ed., 1983, Elsevier Science Publishing, Amsterdam, 1991.
9. P. H. Courty, H. Ajot, C. Marcilly and B. Delmon: *Powder Technol.*, 1973, **7**, 21.
10. D. J. Anderton and F. R. Sale: *Powder Metall.*, 1979, **1**, 14.
11. E. T. Keve and K. L. Bye: *J. Appl. Phys.*, 1975, **46** (2), 810.
12. H. Wang, D. A. Hall and F. R. Sale: *J. Am. Ceram. Soc.*, 1992, **75** (1), 124.

Citrate Gel Route Processing of ZnO Varistors

J. FAN and F. R. SALE
Materials Science Centre, University of Manchester and UMIST, Grosvenor Street, Manchester, M1 7HS, UK

ABSTRACT
ZnO varistors have been prepared using the citrate gel route. The dried precursors have been thermally decomposed to give an oxide powder with submicron primary crystals, the size of which depends upon the temperature of decomposition. Based on a typical composition of ZnO 96.5, Bi_2O_3 0.5, Sb_2O_3 1.0, MnO 0.5, CoO 1.0 and Cr_2O_3 0.5 (all mol.%), it was found that higher breakdown fields ($E_b > 4000$ V/cm) and non-linear exponents ($\alpha > 20$) could be obtained on the sintered gel-derived samples compared with products of identical overall composition produced by a conventional ceramic processing route. The improvements in electrical properties may be explained by the chemically more homogeneous and finer-grained microstructures obtained in the gel-processed materials.

1. INTRODUCTION

ZnO varistors are ZnO-based electroceramic devices, which are widely used in electric and electronic equipment to protect against dangerous electrical surges. About ten to twenty different dopants (such as Bi_2O_3, Sb_2O_3, MnO, CoO and Cr_2O_3, etc.) have been used in the preparation of commercial ZnO varistors and it has been found that the electrical properties of sintered devices are significantly influenced by the distribution of the dopants in this multicomponent material. Consequently, there is a considerable interest in the preparation of ZnO varistor powder of high homogeneity by various chemical and physical methods.[1–5]

The citrate gel process has been employed successfully to prepare a range of substituted perovskite oxides[6,7] superconductors[8] and soft ferrite powders.[9] The process consists of the production of solid amorphous precursors by the dehydration of gels which are produced by the addition of citric acid to a nitrate solution containing salts of the required metals. The solid precursors are subsequently decomposed and oxidised by heating in air to yield the oxide materials. The technical advantages of this method are high chemical homogeneity (on an atomic scale) in the resultant powder, provided that unwanted precipitation does not occur during gel manufacture, and the simplicity of the basic chemistry and the processing needed for the production of the powder. Therefore it is potentially a promising processing route for the preparation of ZnO varistor powder.

The present work has investigated the citrate gel processing of ZnO varistors, based on a typical ZnO varistor composition[1,2] (ZnO 96.5 mol.%, Bi_2O_3 0.5 mol.%, Sb_2O_3 1.0 mol.%, MnO 0.5 mol.%, CoO 1.0 mol.% and Cr_2O_3 0.5 mol.%). To provide materials for comparison the conventional ceramic processing route was also employed to give powder of the same composition. The results for these two materials are compared and discussed.

2. EXPERIMENTAL PROCEDURE

2.1 Powder Preparation

Based on the typical ZnO varistor composition given above, individual solutions containing 'analar grade' citric acid and metal nitrates (except for Sb where a solution of high-purity metal Sb was dissolved in 50% HCl) were mixed to produce solutions for gel production. The amount of citric acid used was that necessary to bind the metal ions if all the NO_3 and Cl ions were to be replaced by citrate groupings. The mixed solution was dehydrated using a rotary evaporator at 75 ± 5°C to produce a gel. The viscous gel was then placed in a vacuum oven at 80 ± 5°C for 12 h to produce the dried precursor. To form oxide powders, the precursor was heated for 2 h at 300°C and then a further 2 h to 500°C or 700°C in air. The two-stage processing was required to control the strongly exothermic decomposition/oxidation reaction which occurs on heating such gels.

The batches of powder prepared for comparison purposes were made using a conventional ceramic processing route. Individual oxides were mixed by ball milling to give the same overall composition as that obtained in the gel powders. The mixed oxides were then calcined for 2 h at 500°C or 700°C.

In summary, four batches of powders were prepared. The gel-derived powders were calcined at 300°C and then at either 500°C or 700°C. These samples are designated G5 and G7 throughout this paper. As indicated above, the mixed oxides were also calcined at 500°C or 700°C. These samples are designated M5 and M7 throughout this paper.

2.2 Compaction and Sintering

Each of the powders, G5, G7, M5 and M7, was pressed into pellets of 10 mm in diameter and 3.5 mm in height at 100 MPa and then sintered at 1200°C for one hour. The heating and cooling rates were 240°C/h.

2.3 Microstructural Observation and Phase Analysis

The morphology of each of the powder samples was studied using a SEM (Philips 505 with EDAX) which was also used to give compositional data on sintered and polished specimens. The sintered and polished specimens were also studied using optical microscopy. The ZnO grain size was determined by the linear intercept method using optical micrographs. The individual phases present in the samples were determined by X-ray diffraction analysis using a Philips PW1710 diffractometer and associated data acquisition and peak matching facilities.

2.4 Electrical Properties

The current–voltage (I–V) characteristics were determined at room temperature using a variable d.c. power supply (Branderburg 475R). From the I–V measure-

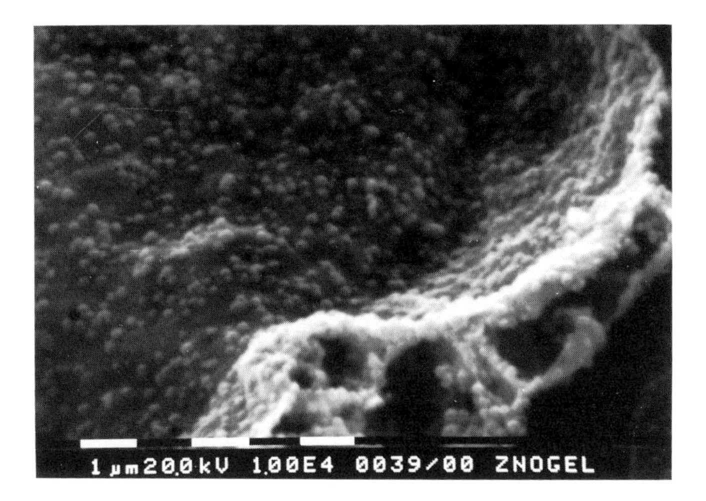

Figure 1 SEM micrograph of gels decomposed at 700°C.

ment it was possible to determine the breakdown field (E_b) and the non-linear exponent (α).

3. RESULTS AND DISCUSSION

3.1 Powder Characterisation

The dried precursors were pale blue in colour with a porous and honeycomb-like structure. After decomposition, submicron primary crystals appeared, the size of which depended upon the temperature of decomposition. Figure 1 shows the submicron crystals obtained by decomposition of the gel at 300–700°C. The crystal size is about 0.2–0.3 μm. At the lower temperature 300–500°C, the crystals were finer and an approximate size of 0.1–0.2 μm was obtained from the SEM micrographs. The X-ray diffraction spectra in Fig. 2 confirm that the crystals shown in Fig. 1 are mainly ZnO. From Fig. 2, it can be seen that the dried precursor was amorphous. During decomposition, as the submicron ZnO particles appear, the crystallinity appears to increase with increase in the temperature (indicated by the height: width ratio of the X-ray diffraction peaks). The presence of the $Zn_7Sb_2O_{12}$ spinel phase was detected in the powders obtained at 700°C (Fig. 2, 700°C curve). This observation confirms previously reported data[10] on the formation of this phase. This phase was also detected in the 700°C powder prepared by conventional ceramic route.

Figure 3 shows the SEM micrographs of sintered samples (Fig. 3a, gel-derived sample G7 and Fig. 3b, conventional route sample M7). Two phases (ZnO and the spinel phase) can be seen in these micrographs. The Bi-rich grain boundary phase has been removed during etching. It is clear that the grain size of the gel-derived

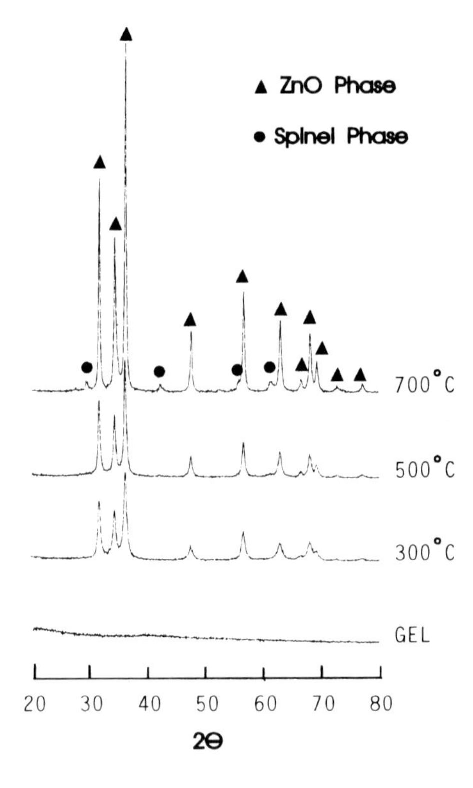

Figure 2 X-ray diffraction spectra of precursors calcined at different temperatures.

sample is smaller than that of the sample prepared by the conventional ceramic route.

3.2 Electrical Characteristics of Sintered Materials

The current–voltage (I–V) characteristics of the gel-derived (G5, G7) and the conventional route (M5, M7) varistor samples are shown in Fig. 4. From the I–V characteristics it is obvious that the gel-derived samples (G series) have higher breakdown fields, E_b, than those prepared by the conventional route (M series). The values of breakdown field determined for the samples are listed in Table 1. In the breakdown region, flatter I–V curves are obtained with the gel-derived samples, which indicates that they have better non-linearities than the samples produced by the conventional route. The non-linear exponents (α) of the various samples are also shown in Table 1. The α values of the conventional route samples are about 13–15, and thc α values of the gel-derived samples are almost double this at about 24–27.

It is well known that the electrical properties of ZnO varistor are attributed to the electrically active barriers in the grain boundaries. The breakdown field (E_b) is

(a)

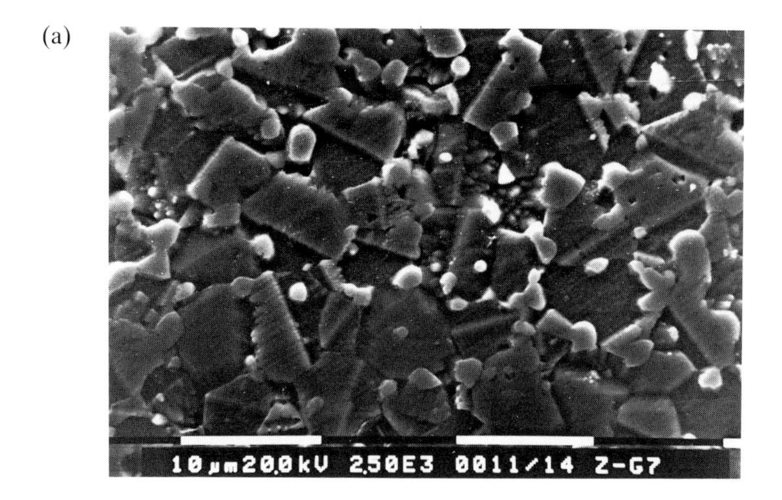

(b)

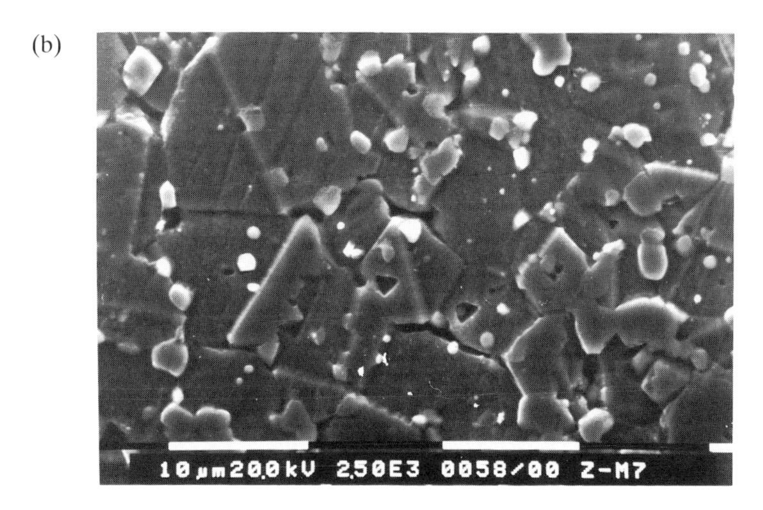

Figure 3 SEM micrographs of sintered samples: (a) gel-derived sample (G7) (b) conventional route sample (M7).

Table 1
Breakdown field and non-linear exponent

Samples	Breakdown field (V/cm)	Non-linear exponent α
M-5	2310	15.3
M-7	2180	13.2
G-5	4430	27.8
G-7	4130	24.3

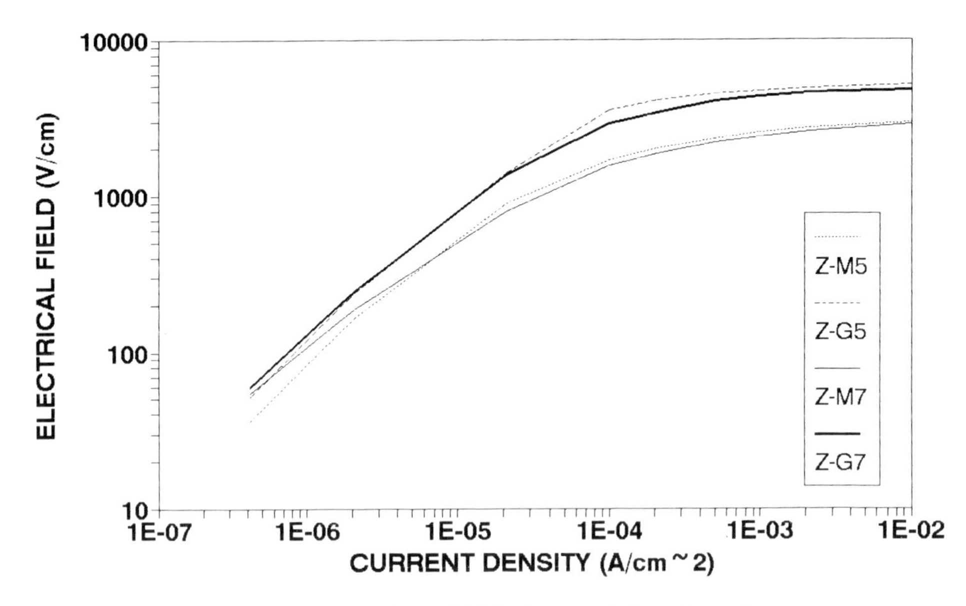

Figure 4 Current–voltage (I–V) characteristics of samples.

Table 2
Average grain size and barrier voltage

Samples	Average grain size (μm)	Barrier voltage (V)
M-5	11.9 ± 1.4	2.75
M-7	12.4 ± 1.5	2.70
G-5	8.5 ± 0.2	3.74
G-7	9.0 ± 0.5	3.63

a statistical value which is selected as a measure of breakdown voltage of such a ZnO varistor. This value is closely related to the average barrier voltage and the average grain size through the following relations:[10]

$$E_b = V_{gb}N \ (\text{V/cm}) \tag{1}$$

$$N \approx G^{-1} \ (\text{cm}^{-1}) \tag{2}$$

where V_{gb} is the average barrier voltage per grain boundary, N is the number of grain boundaries per cm and G is the average grain size (in cm).

From the above equations, it is obvious that either a smaller grain size (G) or a higher barrier voltage (V_{gb}) could result in a higher breakdown field (E_b). Table 2 shows the values of the various average grain size and barrier voltage of samples. For gel-derived samples, both G^{-1} and V_{gb} are higher than those of the conventional route samples. The smaller grain size is a direct result of the finer ZnO particles (~0.3 μm) in the gel-derived powder than those (~0.7 μm) in the commercial ZnO powder which was used to produce M-series samples.

Similarly to the breakdown field (E_b), the average barrier voltage (V_{gb}) and the non-linear exponent (α) are also statistical values which depend upon the properties of the grain boundary barriers. Because the grain boundary barriers result from the defects or second phases present in the grain boundaries after processing, the electrical properties of those barriers are closely related to the distribution of the many dopants which are present in the varistors. Relative to the conventional powder preparation, the citrate gel-processing produces a chemically more homogeneous powder in which the distribution of dopants can be on an atomic scale. Therefore, such highly homogeneous powder should produce smaller deviations in the microstructure and hence the properties of barriers. As a result, it is anticipated that the higher average barrier voltage (V_{gb}) and non-linear exponent (α) of the gel-derived samples may represent more truly the effect of the dopants on the properties of ZnO varistor. A full analytical, microstructural characterisation of the gel-processed materials is presently underway and will be the subject of a future communication. The data obtained for the samples made by the conventional route are reasonable and within the data range of ZnO varistors reported by other investigators,[2,11] so the non-linearities obtained in the gel-derived materials may represent true improvements on existing materials.

4. CONCLUSIONS

(i) Based on a typical ZnO varistor composition, citrate gel-processed ZnO varistors have higher breakdown fields ($E_b > 4000$ V/cm), average barrier voltages ($V_{gb} > 3$ V) and non-linear exponents ($\alpha > 20$) compared with products produced by a conventional ceramic processing route.

(ii) These improvements and differences in electrical properties may be explained by the chemically more homogeneous and finer-grained microstructures obtained in the gel-processed materials.

REFERENCES

1. M. Matsuoka: *Jpn. J. Appl. Phys.*, 1971, **10**, 736.
2. K. Eda: *IEEE Elec. Ins. Mag.*, 1989, **5**, 28.
3. R. J. Lauf and W. D. Bond: *Am. Ceram. Soc. Bull.*, 1984, **63**, 278.
4. E. Sonder, T. C. Quinly and D. L. Kingser: *Am. Ceram. Soc. Bull.*, 1986, **65**, 665.
5. E. Invers-Tiffee and K. Seitz: *Am. Ceram. Soc. Bull.*, 1987, **66**, 1384.
6. D. J. Anderton and F. R. Sale: *Powder Metall.*, 1979, **1**, 14.
7. M. S. G. Baythoun and F. R. Sale: *J. Mater. Sci.*, 1982, **17**, 2757.
8. F. R. Sale and F. Mahloojchi: *Ceram. Int.*, 1988, **14**, 229.
9. F. Mahloojchi and F. R. Sale: *Ceram. Int.*, 1989, **15**, 51.
10. M. Inada: *Jpn. J. Appl. Phys.*, 1980, **19**, 409.
11. T. K. Gupta: *J. Am. Ceram. Soc.*, 1990, **73**, 1817.

Suppression of Zinc Interstitial Ion Migration in ZnO Due to the Presence of Sodium Ions

D. J. BINKS, R. W. GRIMES and D. L. MORGENSTEIN
Davy–Faraday Laboratory, The Royal Institution of Great Britain, 21 Albemarle Street, London, W1X 4BS, UK

ABSTRACT
Atomistic simulation techniques have been used to investigate the solution mechanisms of Na_2O and NaCl in terms of both isolated and clustered defects. The significance of these results to the prevention of varistor degradation is discussed.

1. INTRODUCTION

Zinc oxide varistors are electronic devices that are used in electrical circuits as surge protectors. The current–voltage characteristics of varistors are dramatically non-linear; consequently a small voltage increase can precipitate a massive rise in current flow across the device.[1] Thus, when connected in parallel with a sensitive electronic device any sudden voltage surges will be preferentially channelled through the varistor, leaving the circuit undamaged. Ideally this process should not damage the varistor which is then able to continue protecting against repeated surges. However over a period of time the varistor experiences a gradual reduction of its non-linear characteristics; this process is termed degradation.[1] It has been reported that this degradation is linked with the migration of zinc ions from the grain interior to the grain boundary.[1]

In a previous communication employing the same simulation technique,[2] the experimental observation was confirmed that the addition of sodium oxide can help to slow the rate of degradation.[3] In this paper the investigations are extended by considering the influence of defect clustering on sodium ion migration. In addition, the detrimental effect of chloride ions is discussed.

2. METHODOLOGY

In this study, calculations were performed using the generalised Mott–Littleton methodology available in the CASCADE code.[4] The procedures are based upon a description of the lattice in terms of interionic potentials. We consider interactions due to long-range Coulombic forces, which are summed using Ewald's method and also short-range forces that are modelled using parameterised pair potentials. The short-range terms account for the electron cloud overlap and dispersion interactions which are negligible beyond a few lattice spacings. The Buckingham potential form was chosen to represent the short-range interaction energy, $E(r)$, such that,

$$E(r) = A \exp(-r_{ij}/\rho) - C/r^6 \quad (1)$$

where A, ρ and C are the variable parameters, see Table 1. The total energy of the crystal is then minimised by allowing the ions in the unit cell and the lattice vectors to relax to zero strain.

After the perfect lattice has been energy minimised, to calculate defect energies this lattice is partitioned into two regions: a spherical inner region I at the centre of which the defect is introduced and an outer region II which extends to infinity. In region I, interactions are calculated explicitly so that the response of the lattice to the defect is modelled by relaxing the positions of all ions to zero force using a Newton–Raphson minimisation procedure. The response of region II is treated using the Mott–Littleton approximation.[5,6]

To ensure a smooth transition between regions I and II, we incorporate an interfacial region IIa in which ion displacements are determined via the Mott–Littleton approximation but in which interactions with ions in region I are calculated by explicit summation.

Oxygen ions are treated as polarisable and described by the shell model.[7] In this, a mass-less shell of charge Y is allowed to move with respect to a massive core of charge X; the charge state of each ion is therefore equal to $(X + Y)$. In this study formal charges are used for all ions (e.g. O^{2-}). The core and shell charges are connected by an isotropic harmonic spring of force constant k, see Table 2. The shell charge and force constant were chosen in such a way that the high-frequency dielectric constants of ZnO were correctly reproduced. In line with studies of many other oxides whose cations are significantly less polarisable than oxygen, all cations are modelled as rigid ions.[8]

Discussions of the model parameters and of the methodology generally can be found in recent reviews.[5,6,9,10]

3. RESULTS

3.1 Solution of Na_2O

The solution of sodium oxide can occur by four possible mechanisms, which are presented in Table 3.

The lowest energy solution pathway is mechanism 4, which is a self-compensating process involving the formation of equal numbers of substitutional and interstitial sodium ions. The onset of defect clustering reduces the solution energies considerably, particularly for mechanism 3, where one of the interstitial sodium ions moves into the vacant zinc site to produce a final cluster configuration which is equivalent to that formed by mechanism 4. The binding energy of isolated Na'_{Zn} and $Na^{\bullet}_{i}$ defects to form the $\{Na'_{Zn}:Na^{\bullet}_{i}\}$ cluster is 1.51 eV.

Table 1
Zinc oxide potential parameters

Empirical short-range pair potentials			
Species	A (eV)	ρ (Å)	C (eVÅ^{-6})
O^{2-}—O^{2-}	9547.96	0.2192	32.0
Zn^{2+}—O^{2-}	529.7	0.3581	0.0
Zn^{+}—O^{2-}	470.411	0.3718	0.0
Zn°—O^{2-}	180.815	0.4418	0.0

Table 2
Oxygen shell model parameters

Shell parameters		
Species	Y (e)	k (eVÅ^{-2})
O^{2-}	−2.04	6.3

Table 3
Solution of sodium oxide

Mechanism	Isolated defects (eV)	Clustered defects (eV)
1 $Na_2O \rightarrow 2Na'_{Zn} + V^{\cdot\cdot}_{O} + 2ZnO$	4.75	2.22
2 $Na_2O \rightarrow 2Na^{\cdot}_{i} + O''_{i}$	11.17	7.66
3 $Na_2O \rightarrow 2Na^{\cdot}_{i} + V''_{Zn} + ZnO$	7.06	2.02
4 $Na_2O \rightarrow Na'_{Zn} + Na^{\cdot}_{i} + ZnO$	3.53	2.02

3.2 Solution of NaCl

There are six incorporation mechanisms by which the solution of sodium chloride can occur, see Table 4.

Clearly the most favourable route, if the Na and Cl defects are isolated from each other, is mechanism 2 which produces both sodium and chlorine substitutional ions. If these defects are adjacent to each other during the energy minimisation calculation, the interstitial chlorine and sodium in schemes 1 and 3 respectively are drawn into the lattice where they occupy the vacant anion and cation sites, as appropriate. The solution energy for reactions 1 and 3 is therefore reduced to 0.98 eV, equal to that for reaction 2 where the ions initially occupied the lattice

Table 4
Solution of sodium chloride

Mechanism	Isolated defects (eV)	Clustered defects (eV)
1 $NaCl \rightarrow Na'_{Zn} + Cl'_{i} + V^{\cdot\cdot}_{O} + ZnO$	8.21	0.98
2 $NaCl \rightarrow Na'_{Zn} + Cl^{\cdot}_{O} + ZnO$	1.28	0.98
3 $NaCl \rightarrow Na^{\cdot}_{i} + Cl^{\cdot}_{O} + V''_{Zn} + ZnO$	7.09	0.98
4 $NaCl \rightarrow Na^{\cdot}_{i} + Cl'_{i}$	6.98	6.31
5 $NaCl \rightarrow Na'_{Zn} + Cl'_{i} + Zn^{\cdot\cdot}_{i}$	9.96	5.41
6 $NaCl \rightarrow Na^{\cdot}_{i} + Cl^{\cdot}_{O} + O''_{i}$	8.92	6.50

Table 5
Zinc interstitial migration

Activation energies (eV)		
Migrating species	*a*, *b* Plane	*c* axis
$Zn^{\cdot\cdot}_{i}$	1.36	1.20
$Zn^{\cdot}_{i}$	1.94	1.24
Zn^{X}_{i}	1.35	0.80

sites; indeed, the same defect cluster is formed. The substitutional ions favour nearest neighbour sites in the *a*, *b* plane, the alternative arrangement, in which the ions lie parallel to the *c* axis, results in a slightly higher energy of 1.00 eV.

3.3 *Zinc Interstitial Migration*

The wurtzite structure of the zinc oxide lattice leads to a different interstitial migration pathway in each of the two unique crystallographic directions. The pathway perpendicular to the *a*, *b* plane is via the wide *c* axis channels, whilst parallel to the *a*, *b* plane, migration will occur via a zig-zag pathway as the migrating species moves from one *c* axis channel to one of the six neighbouring *c* axis channels. We have calculated the activation energies for zinc interstitial ions in both directions and at each of the three possible charge states, see Table 5. In each case the activation energies are lower for the pathway along the *c* axis channel, the differences being particularly pronounced for the neutral and singly charged species.

3.4 *Zinc Migration via a Vacancy Mechanism*

When zinc migration is mediated by cation vacancies the wurtzite lattice imposes a different migration pathway for ions moving along the *c* axis compared to

Table 6
Experimental results from Wuensch *et al.*[11]

Zinc migration via a vacancy mechanism		
	a, b Plane	c axis
D_o (10^{-6} cm^2/sec)	4.9	8.2
E_a (eV)	1.75	1.78

migration confined to the a, b plane. The calculated energy for migration in the a, b plane is 1.82 eV. The value resulting in migration along the c axis is 0.91 eV.

The results clearly show a marked preference for migration parallel to the c axis; however, it has been shown experimentally that the zinc migration activation energy is isotropic, see Table 6.

In fact, as shown in Fig. 1, the individual jumps responsible for migration along the c axis, represented by α, involve a component of movement in the a, b plane. Thus a zinc ion can migrate in the a, b plane, either by one β or two α vacancy assisted hops.

The 'two hop' mechanism for a, b plane migration therefore has an activation energy of 0.91 eV. Since the possibility of making two successful jumps is lower than for one, this should be reflected in a lower diffusion coefficient in the a, b plane than along the c axis. Although this model agrees qualitatively with the

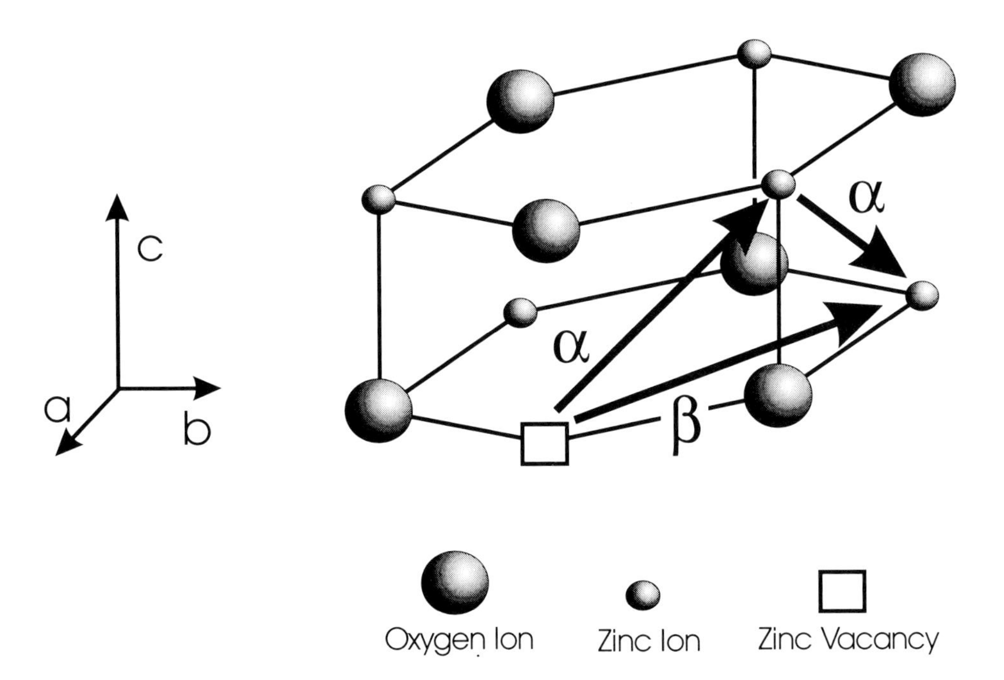

Figure 1 Schematic diagram of ZnO lattice showing vacancy migration pathways.

Table 7
Oxygen interstitial migration

	Activation energies (eV)	
Species	*a*, *b* Plane	*c* axis
$O_i^{\|\|}$	1.62	1.22

Table 8
Sodium interstitial migration

	Activation energies (eV)	
Species	*a*, *b* Plane	*c* axis
$Na_i^{\cdot}$	1.55	1.66

results of Wuensch,[11] in that the *a*, *b* plane and *c* axis activation energies are the same but the diffusion coefficients are significantly different, the experimental activation energy is larger than the calculated value by almost a factor of two. Possible reasons for this discrepancy are presently being investigated, including the possibility that the experimental values include a component of the vacancy formation energy.

3.5 Oxygen Interstitial Migration

The oxygen interstitial migration energies were also calculated, as shown in Table 7.

Again the ions show a clear preference for migration parallel to the *c* axis. The calculated value of 1.22 eV compares well with the experimentally measured value of Robin *et al.*,[12] who determined the activation energy parallel to the *c* axis to be 1.28 eV.

3.6 Sodium Interstitial Migration

The calculated activation energies for sodium interstitial migration are shown in Table 8. The results show that isolated sodium ions exhibit only a slight preference to migrate in the *a*, *b* plane. However, as shown previously in Table 3, the sodium interstitial and sodium substitutional ions will preferentially exist as bound defect pairs. Therefore in order to activate sodium ion migration it is first necessary to overcome the 1.51 eV binding energy of the defect cluster. This has the effect of dramatically increasing the apparent sodium ion migration activation energy.

4. INFLUENCE ON VARISTOR DEGRADATION

The calculations show that the solution of Na_2O results in the presence of sodium interstitial ions (see Table 3) which have a higher activation energy for migration (see Table 8) than do zinc interstitial ions migrating along the *c* axis channel (see Table 5). On a simple level, we can say that the presence of sodium interstitial ions hinders the migration of zinc interstitial ions, effectively increasing the apparent *c* axis migration activation energy of zinc. This result is independent of the charge state assigned to the zinc interstitial ion (by comparing Tables 5 and 8). However, if the effect of clustering is considered, the strong binding of sodium interstitial to sodium substitutional ions significantly increases the activation energy for sodium ion migration, therefore increasing the blocking effect. In addition the singly and doubly charged zinc interstitial ions will bind to the sodium substitutional ions, effectively increasing their migration activation energies.

If sodium is incorporated into the lattice as NaCl, no sodium interstitial ions are formed and zinc interstitial migration is not blocked. In addition, since a strong binding energy is found between the sodium and chlorine substitutional ions, to form a neutral defect cluster, charged zinc ions will not be strongly attracted to the sodium substitutional ion.

ACKNOWLEDGEMENTS

D.J.B. would like to thank the SERC and Harwell Laboratory for financial support. D.L.M. acknowledges the Laura Ashley Foundation for a Summer Fellowship at the Royal Institution of Great Britain.

REFERENCES

1. T. K. Gupta: *J. Am. Ceram. Soc.*, 1990, **73** (7).
2. D. J. Binks and R. W. Grimes: *J. Am. Ceram. Soc.*, 1993, **76** (9).
3. L. M. Levinson ed.: *Advances in Varistor Technology*, vol. 3, American Ceramic Society, Westerville, OH, 1982.
4. M. Leslie: SERC Daresbury Laboratory Report, DL/SCI/TM31T, 1982.
5. C. R. A. Catlow and W. C. Mackrodt eds: *Computer Simulation of Solids*, Springer-Verlag, Berlin, 1982.
6. J. H. Harding: *Rep. Prog. Phys.*, 1990, **53**.
7. B. G. Dick and A. W. Overhauser: *Phys. Rev.*, 1958, **112**, 90–103.
8. M. J. Norgett, C. R. A. Catlow, W. C. Mackrodt and A. M. Stoneham: *Phil. Mag.*, 1977, **35**.
9. C. R. A. Catlow and A. M. Stoneham: *J. Chem. Soc. Faraday Trans.*, 1989, **85** (5).
10. A. H. Harker and R. W. Grimes: *Mol. Sim.*, 1990, **4** (5), *ibid* 1990, **5** (2).
11. B. J. Wuensch, in *Mass Transport in Solids*, F. Beniere and C. R. A. Catlow eds, Plenum Press, New York, 1983, 353.
12. R. Robin, A. R. Cooper and A. H. Hever: *J. Appl. Phys.*, 1973, **44** (8).

Dielectric Properties of *A* and *B* Site Substituted Lead Magnesium Niobate

A. W. TAVERNOR and N. W. THOMAS
School of Materials, University of Leeds, Leeds, LS2 9LT, UK

ABSTRACT
Ceramics of lead magnesium niobate ($Pb(Mg_{1/3}Nb_{2/3})O_3$) have been fabricated with various *A* and *B* site dopants incorporated into the ABO_3 perovskite structure. Lanthanum and bismuth substitutions have been made to the *A* site while zirconium and titanium *B* site dopants have been investigated. Samples have been prepared in a two-stage 'mixed oxide' synthesis, with XRD studies confirming the existence of a solid solution for each composition in the range. Dielectric characterisation, with measurements made of ε_r' and ε_r'' vs. temperature at five different frequencies between 100 Hz and 1 MHz, has shown 'relaxor' ferroelectric properties to be evident in all compositions.

The variation of dielectric response with substituent concentration, x, is described through the use of four empirical parameters, $T(\varepsilon'_{r,max})$, $\varepsilon'_{r,max}$, δ and $\Delta T'$, which are derived from the variation of ε_r' with temperature. A rationalisation of the observed trends is given, employing the 'polarisation cluster' conceptual framework. Consideration of the properties of NbO_6, MgO_6, TiO_6 and ZrO_6 cation coordination polyhedra permits an understanding of the compositional dependence of the 'relaxor' response upon the *B* site. The effects of *A* site additions are described with reference to the 'indirect coupling' model.[1] Taken together, the *A* and *B* site substitution effects provide the building blocks and understanding necessary to design new compositions with desirable dielectric properties.

1. INTRODUCTION

In recent years research based on the class of ferroelectric ceramics known as relaxors has become widespread, both due to their potential in commercial applications and from a purely scientific viewpoint.

Relaxors show several intriguing phenomena which clearly distinguish them from conventional ferroelectrics. Instead of a sharp peak in the permittivity–temperature response (corresponding to a structural phase transition), they show a diffuse maximum in permittivity spreading over a wider temperature interval. The temperature coincident with the peak permittivity $T(\varepsilon'_{r,max})$ shows a strong frequency dependence, being shifted to higher temperatures as the measurement frequency is increased. The temperature at which the maximum in dielectric loss occurs $T(\varepsilon''_{r,max})$ also displays a frequency dispersion but in the opposite direction to that in the permittivity.

Substitution on one or more of the crystal sites in the perovskite structure with foreign cations generally yields many changes to the dielectric response of PMN. The goals of such doping studies may be to search methodically for new materials for a particular application or merely to attain more understanding of the underlying mechanisms of the relaxor dielectric response. Both of these approaches may be used as vehicles for compositional design of relaxor ferroelectrics.

Apart from the obvious application of high-permittivity relaxor ceramics in capacitors, increasing utilisation of these materials is to be found in electrostrictive devices, where their large electric-field-induced polarisations are associated with a mechanical strain. Since the subsequent development of electrostrictive ceramics will embrace both compositional and microstructural factors, the compositionally based approach developed in this paper represents an applicable strategy for optimising the properties of these materials.

Several convenient acronyms are employed in this work. The substituent elements will be referred to not only by chemical symbol but also, for clarity, prefixed by the crystal site on which they reside; bismuth and lanthanum substitutions to the *A* site are therefore denoted by *A*–Bi and *A*–La respectively, while zirconium and titanium on the perovskite *B* sites are referred to as *B*–Zr and *B*–Ti. Note that the universally accepted acronym for the parent composition $Pb(Mg_{1/3}\text{-}Nb_{2/3})O_3$ is PMN.

2. EXPERIMENTAL

2.1 Powder Synthesis

Powders of modified PMN compositions were synthesised from proprietary powders of PbO, Nb_2O_5, $(MgCO_3)_4Mg(OH)_2.5H_2O$, La_2O_3, Bi_2O_3, ZrO_2 and TiO_2 using a 'mixed oxide' route. Two routes were developed, one applicable to *A* site substitutions and one to the *B* site case, both with the common aim of incorporating the substituents homogeneously onto the desired crystal site while also eliminating the parasitic pyrochlore phases known to be problematical in the formation of PMN.[2] These are outlined below.

A Site route

$$(MgCO_3)_4Mg(OH)_2.5H_2O + 5Nb_2O_5 \Longrightarrow 5MgNb_2O_6$$
$$3PbO + MgNb_2O_6 \Longrightarrow 3Pb(Mg_{1/3}Nb_{2/3})O_3$$
$$3A_2O_3 + 0.8[(MgCO_3)_4Mg(OH)_2.5H_2O] + Nb_2O_5 \Longrightarrow 6A(Mg_{2/3}Nb_{1/3})O_3$$
$$[1-x]Pb(Mg_{1/3}Nb_{2/3})O_3 + xA(Mg_{2/3}Nb_{1/3})O_3 \Longrightarrow Pb_{1-x}A_x(Mg_{1+x/3}Nb_{2-x/3})O_3$$

where *A* represents either Bi or La, and $0 \leqslant x \leqslant 0.3$.

X-ray diffraction characterisation of the products lanthanum lead magnesium niobate (PLMN) and bismuth lead magnesium niobate (PBMN) showed them both to be single phase and to possess a rhombohedral lattice as indicated by peak splitting of the *hhh* reflections. The lattice parameter, a_o, was also found to decrease linearly with increasing substituent content, x. Interestingly, the introduction of the aliovalent species onto the Pb^{2+} sites also caused marked ordering of the *B* sites (via doubling of the unit cell size) as detected by X-ray superlattice reflections. These results have been more fully discussed elsewhere.[3]

B Site route

$$(MgCO_3)_4Mg(OH)_2 . 5H_2O + 5Nb_2O_5 \Longrightarrow 5MgNb_2O_6$$

$$3PbO + [1 - x]MgNb_2O_6 + [3x]BO_2 \Longrightarrow 3Pb(Mg_{1-x/3}Nb_{2(1-x)/3}B_x)O_3$$

where *B* represents either Zr or Ti, and $0 \leqslant x \leqslant 0.3$.

X-ray diffraction analysis of the finished ceramics of lead magnesium zirconium niobate (PMNZ) and lead magnesium titanium niobate (PMNT) proved them to be single-phase perovskite. The parent phase ($x = 0$) was found to correspond exactly to the known PMN phase (JCPDS data file 22-1199). Upon increasing substituent levels, all the peaks shifted to slightly higher angles, indicative of a small increase in unit cell size. These results are thoroughly treated elsewhere.[4]

2.2 *Ceramic Fabrication*

A standard pseudo double-ended uniaxial die-pressing technique (10 mm diameter, 80 MPa pressure, 5 min) was used to press the powders into *green* pellets. To aid pressing and green densification, polymeric binders and plasticisers were added (0.25 wt% PVA and 0.75 wt% PEG respectively), in optimum amounts which were determined empirically.[3] These additives were burnt off at 450°C for a period of 4 h. Furnace ramp rates were $\leqslant 2\,Kh^{-1}$, in order to reduce the risk of porosity or lamination caused by sudden burnout. All pellets were densified by sintering at 1220°C for 2 h.

2.3 *Preparation and Measurement of Dielectric Specimens*

Single-phase samples of comparable density (in excess of 98.5% of 'theoretical density') wre selected for dielectric measurements. The sintered pellets were ground and polished down to parallel sided disks of thickness 200 μm and ultrasonically cleaned in acetone. Gold electrodes were sputtered on to both faces. Complex impedance measurements were taken over 5 decades of frequency (100 Hz–1 MHz) from −160°C to 100°C using an HP4284A precision LCR meter in conjunction with a Delta Design 9023 environmental test chamber. The system was operated remotely from a personal computer, using software developed at Leeds.

3. RESULTS

Figure 1 gives the measured variations with temperature and composition of relative permittivity, ε_r', for (a) PLMN ceramics, (b) PBMN ceramics, (c) PMNZ ceramics and (d) PMNT ceramics. These graphs show measurements taken at 100 Hz. Data from the other frequencies have been omitted for clarity although they have been used to calculate values for the parameters δ, γ and ΔT. In order to focus attention on the dependence of the key parameters of the dielectric response,

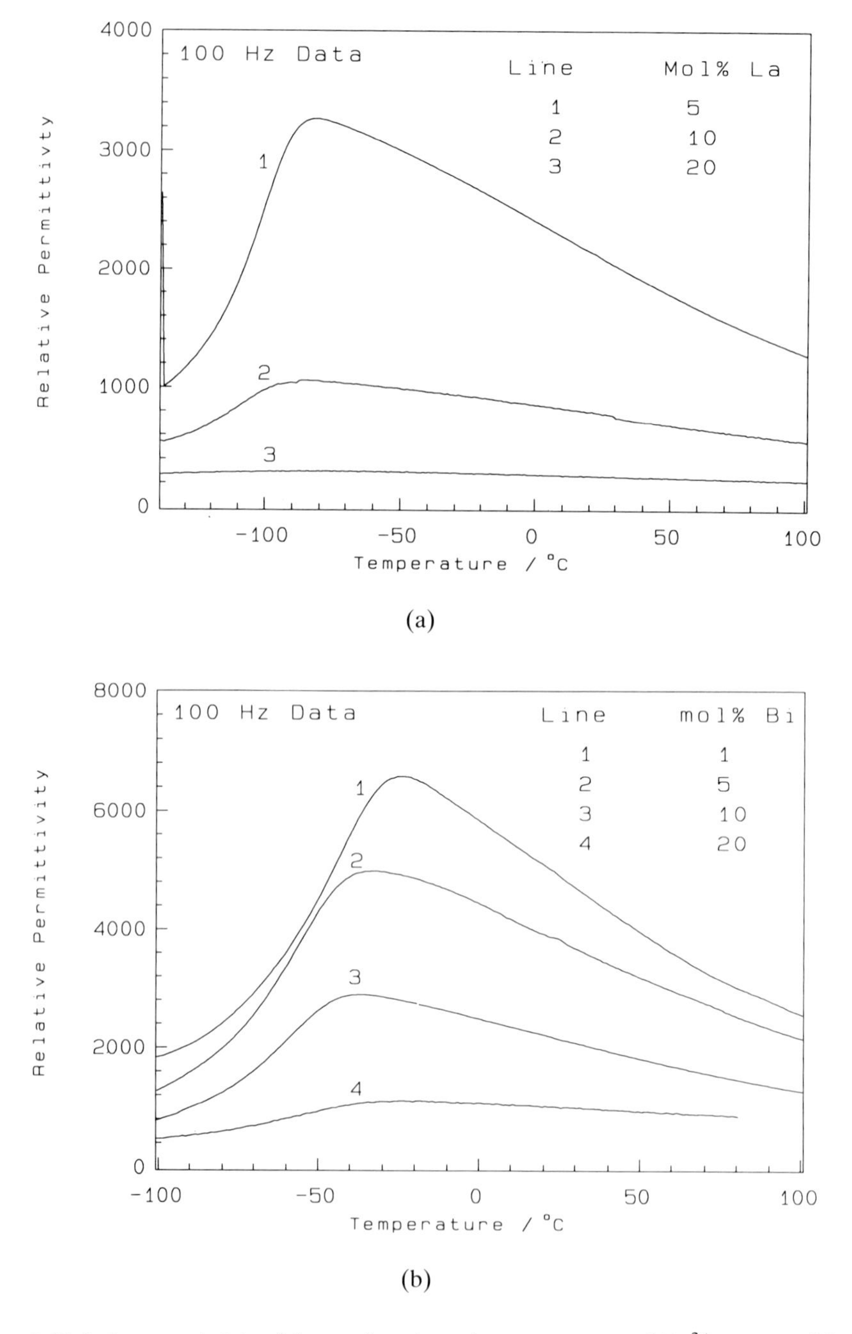

Figure 1 Relative permittivity (a) as a function of temperature and La^{3+} content (b) as a function of temperature and Bi^{3+} content (c) as a function of temperature and Zr^{4+} content (d) as a function of temperature and Ti^{4+} content.

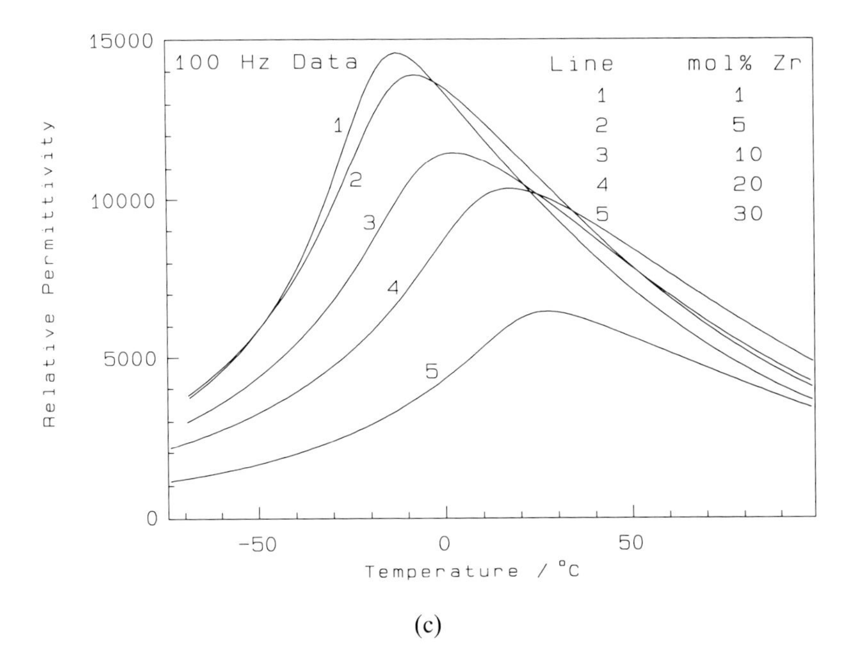

15000
10000
5000
0
Relative Permittivity
100 Hz Data
Line mol% Zr
1 1
2 5
3 10
4 20
5 30
1
2
3
4
5
-50
0
50
Temperature / °C

(c)

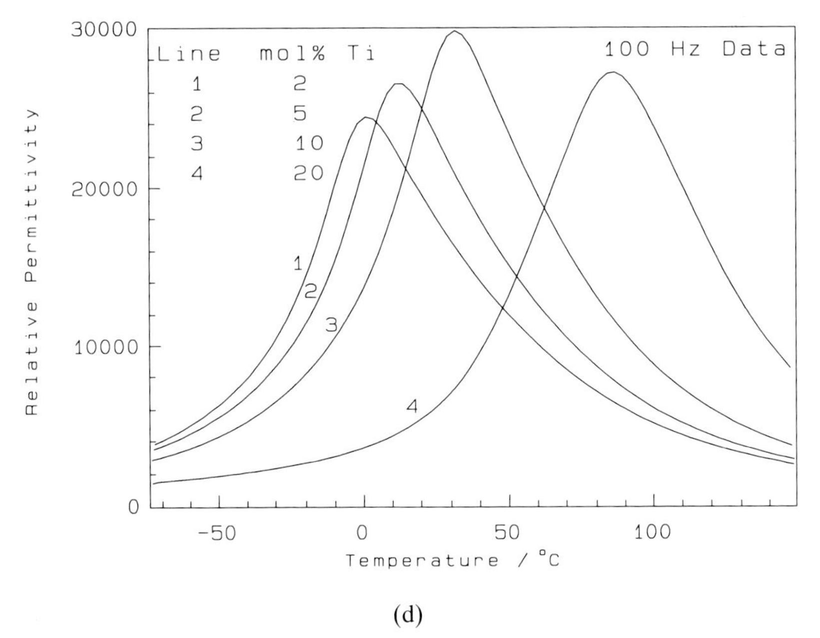

30000
20000
10000
0
Relative Permittivity
Line mol% Ti
1 2
2 5
3 10
4 20
100 Hz Data
1
2
3
4
-50
0
50
100
Temperature / °C

(d)

Fig. 2 gives plots of the variation of the magnitude of the maximum relative permittivity, $\varepsilon'_{r,max}$, with substituent levels, x, for (a) PLMN ceramics, (b) PBMN ceramics, (c) PMNZ ceramics and (d) PMNT ceramics. Figure 3 depicts the measured variation of the temperature coincident with the maximum relative permittivity, $T(\varepsilon'_{r,max})$, for (a) PLMN ceramics, (b) PBMN ceramics, (c) PMNZ ceramics and (d) PMNT ceramics.

Three further parameters, δ, γ and $\Delta T'$, are invoked to quantify the variation of ε'_r with temperature and frequency, with the values of these quoted in Table 1. The first of these, δ, was introduced by Kirillov and Isupov[5] to quantify the 'diffuseness' of a given $\varepsilon'_r(T)$ maximum. According to this parametrisation, the following general relationship applies:

$$\frac{1}{\varepsilon'_r} - \frac{1}{\varepsilon'_{r,max}} = \frac{(T - T_{\varepsilon'_{r,max}})^2}{2\varepsilon'_{r,max}\delta^2}$$

The second parameter, γ, termed the 'critical exponent of non-linearity', relates to an alternative parametrisation proposed by Uchino and Nomura:[6]

$$\frac{1}{\varepsilon'_r} - \frac{1}{\varepsilon'_{r,max}} = \frac{(T - T_{\varepsilon'_{r,max}})^\gamma}{C}$$

where C is a 'Curie like' constant.

The exponent γ lies in the range $1 \leqslant \gamma \leqslant 2$, tending to 1 for a normal (i.e. non-diffuse) ferroelectric and rising towards 2 for a relaxor.

In the current work, δ and γ are regarded as independent indicators of 'diffuseness', with a third parameter, $\Delta T'$, utilised to quantify directly the degree of relaxation behaviour:

$\Delta T'$ quantifies the observable *relaxor* properties explicitly: the extent to which the dielectric maxima are shifted to higher temperatures with increasing frequency.

Table 1 give the calculated values for δ, γ and ΔT for (a) PLMN ceramics, (b) PBMN ceramics, (c) PMNZ ceramics and (d) PMNT ceramics.

4. DISCUSSION

Compositional design of relaxor ferroelectrics depends on the ability to control various key dielectric parameters via chemical changes. This paper focuses on $\varepsilon'_r(T)$ since it is of the most direct importance for the design of capacitor materials. Obviously several other responses are important such as $\varepsilon''_r(T)$ and also the *field tolerance*, $\varepsilon^*_r(E)$; these, however, may be investigated in the same way and will be the subject of future publications.

Before an attempt is made to rationalise the results presented here, it is necessary to define the salient parameters responsible for the observed properties. Previous work[1,3] has provided a set of variables operating within a theoretical framework that allow a description of the observed dielectric response from a

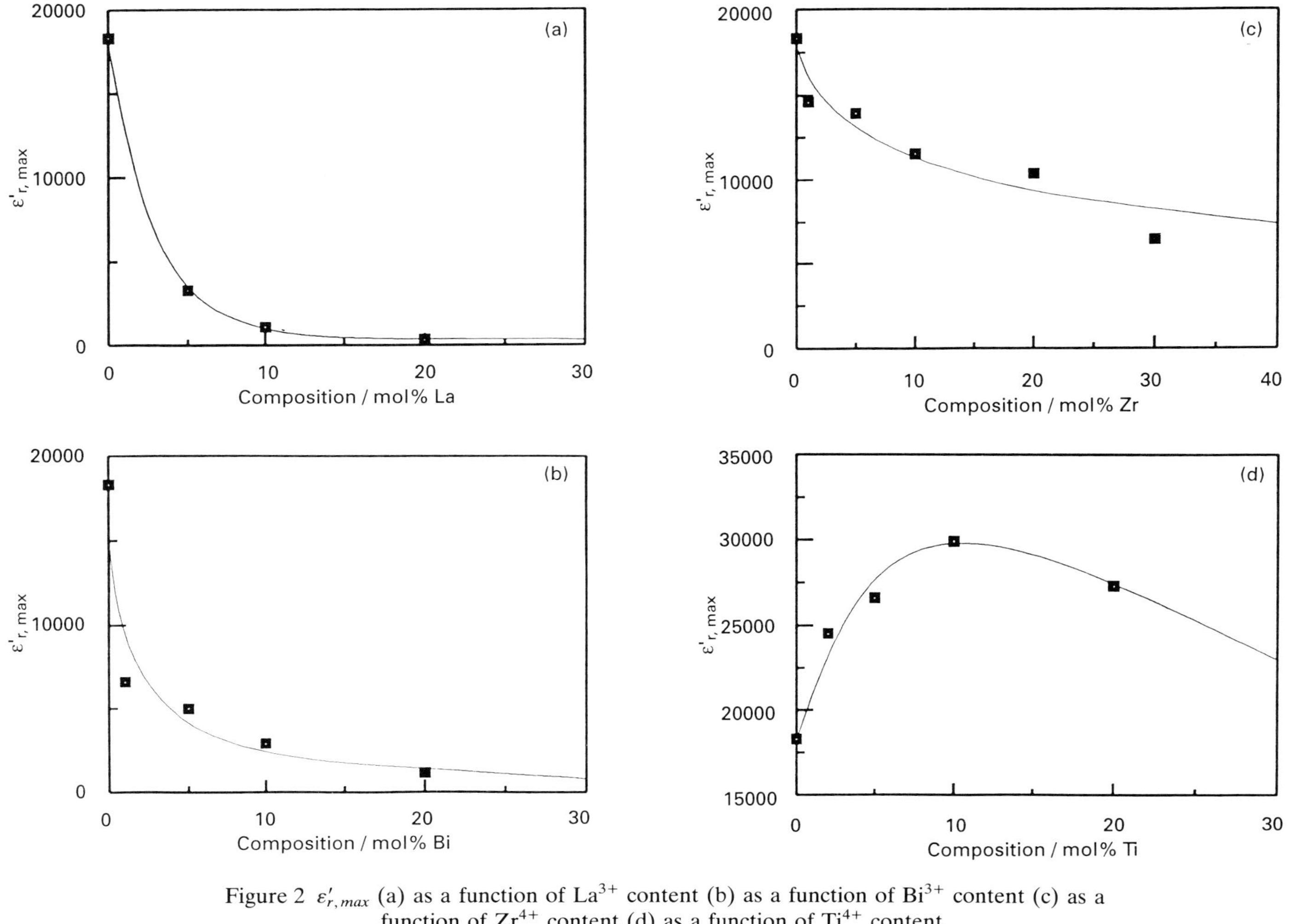

Figure 2 $\varepsilon'_{r,max}$ (a) as a function of La^{3+} content (b) as a function of Bi^{3+} content (c) as a function of Zr^{4+} content (d) as a function of Ti^{4+} content.

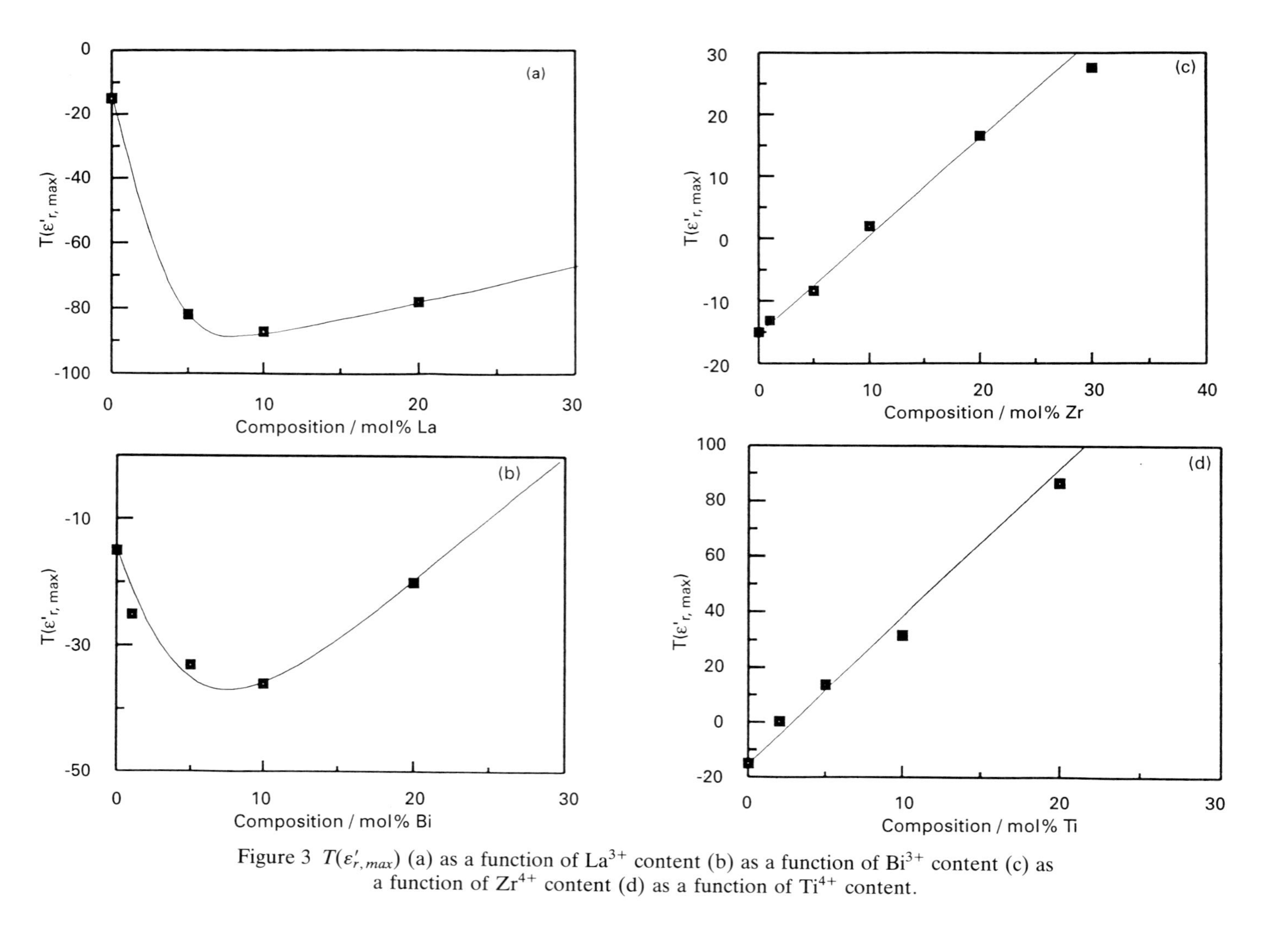

Figure 3 $T(\epsilon'_{r,max})$ (a) as a function of La^{3+} content (b) as a function of Bi^{3+} content (c) as a function of Zr^{4+} content (d) as a function of Ti^{4+} content.

Table 1(a)
δ, γ and ΔT for PLMN ceramics

Composition/x	Diffuseness, δ/K	Critical exponent, γ	ΔT/K
0.00	43.66	1.54	23.7
0.05	105.54	1.51	33.0
0.10	140.12	1.48	36.0
0.20	214.77	1.02	11.0

Table 1(b)
δ, γ and ΔT for PBMN ceramics

Composition/x	Diffuseness, δ/K	Critical exponent, γ	ΔT/K
0.00	43.66	1.54	23.7
0.01	71.45	1.70	28.0
0.05	83.58	1.74	36.0
0.10	87.36	1.47	38.0
0.20	136.11	1.14	60.0

Table 1(c)
δ, γ and ΔT for PMNZ ceramics

Composition/x	Diffuseness, δ/K	Critical exponent, γ	ΔT/K
0.00	43.66	1.54	23.7
0.01	46.37	1.63	27.3
0.05	49.29	1.69	26.37
0.10	52.80	1.78	25.00
0.20	55.30	2.13	19.80
0.30	53.76	1.56	18.21

Table 1(d)
δ, γ and ΔT for PMNT ceramics

Composition/x	Diffuseness, δ/K	Critical exponent, γ	ΔT/K
0.00	43.66	1.54	23.7
0.02	36.06	1.72	21.3
0.05	33.51	1.57	18.4
0.10	31.46	1.68	15.8
0.20	29.42	1.61	7.8

crystal chemical viewpoint. These parameters allow not only a description of an observed dielectric response but also may be used as a predictive tool in the design of new materials.

The activity, a, may be defined as a measure of the contribution of a particular active species towards local electric dipole moment at a given temperature. An ionic species possessing a high activity will, with all other conditions held constant, manifest a greater dipole moment than one with a low activity. Inactive species, such as magnesium in PMN, may be considered to possess an activity of zero. In some cases, however, it is probably more accurate to describe inactive species as those with such a small activity that their contribution to the dielectric response is negligible.

The concentration, C_a, of the active species within a particular material is also an important parameter. Primarily, C_a will act to affect the magnitude of $\varepsilon'_{r,max}$. As the mole fraction of active species present on the B site is increased, the proportion of the material capable of becoming polarised is increased and the relative permittivity is increased.

A secondary, more subtle, effect of concentration alterations may be proposed. In lower concentrations, the inactive regions separating the active islands or *polarisation clusters* will be larger. Following the classical interpretation of Smolenskii,[7] this type of situation would be expected to cause an increase in the diffuseness, δ, and the frequency dispersion ΔT.

The discussion of concentration effects given so far implies that C_a should only be considered when studying B site substitutions to PMN. This is, in fact, not the case since certain A site substituents may also cause subtle alterations to the concentration, for example those ions which are aliovalent with respect to lead.

The introduction of 3+ substituents such as lanthanum or bismuth makes it necessary to adjust the Mg/Nb ratio on the B site to preserve charge neutrality. In such cases, although the activity, a, remains constant (since the identities of the B site occupants are unchanged), the concentration, C_a, will be reduced.

It follows from the above arguments that it is desirable to define a parameter which includes both the activity and the concentration. This term will be denoted the net activity, $\overline{a}$, for the material where:

$$\overline{a} = a . C_a$$

The provision of such a parameter allows more fundamental arguments to be applied to complex systems where both A and B site occupancy is altered.

Thomas[1] has postulated that the coupling between the active species occurs via an indirect interaction in either the $\langle 110 \rangle$ or $\langle 111 \rangle$ crystal directions. Which of these directions, if either, is predominant remains unclear. For the purposes of this paper, $\langle 110 \rangle$ coupling will be assumed to be the dominant mechanism, i.e. the cooperative ferroelectric response of one particular active ion within a unit cell with neighbouring active ions occurs via the A site ions (i.e. Pb^{2+} in PMN) related by $\langle 110 \rangle$ directional vectors in the crystal.

Thomas[1] and Butcher[8] made the assumption of indirect $\langle 110 \rangle$ coupling based on results obtained for the *A*–Ba substituted PMN system. No significant flaws in the hypothesis could be found and the reader is referred to the works in question for a full discussion.

The dependence of coupling strength (termed S_c) on *A* site occupancy is of primary concern to a compositionally based rationale. By substitution of a Pb^{2+} ion with a foreign species, coupling strength will be altered. To date, no observations have been made of materials where substitution of lead for another species has acted to increase ferroelectric coupling. It is therefore postulated that the –Pb– linkage produces the most *ideal* coupling possible in complex perovskites and, furthermore, *A* site substitutions will act to disrupt coupling with the magnitude of the disruption being dependent on the identity of the substituent species.

In summary, the compositional dependence of the coupling strength parameter, S_c, is difficult to quantify accurately. Descriptions of coupling strength alterations may be given in terms of substituent size and/or polarisability. It is likely that this type of visualisation is an oversimplification and the situation is, in fact, considerably more complex.

The degree of ordering of the cations, in relaxors, is postulated to cause alterations to the dielectric response. Several previous studies serve to substantiate this hypothesis.

An ideal example of ordering effects may be found in the lead scandium tantalate (PST) system. In the ordered state PST reveals a sharp, first-order ferroelectric transition while in the disordered state, diffuse relaxor-type behaviour is observed. In terms of the Smolenskii model, this is described in terms of the development of polar microregions in the disordered state. Again, a contrast may be drawn with this by considering aliovalent *A* site substitutions. As 3+ ion concentration on the *A* site is increased, the system develops 1 : 1 (Mg : Nb) ordered regions[9] as a charge compensation mechanism. In such cases, a reduction in ΔT or γ is *not* observed as the material orders. This may be rationalised quite simply since, in some regions, the coupling strength is seriously reduced preventing conventional long-range ferroelectric coupling. Furthermore, it is anticipated that the ordered regions will not contribute to the dielectric response since they are not ferroelectric. The observed behaviour will, therefore, be due to the remaining disordered matrix and, as such, remain a relaxor. Ordering effects in PMN are also seen in the present work to cause alterations to the value of $T(\varepsilon'_{r,max})$.

The traditional compositional fluctuation model of relaxors describes the existence of chemical microregions resulting from regions rich in active species residing within an inactive matrix. Such a model does not allow the volume of these regions of polarisation to change dynamically, i.e. via effects other than concentration.

If, however, the microregion is replaced by the more dynamic concept of a *polarisation cluster*, characterised by its local coherence length l_c, the size of these regions may be altered by variables other than concentration. As the coupling behaviour within a material is changed, the effective sizes of the polarisation clusters may be changed. This rationalisation permits the interpretation of certain alterations to the dielectric response in terms of changes to the coupling strength,

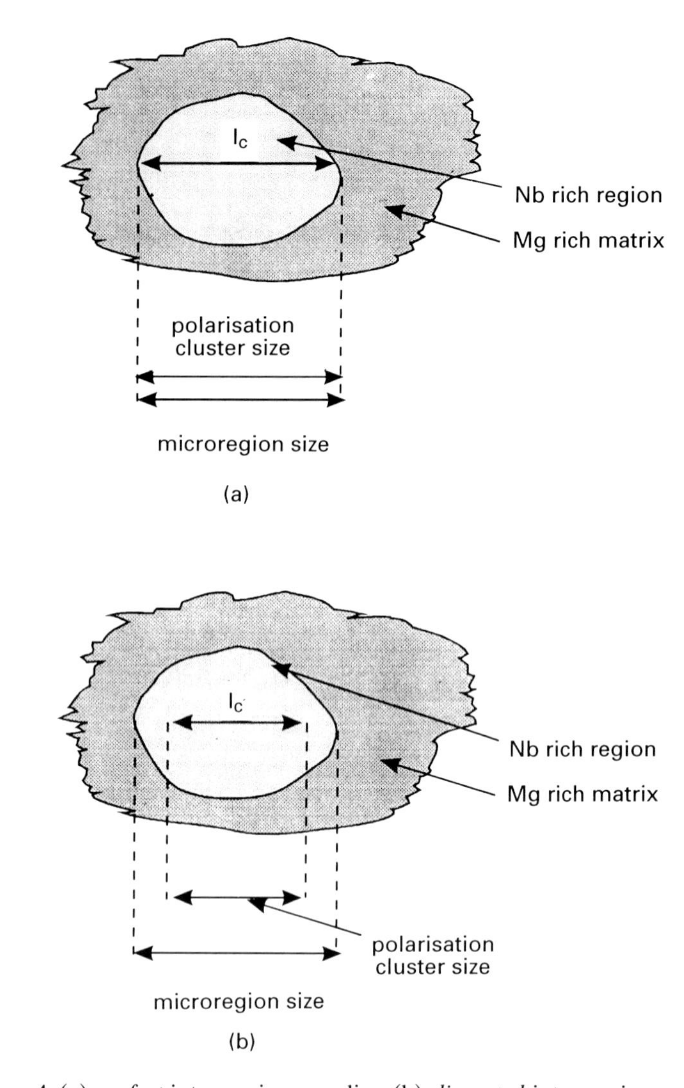

Figure 4 (a) *perfect* intra-region coupling (b) *disrupted* intra-region coupling.

S_c. Figures 4(a) and 4(b) depict schematically a single microregion showing the cases of *perfect* and *disrupted* '*intra-region*' coupling respectively.

Thomas[1] observed that as coupling strength is reduced then $\varepsilon'_{r,max}$ is decreased. Based on the above hypothesis of dynamically sized polarisation clusters it is possible to propose further consequences for the dielectric response from disruption of coupling.

Referring to Fig. 4, it can be seen that, without chemical alteration to the B site, the size of the polarisation clusters may be reduced. If the typical dielectric relaxation observed for relaxors is attributed to relaxation *between* polarisation clusters (as proposed by Smolenskii) then, as the cluster size is reduced (i.e. the inter-cluster spacing is increased), an increase in the relaxation would result and ΔT would be expected to increase.

The experimental results presented will now be discussed in terms of the above parameters.

4.1 A–La

(i) $\varepsilon'_{r,max}$ *behaviour*

Since lanthanum additions are made to the *A* site of the perovskite structure, the activity, *a*, will remain unchanged (recall that the activity is defined as being dependent on the identities of the *B* site occupants). Since the Mg:Nb ratio is altered to preserve charge neutrality, according to the general formula $Pb_{1-x}La_x(Mg_{1+x/3}Nb_{2-x/3})O_3$, then the concentration of the active species present, C_a, varies as a function of lanthanum content. With the activity remaining constant and C_a being reduced, it is expected that $\varepsilon'_{r,max}$ will be reduced as lanthanum content is increased.

The macroscopic coupling strength (denoted here as θ), strongly influences the observed value of $\varepsilon'_{r,max}$.[1,3] From the arguments presented earlier, the presence of lanthanum ions on the *A* site in PMN would be expected to yield a reduction in θ.

Referring to Fig. 2(a) it is apparent that θ does not vary proportionally with lanthanum content. The most significant decrease in $\varepsilon'_{r,max}$ is observed at very low substituent concentrations. Further lanthanum additions cause a progressively smaller decrease per mole. This may be reationalised as follows. As lanthanum is first introduced, and subsequently at very low dopant levels, there is a high possibility that the La^{3+} ion will occupy a Pb^{2+} site surrounded only by other Pb^{2+} ions. Ferroelectric coupling will therefore be disrupted in all the ⟨110⟩ directions. At higher lanthanum contents, there is a higher probability that an individual La^{3+} ion will reside on an *A* site in the crystal that is adjacent to one also occupied by lanthanum. The disruptive effect of this La^{3+} ion is therefore reduced, since fewer active NbO_6 octahedra will become decoupled as a result of its introduction into the structure.

(ii) $T(\varepsilon'_{r,max})$ *behaviour*

The behaviour of $T(\varepsilon'_{r,max})$ as lanthanum concentration is increased is unusual. As may be seen from Fig. 3(a), $T(\varepsilon'_{r,max})$ is shifted to lower temperatures at less than 7 mol.% La. At concentrations in excess of 7 mol.% La, $T(\varepsilon'_{r,max})$ is shifted back towards higher temperatures.

The first clue towards the elucidation of this behaviour lies in the results presented by Butcher[8] concerning *A*–Ba substituted PMN ceramics. Barium additions cause a uniform decrease in $T(\varepsilon'_{r,max})$ with increasing substituent content. It is, therefore, postulated that the subsequent increase in $T(\varepsilon'_{r,max})$ observed in this study is a result of ordering effects caused by the aliovalent substituents introduced.

At substituent levels <7 mol.% La, reductions in the observed $T(\varepsilon'_{r,max})$ may be attributed to a reduction in the macroscopic coupling strength, θ. The positive sign of $dT(\varepsilon'_{r,max})/dx$ at compositions of $x > 0.07$ (7 mol.%) is thought to be a consequence of B site ordering effects. At compositions of $x < 0.07$, it is thought that aliovalent, La^{3+}, ions are incorporated randomly into the structure. At a critical composition (thought to be 7 mol.% from this study), significant 1:1 ordering of the Mg and Nb B site occupants begins to occur. This behaviour has been confirmed by X-ray diffraction evidence.[10–12] The ordered regions will be rich in lanthanum while the remaining matrix comprises predominantly PMN (with an Mg:Nb stoichiometry of 1:2). The net dielectric response may, therefore, be interpreted in terms of this matrix and as such $T(\varepsilon'_{r,max})$ is observed to increase, returning towards the value of −15°C (at 100 Hz).

(iii) Diffuseness and ΔT

Referring to Table 1(a), it is apparent that the degree of relaxor-type behaviour increases with lanthanum content. It is important to note that the materials synthesised in this study were of relatively low dopant levels. At higher substituent concentrations, coupling may be disrupted to such an extent that the material cannot sustain ferroelectricity and hence δ and ΔT become meaningless.

The origins of dielectric relaxation remain unclear despite extensive research over many years. The traditional argument put forward by Smolenskii[7] described relaxation occurring across micro-region *walls*. Smaller microregions could be considered to possess 'thicker' walls and hence display more dielectric relaxation.

The essential features of Smolenskii's argument may be carried across to the notion of polarisation clusters. It has already been suggested that additions of lanthanum to PMN cause reductions in θ and hence to $S_c(T)$. It therefore follows that there will be an associated reduction in the coherence length, l_c, as a result of coupling disruption. With a shorter coherence length the edge-to-edge cluster spacing (i.e. the *wall* thickness) is increased leading to an increase in ΔT.

4.2 A–Bi

(i) $\varepsilon'_{r,,max}$ behaviour

The most noticeable observation pertaining to the compositional dependence of $\varepsilon'_{r,max}$ in the $Pb_{1-x}Bi_x(Mg_{1+x/3}Nb_{2-x/3})O_3$ system it is apparent similarity to the behaviour of the lanthanum doped materials. Comparing Fig. 2(a) and (b), it is noticeable that the reduction in $\varepsilon'_{r,max}$ per mole is less for bismuth than for lanthanum. It is anticipated that the ordering effects due to Bi^{3+} substitutions will be the same as those caused by La^{3+} additions. Assuming that the ordering behaviour is the same for all 3+ substituents, differences may be described in terms of the *effectivness* of a substituent to destroy coupling.

Since bismuth substitutions cause less degradation of $\varepsilon'_{r,max}$ per mole addition, it may be posulated that bismuth possesses a greater ability for indirect ferroelectric coupling than does lanthanum i.e. $\theta_{Pb} \gg \theta_{Bi} > \theta_{La}$. As with the lanthanum substituted system, the decrease of $\varepsilon'_{r,max}$ with increasing x was observed to be non-linear. This suggests that the arguments presented above are equally applicable to bismuth substitutions.

(ii) $T(\varepsilon'_{r,max})$ behaviour
The observed variation of $T(\varepsilon'_{r,max})$ with bismuth content (see Fig. 3a) is strikingly similar to that seen with lanthanum additions. Again, the directionality of the $T(\varepsilon'_{r,max})$ shift is reversed around a composition of 7 mol.% substituent.

Utilisation of the assumption that all 3+ substituents will cause identical ordering behaviour allows the differences in the dielectric data gathered from the lanthanum and bismuth systems to be compared by reference to $S_c(T)$ alone. It has been hypothesised above that the ability of bismuth to disrupt coupling is less than lanthanum. This is corroborated further by the relative alterations to $T(\varepsilon'_{r,max})$ observed for La^{3+} and Bi^{3+} additions. Previous work[3] has shown that the greater the degradation to $S_c(T)$ (via θ), the greater is the reduction in $T(\varepsilon'_{r,max})$ per mole substituent.

It is interesting to compare the behaviour of $T(\varepsilon'_{r,max})$ of the *A*–La and the *A*–Bi systems in the $x > 0.07$ regime. The systems display very similar gradients of the $d\varepsilon'_{r,max}/dx$ curve, suggesting that the ordering mechanisms are closely related.

(iii) Diffuseness and ΔT
The diffuseness, γ, and the frequency dispersion, ΔT, increase upon the addition of *A*–Bi substituents to PMN within the composition range studied. From the earlier assumption that $\theta_{Bi} < \theta_{Pb}$, it follows that $l_{c(Bi)} < l_{c(Pb)}$. Upon the addition of bismuth to PMN, the average coherence length (and thus the polarisation cluster size) will be reduced. As the cluster size is reduced then the inter-cluster edge separation is increased.

The increase in cluster spacing is also magnified by the alteration to the Mg:Nb ratio of the materials studied. To preserve charge neutrality, C_a was reduced and hence the average microregion size was reduced. The inter-cluster edge separation is, therefore, slightly greater than would be expected solely for bismuth substitutions. As was the case for lanthanum, an increase in the inter-cluster spacing is thought to cause an increase in dielectric relaxation and thus of ΔT.

4.3 B–Zr

Substitutional effects of zirconium onto the B'(Mg) and the B''(Nb) sites in PMN may be rationalised in terms of the model presented. In this system, unlike those discussed previously, the most significant parameters are a and C_a. Since no alterations to *A* site occupancy are made then the coupling function $S_c(T)$ may be considered to remain constant.

(i) $\varepsilon'_{r,max}$ behaviour
The magnitude of $\varepsilon'_{r,max}$ for $Pb(Mg_{1-x/3}Nb_{2(1-x)/3}Zr_x)O_3$ ceramics decreases as x increases. The variation of $\varepsilon'_{r,max}$ as a function of x is non-linear, with the most significant reductions occurring at low x.

The reduction of $\varepsilon'_{r,max}$ with *B* site substitution is explained by a reduction in the net activity, $\bar{a}$. The net activity may be reduced by a reduction of a and/or C_a. Zirconium may be considered to possess a lower activity than niobium although higher than magnesium due to its size, i.e. $a_{Mg} < a_{Zr} < a_{Nb}$. By substituting for both Mg and Nb, two competing mechanisms affecting $\varepsilon'_{r,max}$ are operative.

Zirconium substitutions to the niobium sites will cause a reduction in activity whereas substitutions to magnesium sites will cause an increase. The net activity, therefore, undergoes a net reduction as zirconium is substituted into PMN since:

(i) The difference in activity between niobium and zirconium is greater than the difference between magnesium and zirconium.
(ii) Owing to valence constraints, 2 niobium ions are replaced for each magnesium replaced.

The reduction is therefore less pronounced than may be expected, primarily due to the partial replacement of Mg^{2+} by Zr^{4+}; making such materials promising candidates for capacitor materials since the response may be diffuse while still retaining a reasonably high permittivity.

(ii) $T(\varepsilon'_{r,max})$ behaviour

Since $T(\varepsilon'_{r,max})$ is increased upon the substitution of zirconium in PMN, it is clear that the zirconium ion, Zr^{4+} is effective in stabilising the polarisation within the 'polarisation clusters' of the relaxor to a higher temperature than in the parent composition. This may be attributed to the probable distortion of the oxygen octahedron coordinating the zirconium ions.[4] This type of distortion, in which the upper triangular face of oxygen ions becomes larger than the lower triangular face, is commonly found in rhombohedral perovskites.[13] These distortions permit a relatively large ion, such as Zr^{4+}, to be displaced from the centre of coordinates of the O_6 octahedral cage towards the larger face, thereby creating an electric dipole moment.

Such ZrO_6 dipoles could act as structural units which would be stable to higher temperatures than their NbO_6 counterparts. Furthermore, these ZrO_6 octahedra would favour the parallel (ferroelectric) alignment of neighbouring NbO_6 dipoles, through electrostatic interactions. The persistence of the polar ZrO_6 octahedra to higher temperatures would thus provide a mechanism for stabilising neighbouring NbO_6 dipoles to higher temperatures than in the parent composition. The associated polarisation clusters would consequently be stable to higher temperatures, with a predicted increase in $T(\varepsilon'_{r,max})$ with x.

(iii) Diffuseness and ΔT

The replacement of both Mg^{2+} and Nb^{5+} by Zr^{4+} will reduce both the activity, a, and the concentration C_a resulting in a reduction of $\bar{a}$. The chemical region size will, therefore, be smaller than for pure PMN and the volume fraction of ferroelectrically inactive matrix will be greater.

The increase in diffuseness, δ, reflects the greater range in the chemical compositon of the polarisation clusters compared to the undoped PMN. Clusters that contain a relatively greater Zr^{4+} concentration will give a larger contribution to the relative permittivity at high temperatures than those with lower concentrations.

It is also worth mentioning that since *A* site occupancy is 100% Pb^{2+}, coupling will be unperturbed and the polarisation cluster size will be equal to the chemical region size. In addition, the reduction of concentration of Mg^{2+} ions will reduce the degree to which the MgO_6 octahedra break up the long-range dipolar coupling and to which ΔT will be reduced. This effect provides materials designers with a useful mechanism with which to increase the temperature stability of devices without incurring huge frequency relaxations.

4.4 B–Ti

In the discussion of *B*–Ti substitutions, many of the assumptions and findings presented for the *B*–Zr system are applicable and, therefore, are utilised.

(i) $\varepsilon'_{r,max}$ behaviour

The addition of titanium to PMN produces an increase in the maximum relative permittivity up to approximately 35 mol.% substitution. Upon addition of Ti^{4+} to both the Mg^{2+} and the Nb^{5+} *B* sites, according to the formula $Pb(Mg_{1-x/3}Nb_{2(1-x)/3}Ti_x)O_3$, both the activity, a, and the concentration, C_a, are increased. These alterations cause a marked increase in $\overline{a}$ and so of $\varepsilon'_{r,max}$.

Associated with an increase in C_a will be an increase in the chemical microregion size and so the polarisation cluster size. As cluster size is increased, the inter-cluster spacing is decreased and an associated decrease in ΔT is observed together with an increase in $\varepsilon'_{r,max}$.

At a critical composition (corresponding to the rhombohedral to tetragonal morphotropic phase boundary), which is hereby defined as x_c, the polarisation clusters will become so large that they begin to coalesce. Such a situation will produce many changes to the dielectric response including a reduction in $\varepsilon'_{r,max}$.

Relaxors are thought to possess generally higher permittivities than normal ferroelectrics owing to their lack of long-range 'mutual' interactions. By increasing the concentration of the active species, C_a, towards the *critical coalescence composition*, x_c, such long-range interactions become operative and at x_c the coherence length, l_c, becomes infinite. In such cases the relative permittivity is reduced towards the normal ferroelectric response of the end member $PbTiO_3$.

(ii) $T(\varepsilon'_{r,max})$ behaviour

The behaviour of $T(\varepsilon'_{r,max})$ as *B*–Ti substitutions are made to PMN is closely analogous to the alterations seen in the *B*–Zr system. Distortions of the coordinating oxygen octahedra leading to enhanced high-temperature stability will result in a positive temperature shift of $T(\varepsilon'_{r,max})$ compared to undoped material.

(iii) Diffuseness and ΔT

The diffuseness and frequency dispersion of the *B*–Ti substituted PMN ceramics were reduced with increasing titanium concentration. In the composition range $0 \leqslant x \leqslant 0.35$, the diffuseness, δ, and ΔT are continually reduced as a consequence of the increase in size of the polarisation clusters.

At the critical coalescence composition, x_c, ΔT becomes zero as *relaxor ferroelectricity* is lost and long-range interactions become predominant. Further increases in titanium content cause a progressive sharpening of the phase transition as the system tends towards the normal ferroelectric end member $PbTiO_3$.

CONCLUSION

The systems studied in this work have provided many varied effects on the real part of the dielectric response for lead magnesium niobate. These results, together with the conceptual framework proposed offer several opportunities for the materials designer to achieve a *tailored* dielectric response with reasonable accuracy.

In addition to the description of observed properties with various substituents, the notion of dynamically sized polarisation clusters has been introduced, whose properties are functions of both *A* and *B* site occupancy. This type of approach allows accurate predictions to be made of considerably more complex (multi-doped) systems that are likely to be of greater commercial interest.

REFERENCES

1. N. W. Thomas: *J. Phys. Chem. Solids*, 1990, **51** (12), 1419.
2. S. L. Swartz and T. R. Shrout: *Mater. Res. Bull.*, 1982, **17**, 1245.
3. A. W. Tavernor: Ph.D. Thesis, The University of Leeds, 1992.
4. A. W. Tavernor and N. W. Thomas: *J. Eur. Ceram. Soc.*, in press.
5. V. V. Kirillov and V. A. Isupov: *Ferroelectrics*, 1973, **5**, 3–9.
6. K. Uchino and S. Nomura: *Ferroelectrics Lett.*, 1982, **44**, 55–61.
7. G. A. Smolenskii, V. A. Isupov, A. I. Agranovskaya and S. N. Popov: *Soviet Phys. Solid State*, 1961, **2**, 2584.
8. S. J. Butcher: Ph.D. Thesis, The University of Leeds, 1989.
9. H. B. Krause and D. L. Gibbon: *Z. Kristallogr.*, 1971, **134**, 44.
10. A. D. Hilton, C. A. Randall, D. J. Barber and T. R. Shrout: *Ferroelectrics*, 1989, **93**, 379.
11. L. J. Lin and T. B. Wu: *J. Am. Ceram. Soc.*, 1990, **73** (5), 1253.
12. L. J. Lin and T. B. Wu: *J. Am. Ceram. Soc.*, 1991, **74** (6), 1360.
13. N. W. Thomas and A. Beitollahi: submitted to *Acta Cryst.*, **B**.

The Effect of Hot Isostatic Pressing on the Microstructure of Hydrothermally Processed $PbTiO_3$ Ceramics

C. E. MILLAR,† W. WOLNY,† J. RICOTE,‡ C. ALEMANY‡ and L. PARDO‡

†Ferroperm A/S, Hejreskovvej 6, 3490 Kvistgård, Denmark

‡Instituto de Ciencia de Materiales (Sede A). CSIC, Serrano, 144, 28006 Madrid, Spain

ABSTRACT

Fine grain samarium modified lead titanate ceramics were fabricated from hydrothermally prepared powders. Powders were sintered between 1100 and 1200°C for 2 to 10 h. Additionally, some of the sintered samples were hot isostatically pressed (HIP) in an argon–oxygen atmosphere at temperatures of 1000–1200°C a pressure of 200 MPa. The evolution of the microstructures of the resulting ceramics was studied by optical and electron microscopy. The grain and pore size distributions were determined using computerised image analysis. The dielectric and electromechanical properties of the poled ceramics were obtained from their resonances and related to their microstructures. Finally, the suitability of these ceramics for high-frequency (>20 MHz) devices is assessed.

1. INTRODUCTION

Earlier results[1] have shown that ceramics produced from hydrothermally processed powders of modified lead titanate have high density (4% porosity), fine grain size and properties suitable for high-frequency bulk devices, for example for medical imaging applications. However, as the thickness required for operation at higher frequencies decreases, the risk of electrical breakdown during poling due to residual porosity increases.[2] Therefore ceramics with close to 100% of theoretical density are desired.

Hot isostatic pressing (HIP) has been shown to be a successful method of eliminating porosity.[3,4] Here, hot isostatic pressing, carried out as a post-sintering treatment on the above ceramics, is investigated. The study examines the effect of sintering and HIP temperature and time on the microstructure and properties of ceramics prepared from hydrothermally synthesised powders. The properties of the HIP ceramics are compared with those of the sintered ceramics, and evaluated for high-frequency applications.

2. EXPERIMENTAL PROCEDURE

A modified lead titanate powder, with composition $(Pb_{0.88},Sm_{0.08})(Ti_{0.98},Mn_{0.02})O_3$ was processed by hydrothermal synthesis as described previously.[1] The powder was pressed into discs using a pressure of 98 MPa. The discs were sintered at 1100, 1150 and 1200°C for 2 h and at 1100°C for 10 h.

Samples sintered under each of the conditions described above were hot isostatically pressed at 1000°C, with a pressure of 200 MPa for 1 h (HIP 1) without encapsulation. The atmosphere was mixture of 20% O_2 and 80% Ar. Heating and cooling rates were 600°C/h. To examine the effect of HIP temperature and time, samples sintered at 1150°C for 2 h were also hot isostatically pressed at 1100°C (HIP 3) and 1200°C (HIP 4) for 1 h, and 1000°C for 16 h (HIP 2). All other conditions, pressure, atmosphere, heating and cooling rates were the same as those for HIP 1.

The microstructures of as-sintered and HIP 1 to HIP 4 ceramics were examined using the following procedure. The ceramics were polished, metallised with gold and examined using either optical or scanning electron microscopy with computerised image analysis. Porosity measurements were made on polished surfaces and the grain size distribution was determined from thermally etched samples. Measurements were made using software based on IMCO 10 system (Kontron Elektronic GmbH 1990) on, typically, 10 images for porosity measurements and on a total number of 1000 grains for grain size determination. The grain size was calculated as equivalent diameters to a circular shape from the measured areas. All samples were examined by X-ray diffraction and the tetragonal distortion (c/a) of the perovskite structure calculated.

For characterisation of the dielectric and piezoelectric properties, the discs were lapped to give a diameter:thickness ratio >20 and then electroded. The discs were poled in a heated oil bath at 130°C with a field of 70 kV/cm with the field applied during cooling. The relative permittivity, ε_{33}^T and dielectric loss at 1 kHz were measured using an LCR meter. Thickness and planar electromechanical coupling coefficients, k_t and k_p, were calculated according to IEEE Standards[5] from resonance data. The piezoelectric d_{33} coefficient was measured using a Berlincourt meter. All measurements were made at least 24 h after poling.

After measuring the properties, the electrical breakdown strengths of the samples were measured by increasing the poling field in increments of 10 kV/cm, until the samples broke.

3. RESULTS

The results of the microstructural characterisation for the as-sintered and HIP 1 samples are shown in Table 1. Looking first at the effect of sintering temperature, it can be seen that a low porosity and fine grain size were obtained for samples sintered at 1100°C for 2 h. Increasing the time or the temperature led to an increase in the grain size and, apart from samples sintered for 10 h, an increase in the porosity.

The effect of the HIP process on the microstructures was to reduce the porosity content to <1%. The lowest porosity content of 0.22% was obtained in samples sintered at 1100°C for 2 h. The porosity content of the HIP 1 samples increased

Table 1
Average porosity, pore area and grain size of as-sintered and HIP 1 ceramics

	1100°C – 2 h		1100°C – 10 h		1150°C – 2 h		1200°C – 2 h	
Material	Sinter	HIP 1	Sinter	HIP 1	Sinter	HIP 1	Sinter	HIP 1
Porosity (%)	1.78	0.22	1.4	0.4	2.2	0.5	3.0	0.8
Av. pore area (μm^2)	0.8	0.2	2.0	0.5	2.0	0.5	2.0	2.0
Av. grain size (μm)	0.6	0.6	0.9	0.8	0.8	0.8	1.2	1.2

Table 2
Average porosity, pore area and grain size of ceramics as sintered (1150°C for 2 h) and various HIP conditions

	1150°C – 2 h				
Material	Sintered	HIP 1	HIP 2	HIP 3	HIP 4
Porosity %	2.2	0.5	0.14	0.09	0.1
Av. pore area (μm^2)	2.0	0.5	0.3	0.2	0.2
Av. grain size (μm)	0.8	0.8	0.8	0.8	1.0

with increasing sintering temperature and time to a maximum of 0.8% for samples sintered at 1200°C for 2 h. It should also be noted that the grain size of the ceramics was not increased significantly during HIP 1.

Table 2 shows the effect of HIP conditions on the microstructure of ceramics sintered at 1150°C for 2 h. It can be seen that increasing the time to 16 h (HIP 2) or temperature to 1100°C (HIP 3) decreased the porosity content, without increasing the grain size. Whereas, increasing the temperature to 1200°C (HIP 4) increased the grain size from 0.8 to 1 μm.

Examples of microstructures of the ceramic sintered at 1150°C for 2 h and after HIP 3 are presented in Figs 1 to 4. Figures 1 and 2 show typical polished surfaces used to characterise the porosity content of the as-sintered and HIP 3 samples respectively. The reduction in percentage of porosity, as well as pore size, of the HIP 3 sample compared with the sintered sample can be observed. Figures 3 and 4 show SEM micrographs of thermally etched surfaces of the as-sintered ceramic and after HIP 3 respectively. It can be seen that the grain size of both materials is similar.

A comparison of the structural, dielectric and piezoelectric properties of the as-sintered and HIP 1 samples is given in Table 3. The tetragonal distortion (c/a) was always lower for the HIP 1 samples. As expected, the relative permittivity is higher in all of the HIP 1 samples corresponding to the reduction in porosity. The

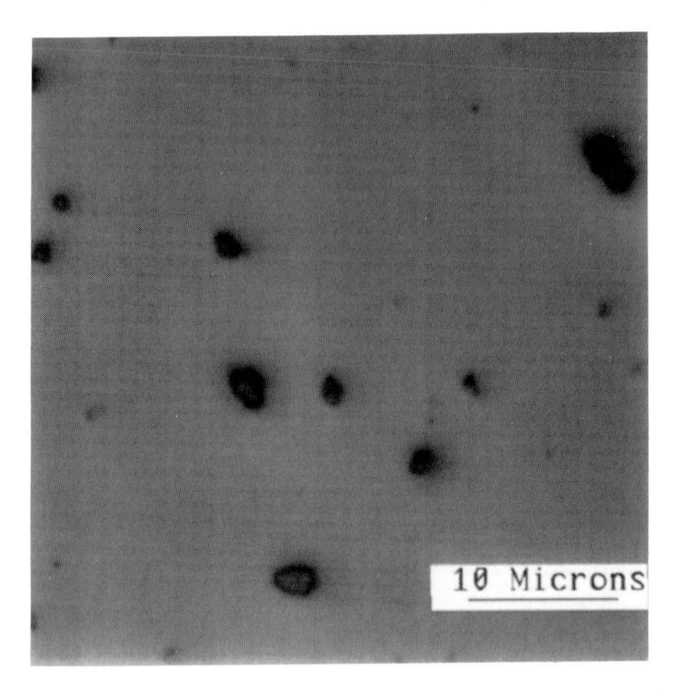

Figure 1 Polished surface of ceramic sintered at 1150°C for 2 h.

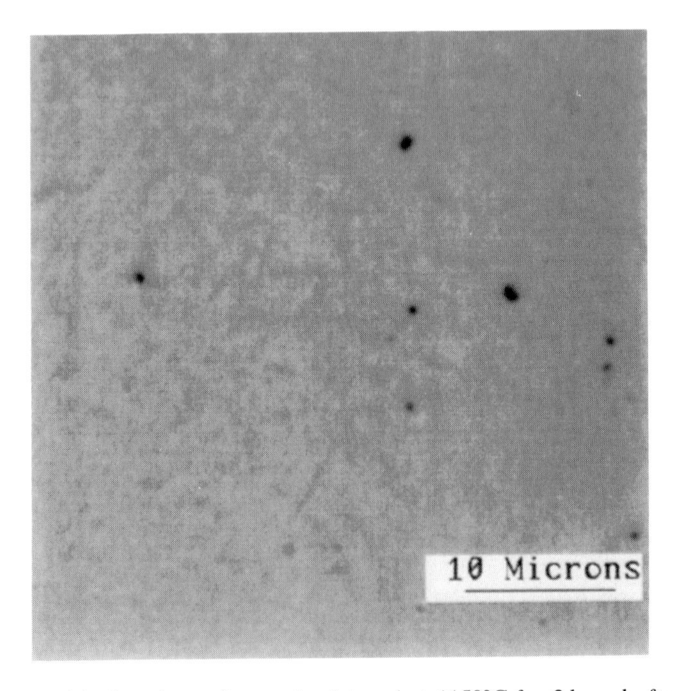

Figure 2 Polished surface of ceramic sintered at 1150°C for 2 h and after HIP 3.

Figure 3 SEM micrograph of thermally etched surfaces of ceramic sintered at 1150°C for 2 h.

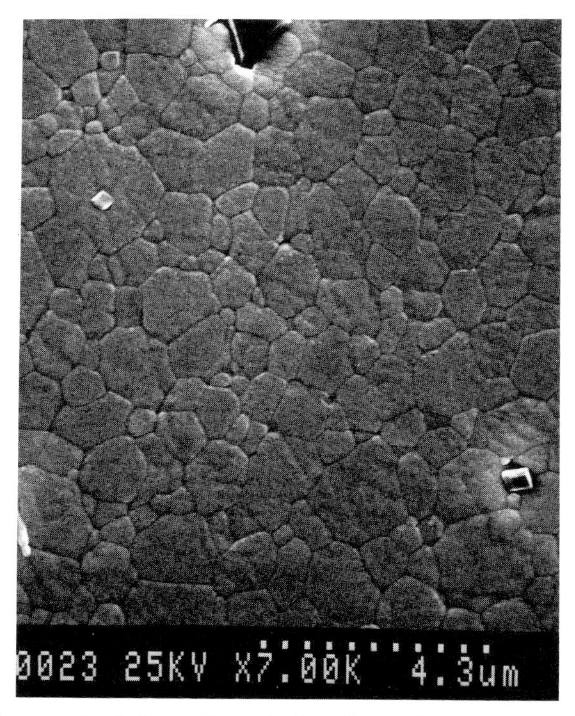

Figure 4 SEM micrograph of thermally etched surfaces of ceramic sintered at 1150°C for 2 h and after HIP 3.

Table 3
Structural, dielectric and piezoelectric properties and breakdown strength (E_B) of as-sintered and HIP 1 ceramics

	1100°C – 2 h		1100°C – 10 h		1150°C – 2 h		1200°C – 2 h	
Material	Sinter	HIP 1	Sinter	HIP 1	Sinter	HIP 1	Sinter	HIP 1
c/a	1.044	1.037	1.045	1.038	1.042	1.038	1.045	1.039
ε_{33}^T	195	192	189	205	192	201	177	190
$\tan\delta$	0.016	0.013	0.026	0.026	0.021	0.015	0.013	0.014
k_t (%)	43.8	47.5	45.7	44.2	45.4	45.7	43.4	46.1
k_p (%)	5.6	7.1	3.9	5.6	4.1	5.6	2.69	3.9
k_t/k_p	7.8	6.7	11.7	7.9	11.0	8.2	16.1	11.8
N_t (kHz mm)	2113	1952	2101	2189	2135	2148	2092	2146
d_{33}(pCN^{-1})	56	60	57	56	59	58	56	58
E_B (kV cm^{-1})	80	90	80	70	70	60	80	70

Table 4
Structural, dielectric and piezoelectric properties and breakdown strength (E_B) of HIP 1 to HIP 4 ceramics

	1150°C – 2 h				
Material	Sintered	HIP 1	HIP 2	HIP 3	HIP 4
c/a	1.042	1.038	1.039	1.040	1.040
ε_{33}^T	192	201	217	214	214
$\tan\delta$	0.021	0.015	0.006	0.007	0.008
k_t (%)	45.4	45.7	46.8	46.3	47.1
k_p (%)	4.1	5.6	9.2	8.6	7.6
k_t/k_p	11.1	8.2	5.1	5.4	6.2
N_t (kHz mm)	2135	2148	2169	2196	2243
d_{33} (pCN^{-1})	59	58	67	65	63
E_B (kV cm^{-1})	70	60	70	90	70

dielectric loss did not change significantly, indicating that at this level the porosity does not contribute to the loss value.

In almost all cases the piezoelectric properties were improved. However, the planar coupling coefficient k_p increased more than the thickness coupling coefficient k_t leading to a decrease in the $k_t : k_p$ ratio. Finally, the breakdown field tends to be reduced for the HIP 1 samples.

The structural, dielectric and piezoelectric properties of the ceramics after different HIP treatments are presented in Table 4. The tetragonal distortion (c/a) was lower for all of the HIP samples, however, it was slightly higher for HIP 3 and

HIP 4 samples. Again, it can be seen that the relative permittivity of all of the HIP ceramics was higher than for the as-sintered ceramics. Here, it is interesting to observe that the dielectric loss was reduced when the HIP treatment was carried out at higher temperature or for longer time.

The piezoelectric properties improved slightly with increasing HIP temperature or time. Again the increase in k_p was higher than k_t, leading to a decrease in $k_t:k_p$ ratio. For these samples the breakdown strength depended on the HIP conditions used and was highest for HIP 3.

4. DISCUSSION

Using a post-sintering HIP process it was possible to reduce the porosity of modified lead titanate ceramics without increasing the grain size. The final porosity and grain size obtained depended on both their initial values and the HIP conditions used. A HIP treatment at 1000°C for 1 h with a pressure of 200 MPa (HIP 1) reduced the porosity from 2.2% in the as-sintered ceramic (1150°C – 2 h) to 0.5%. Increasing the HIP temperature to 1100°C (HIP 3), reduced the porosity further to 0.09%, indicating that the densification rate was faster at the higher temperature. However, if the temperature was increased to 1200°C (HIP 4), the reduction in porosity was accompanied by an increase in grain size. This grain growth, although undesirable, was still moderate. Alternatively, keeping the temperature at 1000°C and increasing the time to 16 h (HIP 2), reduced the porosity to 0.14%. These results are in agreement with those reported by Härdtl,[3] who found that close to 1000% density could be achieved using a HIP treatment for $BaTiO_3$ and $SrTiO_3$ ceramics at <100°C below their sintering temperatures for 1 h with pressures of 200 MPa. At 100°C lower, however, at least 10 h was required.

The effect of the HIP process was to increase the relative permittivity, and piezoelectric properties of the ceramics. The increase in the permittivity, resulting from the decrease in porosity, is as expected.[6]

The improvements in piezoelectric properties are more difficult to explain as their increase may be caused by, either or both, the reduction in porosity, or the increase in the extent of poling, due to the decrease in $c:a$ ratio. Here, the k_t and d_{33} value increased by up to 4 and 13% respectively. Whereas the k_p value increased by more than 100%, indicating that the k_p is more sensitive to changes in the microstructure. To examine the effect of porosity on the anisotropy of the materials, $k_t:k_p$ was plotted as a function of porosity for all the sintered and HIP ceramics measured here (Fig. 5). It can be seen that the anisotropy decreases as the porosity content was reduced. This is consistent with previous results.[7] This shows that the decrease in anisotropy found could not be accounted for by the reduction in c/a ratio alone. Indeed, samples with the lowest c/a ratios did not have the lowest k_t/k_p ratio.

Looking next at the decrease in the dielectric loss, which was obtained for samples HIP treated at the higher temperatures (1100°C and 1200°C) or longer

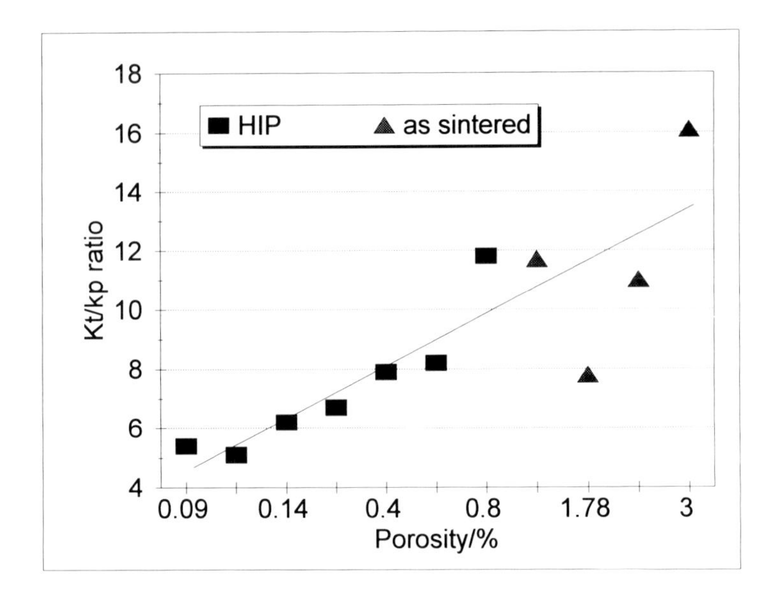

Figure 5 $k_t:k_p$ as function of porosity for all the sintered and HIP ceramics.

time (16 h). As there was little change in the values for HIP 1 samples, it seems unlikely that this decrease is related to the reduction in porosity alone. The decrease in dielectric loss may be due to a change in the oxidation state of the Mn ions, similar to that described for $BaTiO_3$ by Huybrechts *et al.*,[8] or to a change in oxygen content.

Considering the breakdown strengths of the ceramics, it was foreseen that the close packing of the grains in the HIP ceramics might not allow accommodation of the high strains occurring during poling, which arise from reorientation of 90° in ceramics with high anisotropy. However, due to the reduction in the tetragonal distortion the breakdown fields were similar in the sintered and HIP treated ceramics.

For high-frequency devices, there are advantages to be gained by using the HIP process: reduced porosity, higher sensitivity (due to higher k_t value), higher N_t (leading to an increase in thickness for a particular resonant frequency) and lower tan δ. The $k_t:k_p$ ratio is still high enough for most high-frequency applications.

5. CONCLUSIONS

The HIP process was used to reduce the porosity of fine-grained modified lead titanate ceramics to 0.09% without increasing the grain size. The porosity, grain size and densification rate all depended on the HIP conditions used. Of the conditions studied here, the best results were obtained for samples HIP treated at 1100°C for 1 h with a pressure of 200 MPa (HIP 3).

In addition to improving the microstructure of the ceramics, the HIP process also altered properties. For example, the relative permittivity and the piezoelectric properties were improved for all HIP samples, whereas the dielectric loss was reduced only in materials subjected to HIP treatments at higher temperature or longer time. Thus the HIP process gave an overall enhancement of properties for high-frequency applications.

ACKNOWLEDGEMENTS

This work was funded by Brite/Euram project, contract number BREU 0504. The authors would like to acknowledge D. Gomez from CNM-CSIC (Madrid) for SEM work.

REFERENCES

1. C. E. Millar, W. W. Wolny and L. Pardo: Proc. 8th ISAF, 1992, 59–62.
2. R. Gerson and T. C. Mashaet: *J. Appl. Phys.*, 1959, **30**, 1650–1653.
3. K. H. Härdtl: *Philips tech. rev.*, 1975, **35**, 65–72.
4. L. J. Bowen, W. A. Schulze and J. V. Biggers: *Powder Met. Int.*, 1980, **12**, 92–95.
5. IEEE Standards on Piezoelectricity, ANSI/IEEE Std 176, 1987.
6. S. Yomura, K. Nagatsuma and H. Takeuchi: *J. Appl. Phys.*, 1981, **52**, 4472.
7. L. Pardo, J. Ricote, C. Alemany, B. Jimenez and C. E. Millar: Proc. 8th ISAF, 1992, 512–514.
8. B. Huybrechts, K. Ishizaki and M. Takata: Presented at INT. Conf. HIP 93, Antwerp, April 1993.

Hot Isostatic Pressing of Aurivillius Compounds for High-Temperature Device Applications

S. G. BRODIE, J. RICOTE* and C. E. MILLAR
Ferroperm A/S, Hejreskovvej 6, 3490 Kvistgård, Denmark
**Instituto de Ciencia de Materiales (Sede A) CSIC, Serrano, 144, 28006 Madrid, Spain*

ABSTRACT
Aurivillius compounds, with their high Curie temperatures, should be ideal candidates for high temperature device applications. However, present materials based on these compounds have low piezoelectric sensitivity and, at elevated temperatures, low electrical resistivity. Changing the microstructure and crystal structure can improve the properties. The aim of this research was to examine the effect of hot isostatic pressing (HIP) on the microstructure and properties of two Aurivillius compounds, $SrBi_4Ti_4O_{15}$ and $Na_{0.5}Bi_{4.5}Ti_4O_{15}$.

1. INTRODUCTION

The majority of Aurivillius compounds are ferroelectric and possess relatively high Curie temperatures, in some cases as high as 800°C. For this reason these materials are suitable for high-temperature piezoelectric sensors and devices. However, their use is limited by their low piezoelectric sensitivity and low electrical resistivity at elevated temperatures. Both of these problems have been addressed in recent years[1,2] but, as yet, no significant improvement has been obtained.

Aurivillius ceramics consist of a plate-like grain morphology, which, when prepared by conventional methods, contain a high percentage of porosity. The porosity content can be reduced, or even eliminated, by hot pressing or hot forging. However, both these techniques result in grain-oriented materials.

In this research, hot isostatic pressing (HIP), as a post-sintering technique, was used to increase the density of Aurivillius compounds to between 98% and 100% of their theoretical density. The two compositions studied were $SrBi_4Ti_4O_{15}$ (SBT) and $Na_{0.5}Bi_{4.5}Ti_4O_{15}$ (NBT). The effect of the HIP process and atmosphere on the microstructure, dielectric and piezoelectric properties of these materials was investigated.

2. EXPERIMENTAL

Ceramic samples were produced using a conventional mixed oxide technique. After, mixing the raw materials, the powders were calcined in alumina crucibles at 900°C for 4 h. The powder was then milled to reduce the particle size and a binder added. Samples were pressed at 100 MPa. The pellets were then sintered in air for 2 h at the temperatures shown in Table 1.

Table 1
Sintering temperatures for SBT and NBT ceramics

Composition	Sintering temperature °C		
SBT	1180	1200	1220
NBT	1100	1120	1140

After sintering, some of the pellets hot isostatically pressed without encapsulation. Two HIP treatments were carried out, the first with an 80% Ar/20% O_2 atmosphere (HIP-O) and the second with an atmosphere of pure Ar (HIP-Ar). Both treatments were carried out at 1000°C for 1 h at a pressure of 200 MPa. All the disks were lapped to 0.5 mm thickness and their densities measured. Samples were examined by X-ray diffraction (XRD).

Polished and thermally etched surfaces of NBT ceramics, as sintered, and NBT–HIP-O ceramics were examined by scanning electron microscopy (SEM). Two of the obtained microstructures, NBT (1100°C) and NBT–HIP-O (1100°C), were characterised in detail using computerised image analysis of the SEM micrographs. Images of polished surfaces were used for porosity measurements, and thermally etched surfaces for grain size determination. In both cases, 500 to 1000 features were examined.

In order to measure the electrical characteristics of these materials, electrodes of either Ag or Au–Pd were screen printed on to the faces of the ceramics. All samples were poled at 200°C with a poling field of 6 kV/mm, for 5 min. The dielectric properties were measured before and after poling at 100 kHz. Measurement of the piezoelectric d_{33} coefficient was performed using dynamic test equipment similar to a Berlincourt meter, at least 24 hours after poling.

3. RESULTS

The densities, as a function of sintering temperature, of the SBT and NBT ceramics are shown in Fig. 1(a) and (b) respectively. The HIP process increased the density of both materials to >98%, regardless of the HIP atmosphere. For SBT, the HIP-O ceramics had higher density than the HIP-Ar ceramics, whereas for NBT there was little difference between the densities of HIP-O and HIP-Ar ceramics.

Representative microstructures for NBT, sintered at 1100°C for 2 h and after HIP-O, are presented in Fig. 2. In all of the samples examined it could be seen that the porosity content and pore size were significantly reduced by either of the HIP processes.

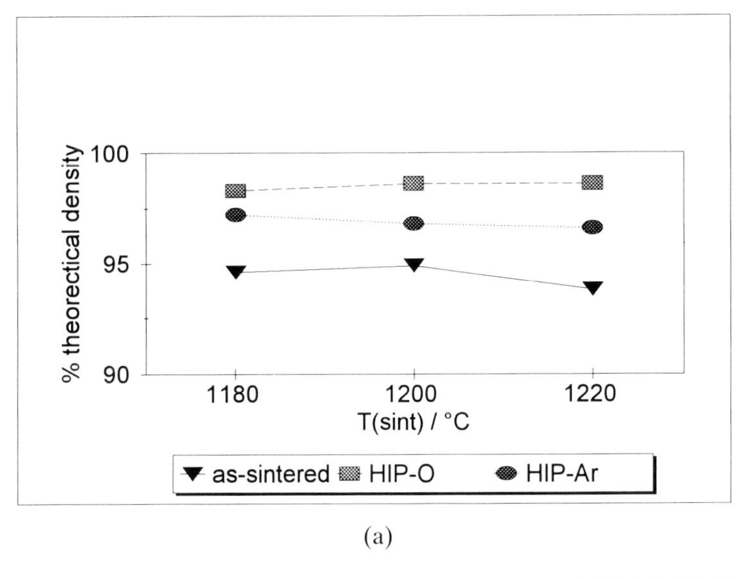

(a)

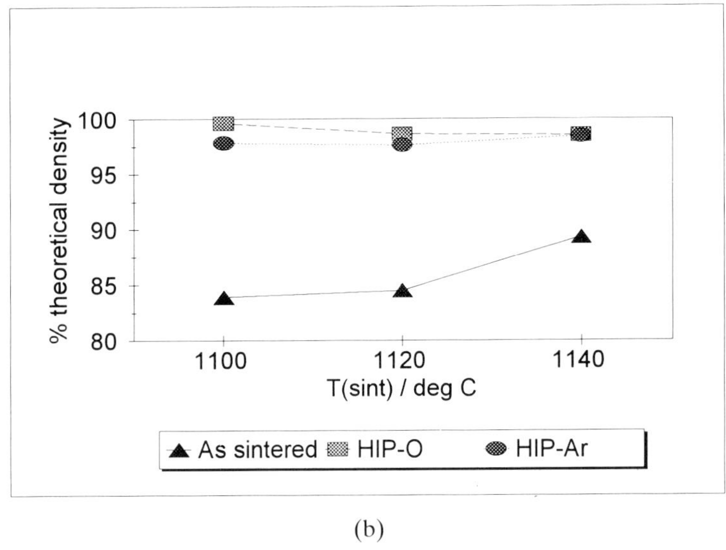

(b)

Figure 1 Density as a function of sintering temperature for (a) SBT and (b) NBT ceramics.

The image analysis data obtained are given in Table 2. The results show that the porosity and pore area of the as-sintered NBT (1100°C) ceramic were drastically reduced by the HIP-O treatment, from 20% and 3 μm^2 to 0.8% and 0.3 μm^2 respectively. No significant change in the grain dimensions was observed. To confirm that the samples were not oriented during the HIP process sections parallel to the pressing direction have also been examined by SEM. No orientation differences were found between the as-sintered samples and the HIP samples.

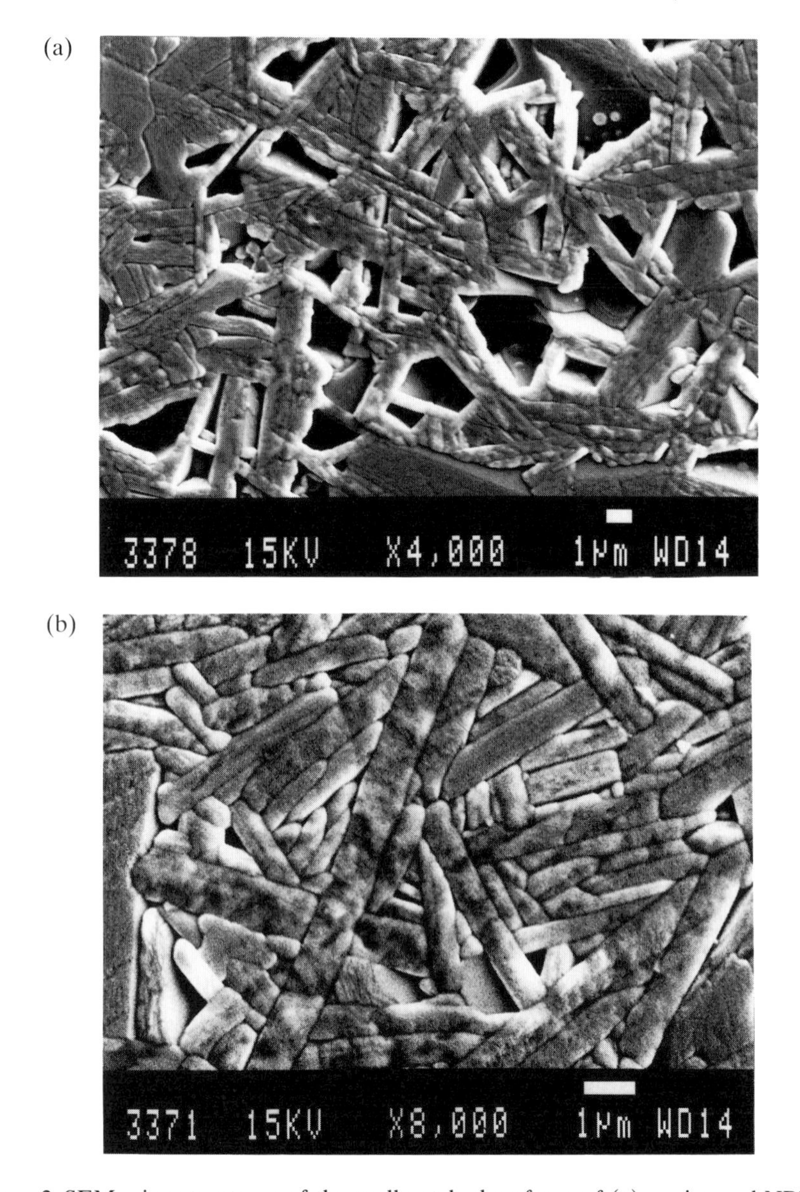

Figure 2 SEM microstructures of thermally etched surfaces of (a) as-sintered NBT and (b) hip processed NBT ceramics.

XRD data showed that the samples were single phase. No difference could be detected in the XRD patterns between the as-sintered ceramics and the HIP ceramics.

The relative permittivity and dielectric loss at 100 KHz for SBT and NBT ceramics plotted as a function of sintering temperature are shown in Figs 3 and 4. The permittivity was higher for both materials after either of the HIP processes,

Table 2
Image analysis results of porosity and grain diameter, *D*, for NBT (1100°C) and NBT (1100°C)-HIP-O ceramics

	NBT (1100°C)		NBT-HIP-O (1100°C)	
	Average	Stand. dev.	Average	Stand. dev.
Porosity (%)	21	3	0.8	0.2
Pore area (μm^2)	3	5	0.3	0.5
D_{min} (μm)	0.7	0.5	0.5	0.3
D_{max} (μm)	1.7	1.6	1.3	1.2
D_{min}/D_{max}	0.9	0.7	0.6	0.4

owing to the reduction in porosity. However, the values of the HIP-Ar ceramics were a little higher than the HIP-O ceramics. The dielectric loss increased after the HIP-Ar process, whereas after HIP-O, ceramics had values comparable to those of the as-sintered ceramics.

Graphs of the piezoelectric charge coefficient d_{33} for SBT and NBT ceramics as a function of sintering temperature are presented in Fig. 5(a) and (b). In all of the HIP ceramics the d_{33} was lower than for the as-sintered ceramics. In the SBT ceramics, there was little difference between values for either of the HIP processes, whereas for the NBT ceramics the HIP-Ar samples had lower d_{33} coefficients than HIP-O samples. By poling HIP-O ceramics at higher poling fields, d_{33} values similar to those of the as-sintered ceramics could be obtained.

4. DISCUSSION

The HIP process reduced the porosity content of Aurivillius ceramics from approximately 20% to <1%, without significantly affecting grain size or orientation. Such a large reduction in porosity is remarkable, when obtained by a non-encapsulated process. Most reports indicate that to achieve close to 100% density in materials with equiaxed grains, starting materials must contain <4% porosity, which is closed.[3,4] This indicates that the ceramics studied here may have closed porosity, resulting from the packing of the elongated grains.

The reduction of porosity led to an increase in permittivity, which is in agreement with the mixture rules for dielectric materials which contain porosity.[5] The HIP-Ar samples had the highest permittivity despite having a lower density than the HIP-O samples, indicating that the increase was not due to the reduction in porosity alone. The HIP-O ceramics had similar dielectric loss values to those of the as-sintered ceramics, whereas, for the HIP-Ar ceramics, the loss was increased. This was probably due to oxygen loss during the process.

A reduction in the piezoelectric coefficient d_{33} was observed for all HIP materials. This reduction is most likely due to the difficulty in poling high-density

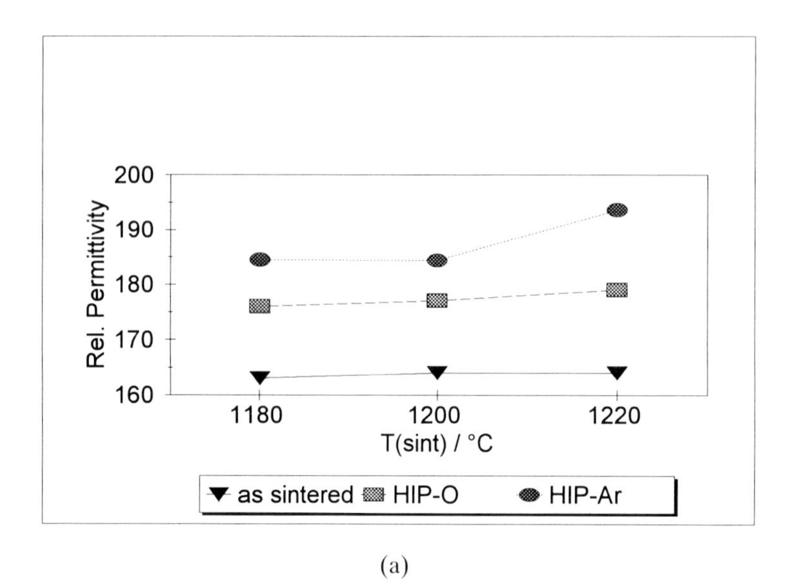

(a)

Dielectric loss
0.5
0.4
0.3
0.2
1180
1200
1220
T(sint) / °C
as sintered HIP-O HIP-Ar

(b)

Figure 3 Relative permittivity and dielectric loss as a function of sintering temperature for SBT ceramics.

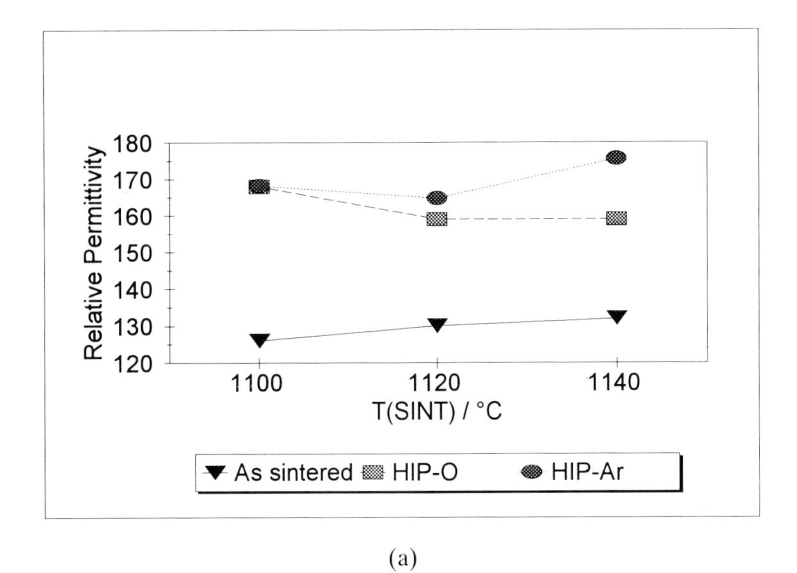

(a)

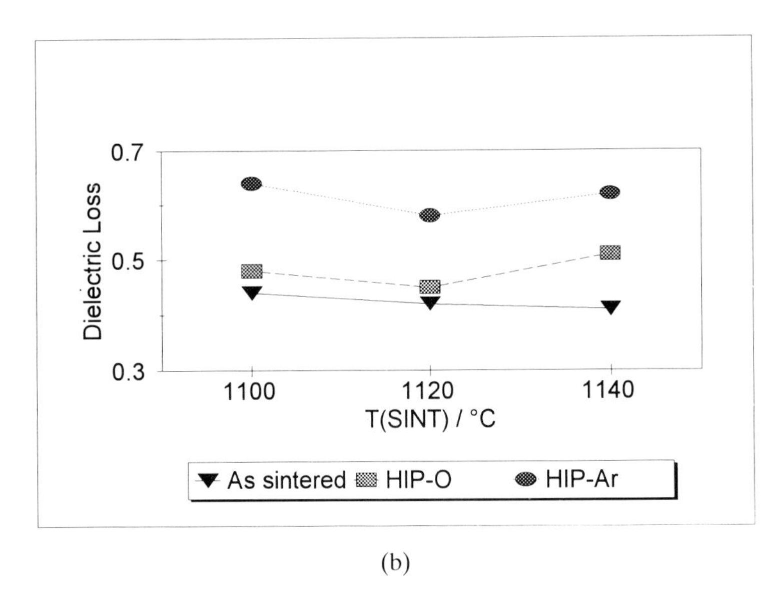

(b)

Figure 4 Relative permittivity (a) and dielectric loss (b) as a function of sintering temperature for NBT ceramics.

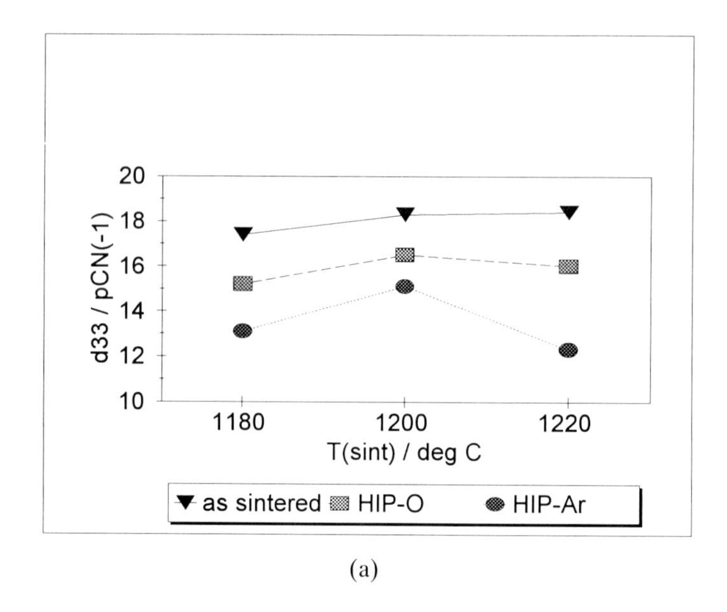

(a)

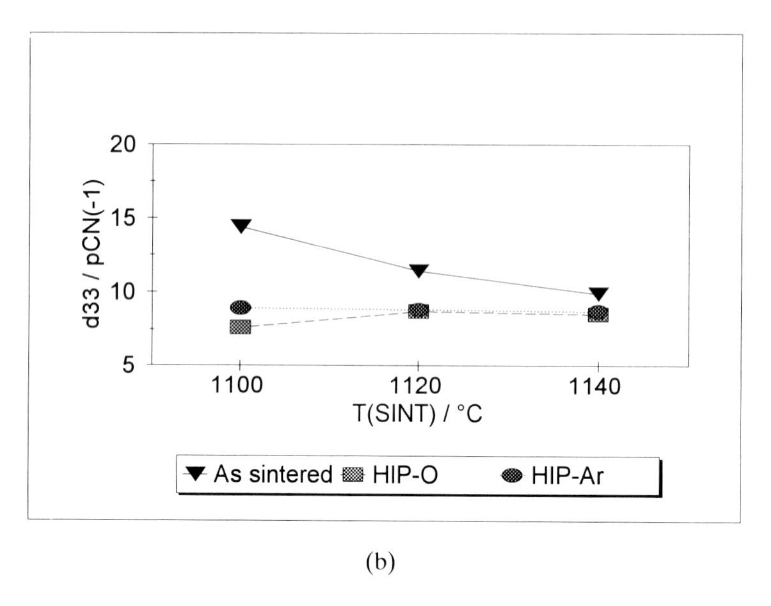

(b)

Figure 5 Piezoelectric sensitivity d_{33} as a function of sintering temperature for (a) SBT and (b) NBT ceramics.

materials. In these types of materials, the reorientation of domains during poling leads to high stress, which can more easily be accommodated when there is porosity present. Thus higher poling fields are required to achieve similar properties.

5. CONCLUSIONS

The HIP process increased the density of both SBT and NBT ceramics to >98% of theoretical density, without appearing to change the grain orientation. The resulting materials had higher permittivity. The HIP-Ar samples had the higher dielectric losses, whereas the HIP-O samples had losses similar to those of the as-sintered ceramics.

Samples from HIP-O and HIP-Ar had lower piezoelectric coefficient d_{33}, owing to the reduction in porosity. A certain amount of porosity is required to accommodate volume changes due to domain reorientation during poling.

REFERENCES

1. T. Takenaka and K. Sakata: *Sensors and Materials*, 1988, **1**, 35–46.
2. M. P. McNeal, J. Dougherty and T. R. Shrout: ISHM-92, Proc. Int. Symp. Microelectronics, **1847**, 134–139, 1992.
3. K. H. Hardtl: *Philips Tech. Rev.*, 1975, **35** (2/3), 65–72.
4. L. J. Bowen, W. A. Schulze and J. V. Biggers: *Powder Met. Int.*, 1980, **12** (2), 92–95.
5. W. Wersing, K. Lubitz and J. Mohaupt: *Ferroelectrics*, 1986, **68**, 77–97.

Investigation of High-Temperature Piezoelectric Ceramics

J. M. WILLIAMS and N. W. THOMAS
School of Materials, University of Leeds, Leeds, LS2 9JT, UK

ABSTRACT
The development of piezoelectric materials capable of operating at 800°C ideally requires ferroelectric systems with a Curie point in excess of 1000°C. A candidate system is reviewed, $Sr_2Nb_2O_7$, with a discussion of the appropriate ceramic fabrication techniques. The results of a transmission electron microscope (TEM) study are presented (selected area diffraction and lattice imaging). Since the idealised structure corresponds to a layered perovskite, the effects of metal and oxygen non-stoichiometry on the formation of the layers are monitored and rationalised. Attention is also given to the occurrence of alternative phases in the SrO–Nb_2O_5 phase diagram, in particular $SrNb_2O_6$ and $Sr_4Nb_2O_9$.

1. INTRODUCTION

Strontium niobate ($Sr_2Nb_2O_7$) and lithium niobate ($LiNbO_3$) are well known as ferroelectric materials and are utilised by the Electroceramics Industry for their piezoelectric and electro-optic properties.

$Sr_2Nb_2O_7$, a member of the $A_2B_2O_7$ family of oxides, has one of the highest ferroelectric phase transition temperatures T_c = 1342°C, which is known to be heavily dependent on the ionic radius of dopant cations substituted on to the Sr *A* sites. The ferroelectric phase transition temperature shows a marked increase when dopant atoms of relatively small ionic radius such as Ca occupy some of the *A* sites ($T_c >$ 1400°C), whereas substituting with larger Pb or Ba ions reduces T_c (T_c = 825°C for $(Sr_{0.6},Ba_{0.4})_2Nb_2O_7$). It is clear, therefore, that appropriate substitution of dopant cations onto the Sr *A* sites can be used to tailor the properties of a ferroelectric material. However, less well documented is the effect of partial cation substitution on the Nb *B* sites on the ferroelectric properties of this material. As a first step to investigating this effect, the structural changes in $Sr_2Nb_2O_7$ produced by variations in the Sr:Nb ratio have been investigated by XRD and TEM.

Initial results on the partial substitution of Nb^{5+} by Ti^{4+} on the $Sr_2Nb_2O_7$ structure are also presented.

2. CRYSTALLOGRAPHY OF STRONTIUM NIOBATE $Sr_2Nb_2O_7$

The structure of $Sr_2Nb_2O_7$ has been found to consist of distorted perovskite-type slabs parallel to (010) composed of NbO_6 octahedra and Sr atoms.[1] The symmetry is orthorhombic $Cmc2_1$ with a = 0.3933(1), b = 2.6726(7), c = 0.5683(4) nm,

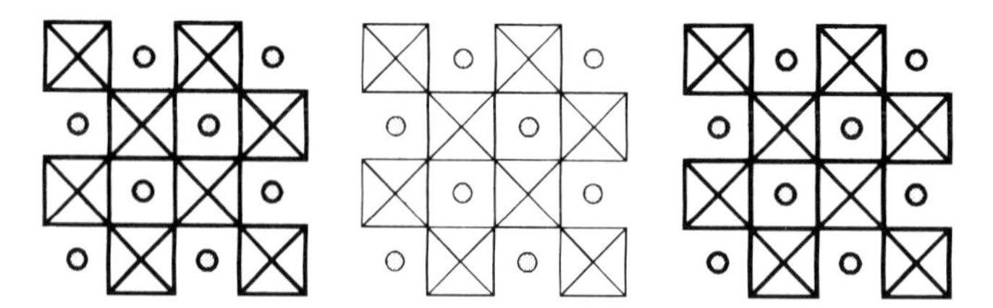

Figure 1 Structure of $Sr_2Nb_2O_7$ viewed down (100) axis. Perovskite slabs are idealised. Slabs drawn in thick lines are shifted with respect to the others by $a/2$.[1]

$z = 4$ and $D_x = 5.26\,\mathrm{g\,cm^{-3}}$. The structure is shown schematically in Fig. 1. Each slab consists of two double layers of NbO_6 octahedra and can be described as an $n = 2$ type material of general formula $A_nB_nO_{3n+1}$, where n is the number of double layers in each slab. When determining the effect of dopant cations on the structure, valence considerations for charge neutrality give

$$nV_A + nV_B = (3n+1)\,V_O$$

$$V_A + V_B = 6 + 2/n$$

$$n = 2/(V_A + V_B - 6)$$

Hence we may predict how n, the number of double layers of NbO_6 octahedra in each slab, will be affected by compositional variations induced by doping with atoms of differing valence. For example, Ti^{+4} substituted onto Nb^{+5} B sites will give $V_A + V_B < 7$, whereas $V_{Sr} + V_{Nb} = 7$. Therefore, Ti substitution should produce a structure tending towards $n = 3$, the degree to which the structure evolves from $n = 2$ to $n = 3$ being dependent on the level of doping. Similarly, if $Sr_2Nb_2O_7$ is doped by substitution on the B sites with a cation of valence greater than +5 (for example W^{+6}) then $V_A + V_B > 7$, which implies that the structure would evolve toward $n = 1$. As before, the degree of structural evolution from $n = 2$ to $n = 1$ would depend on the level of dopant cation substitution.

Implicit in this model is the concept of non-integer values of n. This does not require the formation of materials with non-integer numbers of NbO_6 octahedra (an idea that is difficult to visualise crystallographically), but instead relies upon the formation of incommensurate structures. The existence of incommensurability in strontium niobate $Sr_2Nb_2O_7$ can be demonstrated using TEM, a technique ideally suited to such a task.

3. EXPERIMENTAL

The samples were prepared by normal ceramic methods from Johnson Matthey specpure grade $SrCO_3$ are Johnson Matthey Alfa Nb_2O_5 (99.5 + %) and TiO_2 (99.995%). The component oxides were ground together in a McCrone micronising mill for 30 min under isopropanol and the resulting mixture was dried to give a homogeneous powder which was precalcined at 1400°C for 10 h. The reacted powder was sieved through 300 μm gauze and micronised for 15 min in isopropanol. Pellets were pressed and sintered on platinum foil at 1400°C for 7 days. After reaction the pellets were quenched in air.

The samples were characterised by X-ray powder diffraction using a Philips diffractometer. Electron microscope specimens were prepared by crushing small fragments of the sintered pellet under *n*-butanol in an agate mortar and then allowing a drop of the resulting suspension to dry on a porous carbon film, prior to examination in a Philips EM430 transmission electron microscope. The microscope was operated at 300 kV and fitted with a Super-Twin pole-piece and a double tilt eucentric goniometer stage.

4. RESULTS

4.1 The SrO–Nb_2O_5 System

Samples were prepared with compositions according to the formula

$$x\mathrm{SrCO_3} + (1-x)\mathrm{Nb_2O_5} \rightarrow \mathrm{Sr}_x\mathrm{Nb}_{2(1-x)}\mathrm{O}_{5-4x} + x\mathrm{CO_2}$$

where $0.5 \leqslant x \leqslant 0.8$, with compositions as in Table 1.

X-ray diffraction

The X-ray diffraction pattern of sample S1 ($x = 0.5$) was in broad agreement with the published X-ray powder data for $SrNb_2O_6$ (JCPDS No. 28-1243). No other phase was detected.

The X-ray diffraction pattern of sample S6 ($x = 0.667$) was in good agreement with the published X-ray powder diffraction data for $Sr_2Nb_2O_7$ (JCPDS No. 28-1246). No other phase was detected.

Table 1
Compositions of samples prepared in the SrO-Nb_2O_5 system and the phases present, as determined by X-ray diffraction

Sample	x	Phases present
S1	0.500	$SrNb_2O_6$
S2	0.533	$SrNb_2O_6 + Sr_2Nb_2O_7$
S3	0.567	$SrNb_2O_6 + Sr_2Nb_2O_7$
S4	0.600	$SrNb_2O_6 + Sr_2Nb_2O_7$
S5	0.633	$SrNb_2O_6 + Sr_2Nb_2O_7$
S6	0.667	$Sr_2Nb_2O_7$
S7	0.700	$Sr_2Nb_2O_7 + Sr_4Nb_2O_9$
S8	0.733	$Sr_2Nb_2O_7 + Sr_4Nb_2O_9$
S9	0.767	$Sr_2Nb_2O_7 + Sr_4Nb_2O_9$
S10	0.800	$Sr_4Nb_2O_9$

The major lines in the X-ray diffraction pattern of sample S10 ($x = 0.800$) were in good agreement with published X-ray powder diffraction data for $Sr_4Nb_2O_9$ (JCPDS No. 38-1029). A number of additional reflections were observed, possibly indicative of a long-range supercell. No other phase was detected.

Samples S2, S3, S4 and S5 all contained a mixture of $SrNb_2O_6$ and $Sr_2Nb_2O_7$. As determined from the relative intensities of the major peaks in the X-ray diffraction patterns, S2 contains a significant amount of $Sr_2Nb_2O_7$ even though the stoichiometry of the sample ($x = 0.53$) is close to that of $SrNb_2O_6$. The data indicate that samples S3, S4 and S5 contain progressively higher proportions of $Sr_2Nb_2O_7$, with only trace amounts of $SrNb_2O_6$ being observed in S5 ($x = 0.633$). No phases other than $SrNb_2O_6$ and/or $Sr_2Nb_2O_7$ were observed in samples S1 to S6.

Samples S7, S8 and S9 all contained a mixture of $Sr_2Nb_2O_7$ and $Sr_4Nb_2O_9$. As determined from the relative intensities of the major peaks in the X-ray pattern, the ratio of $Sr_4Nb_2O_9$ to $Sr_2Nb_2O_7$ increases from S7 to S9. No phases other than $Sr_2Nb_2O_7$ and/or $Sr_4Nb_2O_9$ were observed in samples S6 to S10.

Transmission electron microscopy

Selected area electron diffraction in the TEM confirmed the presence of the bulk phases in the samples prepared, as determined by X-ray powder diffraction. In sample S1 ($x = 0.500$) only those diffraction patterns corresponding to $SrNb_2O_6$ were obtained, a typical example of which is shown in Fig. 2a. The crystal fragment has been oriented with the [010] zone axis parallel to the electron beam to give a diffraction pattern of the a^*c^* projection of the monoclinic pseudo-orthorhombic modification of $SrNb_2O_6$ ($a = 1.102$ nm, $b = 0.773$ nm, $c = 0.560$ nm, $\gamma = 90°2'$) as determined by Brusset *et al.*[2] The reflections are sharp with no sign of diffuse scattering or streaking, indicating that the material is well ordered. Figure 2b shows the corresponding lattice image, which is well ordered over a large area. The spacing of the dark fringes corresponds to the unit cell parameter in the a direction, whilst the separation of the pairs of white dots correlates to the c axis. The pairs of white dots are interpretable as the pairs of edge-sharing NbO_6 octahedra present in the $SrNb_2O_6$ structure.

In sample S6 ($x = 0.667$), only those diffraction patterns corresponding to $Sr_2Nb_2O_7$ were obtained. Figure 3a shows an electron diffraction pattern of the a^*b^* projection of the crystal structure of $Sr_2Nb_2O_7$. The reflections are sharp with no sign of diffuse scattering or streaking indicating a high degree of order, particularly along the b axis with its long unit cell parameter. This particular diffraction pattern is incommensurate in the (100) direction, with the ($2k0$) reflections forming a zigzag pattern along the (010) direction. Only some of the diffraction patterns obtained from this sample exhibit incommensurability in this manner, the majority being found to be commensurate.

Figure 3b shows the lattice image corresponding to the electron diffraction pattern of Fig. 3a. The crystal is seen to be highly ordered, with the fringes forming a regular repeating pattern throughout the area visible. The unit cell is delineated in the b direction by the long dark fringes separating the blocks of perovskite-type

(a)

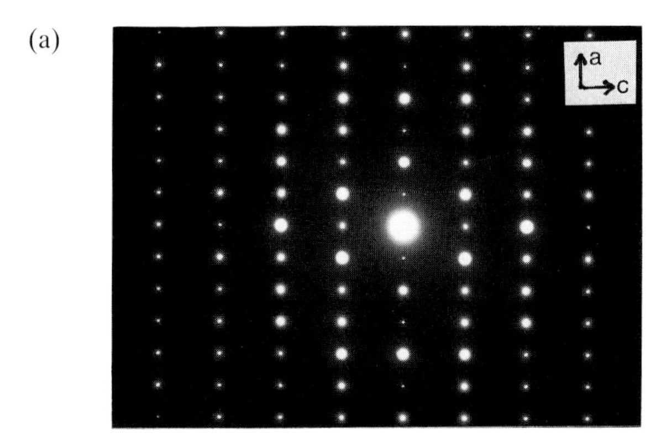

(b)

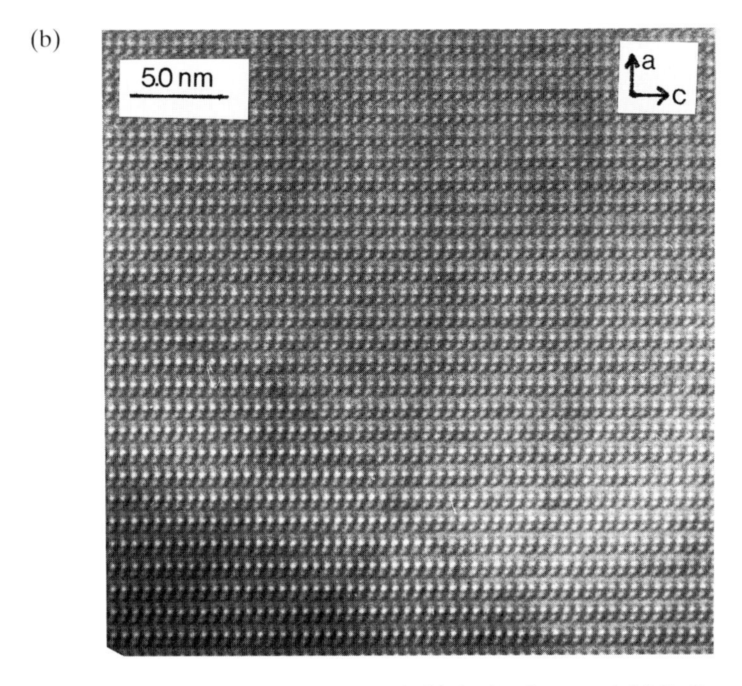

Figure 2 (a) electron diffraction pattern and (b) lattice image of $SrNb_2O_6$ projected down *b* axis. ×2 750 000

material with a separation *b*/2. Between each pair of dark fringes rows of white dots can be resolved. Each row consists of four white dots which can be interpreted as each representing one NbO_6 octahedron. On crossing the dark fringes these rows of NbO_6 octahedra are seen to shift by *a*/2 which is why there are two slabs of perovskite-type material in each unit cell of $Sr_2Nb_2O_7$.

Figure 3c shows a different area of the same crystal fragment as Fig. 3b, but with a significant degree of disorder induced by beam damage. The crystal now consists of blocks of varying widths with the dark fringes that separate them less clearly

(a)

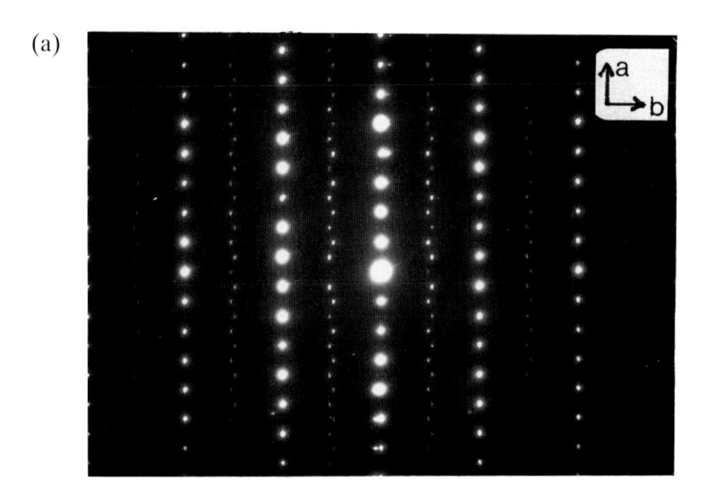

(b)

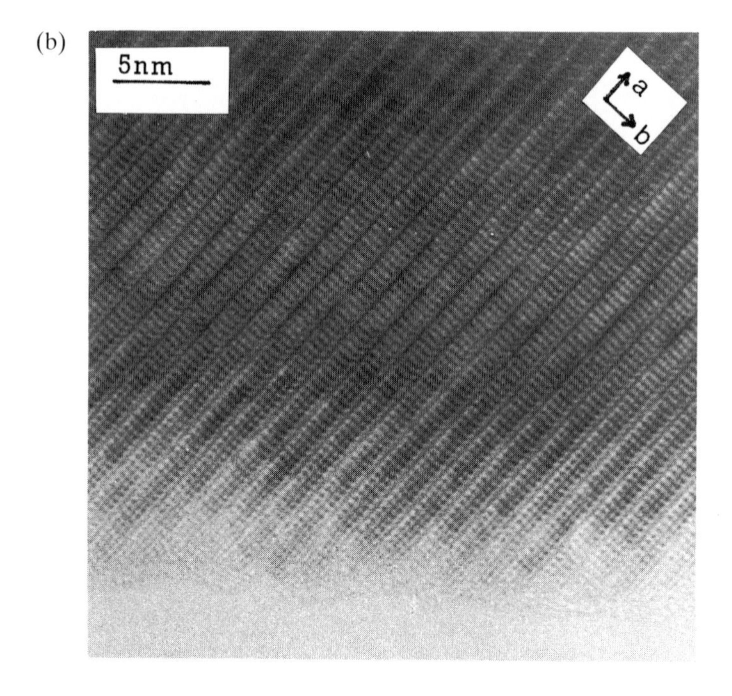

Figure 3 (a) electron diffraction pattern and (b) lattice image of $Sr_2Nb_2O_7$ projected down *c* axis, showing a well-ordered crystal structure. (c) different area of the same crystal fragment showing disorder induced by beam damage effects. ×2 750 000

(c)

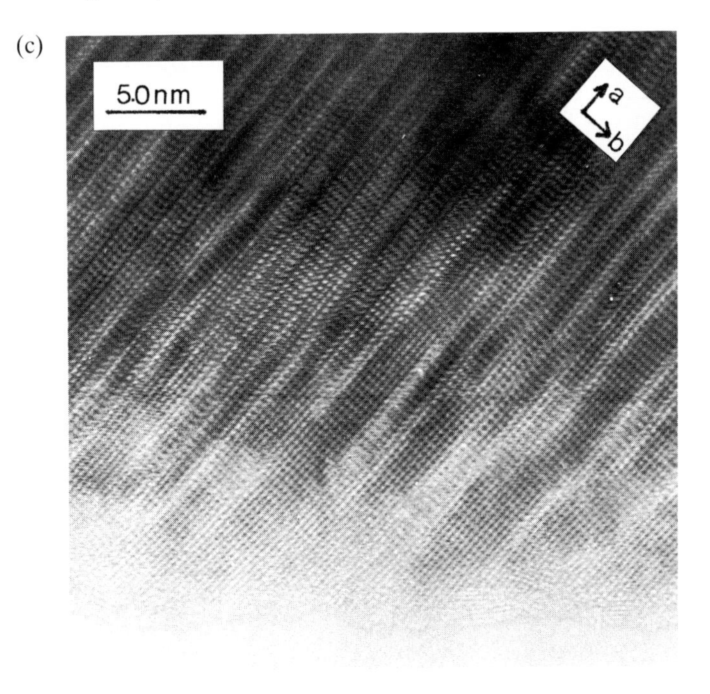

Figure 3—*contd.*

defined. Indeed the dark fringes have disappeared from large parts of the crystal, particularly near the edge, where the crystal is thinner and the heating effect of the electron beam does more damage. Whereas Fig. 3(b) consisted solely of well-ordered blocks of material four NbO_6 octahedra wide, the beam damaged material of Fig. 3(c) consists of blocks 4, 8 or 12 octahedra in width, corresponding to the formation of $n = 2$, $n = 4$ and $n = 6$ type material respectively, along the thin edge of the crystal.

In sample S10 ($x = 0.800$), no diffraction patterns corresponding to $Sr_2Nb_2O_7$ were obtained. Instead diffraction patterns from a complex structure with a large unit cell were identified, a typical example of which is shown in Fig. 4a. The reflections are sharp with no diffuse scattering, though there is evidence of some streaking along one of the crystallographic directions, indicating disorder. The spacing of the diffracted spots in the reciprocal lattice corresponds to unit cell dimensions of 1.19 nm and 1.39 nm with an interaxial angle of 75°. Figure 4b shows the corresponding lattice image, which appears to be well ordered. The structure is block-like in nature and a possible projected unit cell corresponding to the parameters derived from the diffraction pattern of Fig. 4a is indicated. A precise match for these data with a published crystal structure has yet to be achieved, but it may be closely related to the perovskite-type structure with a large orthorhombic supercell proposed for the high-temperature form of $Sr_4Nb_2O_9$ by Hervieu *et al.*[3]

Table 2
Compositions of samples prepared in the SrO-Nb_2O_5-TiO_2 system

Sample	y	$\%TiO_2$
S6	2.00	0.00
ST1	2.05	1.61
ST2	2.15	4.55
ST3	2.25	7.1
ST4	2.50	12.50
ST5	2.75	16.67
ST6	3.00	20.00

Samples S2, S3, S4 and S5 all gave diffraction patterns and lattice images which corresponded either to $SrNb_2O_6$ or $Sr_2Nb_2O_7$. Diffraction patterns from both phases were obtained in all of these samples and were sharp and well ordered. There was no evidence of any structural changes in either of the two phases present that could be related to variations in the chemical compositions of the samples.

Samples S7, S8 and S9 all gave diffraction patterns and lattice images which corresponded either to $Sr_2Nb_2O_7$ or to the complex phase observed in sample S10 with nominal composition $Sr_4Nb_2O_9$. Diffraction patterns from both phases were obtained in all of these samples and were usually sharp and well ordered. There was no evidence of any structural changes in either of the two phases present that could be related to variations in the chemical compositions of the samples.

4.2 The SrO–Nb_2O_5–TiO_2 System

Samples were prepared with composition according to the formula

$$y\mathrm{SrCO_3} + \mathrm{Nb_2O_5} + (y-2)\,\mathrm{TiO_2} \rightarrow \mathrm{Sr}_y\mathrm{Nb_2Ti}_{(y-2)}\mathrm{O}_{3y+1} + y\mathrm{CO_2}$$

where $2 \leqslant y \leqslant 3$, with compositions as in Table 2.

X-ray diffraction

The X-ray powder diffraction data from sample ST1 ($y = 2.05$) is very similar to that of samples S6 ($x = 0.667$, $y = 0$) — i.e. the data correspond closely to that of $Sr_2Nb_2O_7$ (JCPDS No. 28-1246), with some differences observed in the intensities of some of the weaker Bragg reflections around $2\theta = 32°$. In sample ST2 ($y = 2.15$) these weaker Bragg reflections have increased in intensity to be comparable to the main $Sr_2Nb_2O_7$ reflection occurring at $2\theta = 27°$. In ST3 ($y = 2.25$), the main $Sr_2Nb_2O_7$ reflection at $2\theta = 27°$ is no longer discernible and the X-ray diffraction data are relatively unchanged as the composition evolves from ST3 ($y = 2.25$) to ST6 ($y = 3.00$).

(a)

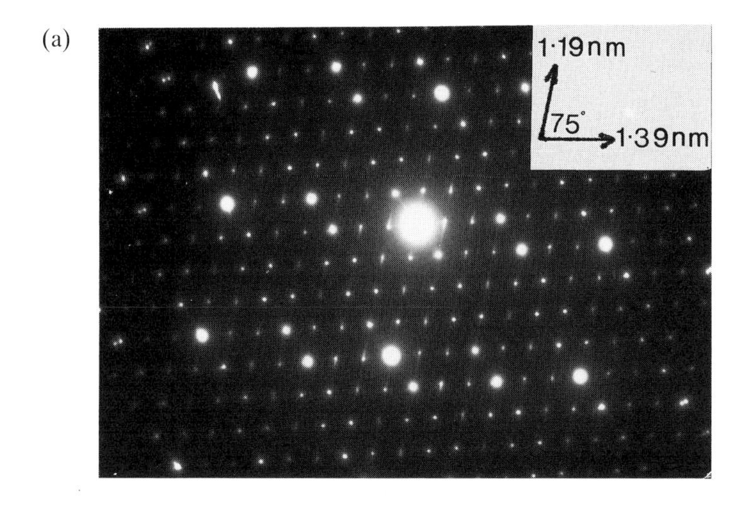

(b)

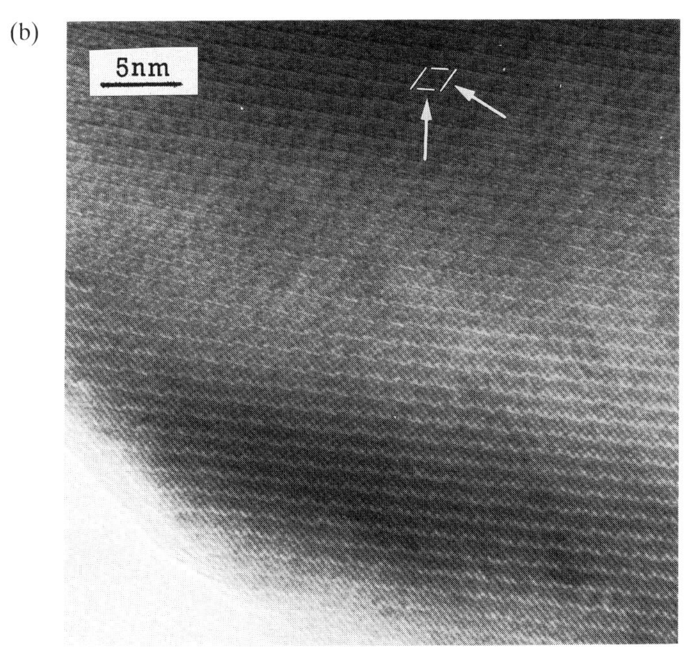

Figure 4 (a) electron diffraction pattern and (b) lattice image of a crystal fragment from a sample of overall composition $Sr_4Nb_2O_9$. A possible projected unit cell is indicated.

Transmission electron microscopy

All the selected area electron diffraction patterns that were obtained from the samples ST1–ST6, were identified as arising from the $Sr_2Nb_2O_7$-type structure. Diffraction patterns from no other phases were identified. Careful measurement of high-magnification prints of the diffraction data exhibited a progressive evolution with composition of the spot spacings along the b^* axis of the diffraction patterns,

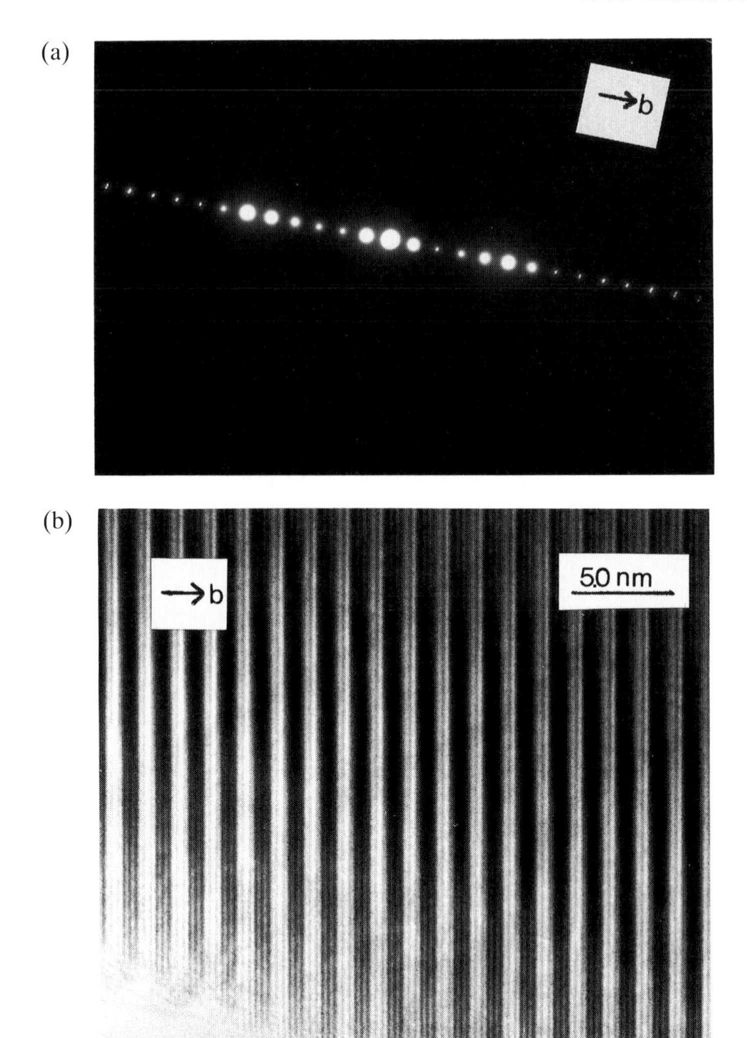

Figure 5 (a) electron diffraction pattern and (b) lattice image of a crystal fragment from sample ST5, with overall composition $Sr_{2.75}Nb_2Ti_{0.75}O_{9.25}$. ×2 750 000

from a value corresponding to 2.67 nm ($n = 2$ type material) for ST1 to a value corresponding to 3.88 nm ($n = 3$ type material) for ST6. No other changes that could be related to variations in composition were observed in the patterns. Although many of the patterns recorded corresponded to $n = 2$ and $n = 3$ type material, a large number were measured and found to correspond to non-integer n-values, the value of n increasing with Ti content.

Figure 5a shows a diffraction pattern of a crystal fragment from sample ST5, with a b axis lattice parameter $b = 3.27$ nm, corresponding approximately to $n = 2.5$ type material. The spots in the diffraction pattern are sharp, with no sign of diffuse scattering or streaking, indicating a high degree of order. The corresponding lattice image in Fig. 5b confirms that the crystal is highly ordered, with all of the perovskite-type slabs apparently of identical width. At the top of the figure, in the thicker part of the crystal, the dark fringes denoting the edges of these slabs are readily apparent, but on moving to the thinner regions at the edge of the fragment, the fringes become more difficult to identify. This is due to changes in thickness and defocus and possibly beam damage effects similar to those shown in Fig. 3c.

A number of the diffraction patterns obtained exhibited a similar type of incommensurability to that observed in Fig. 3a, with ($2k0$) reflections forming a zig-zag pattern parallel to the b^* direction. Some of these diffraction patterns also exhibited diffuse streaking parallel to (010), indicating disorder in the widths of the blocks of perovskite-type material.

5. DISCUSSION

Phase analysis by X-ray powder diffraction indicates that there are three major phases in the $SrO–Nb_2O_5$ system in the region $50\%SrO:50\%Nb_2O_5$ to $80\%SrO:20\%Nb_2O_5$. These are the two boundary phases $SrNb_2O_6$ and $Sr_4Nb_2O_9$ and the perovskite-related $Sr_2Nb_2O_7$, all other compositions in this region producing a mixture of two of these phases. Transmission electron microscopy has provided no evidence of any structural evolution of these three major phases with composition and it is therefore concluded that they correspond to well-defined compounds, with extremely narrow regions of solid solubility. Doping of $Sr_2Nb_2O_7$ by substituting with Ti^{4+} for Nb^{5+} produced a range of different $Sr_2Nb_2O_7$-related structures coresponding to material with $2 \leqslant n \leqslant 3$, which is in good agreement with the structural mode based on valence considerations described earlier. This model will form the basis of further study.

REFERENCES

1. N. Ishizawa, F. Marumo, T. Kawamura and M. Kimura: *Acta Cryst.*, 1975, **B31**, 1912–1915.
2. H. Brusset, Mme Gillier-Pandraud and S. D. Voliotis: *Mat. Res. Bull.*, 1971, **6**, 5–14.
3. M. Hervieu, B. Raveau, J. Lecomte and J. P. Loup: *Chemica Scripta.*, 1985, **25**, 206–211.

0–3 Piezoceramic–Thermoplastic Polymer Composites

M. A. WILLIAMS, D. A. HALL and A. K. WOOD
Materials Science Centre, University of Manchester and UMIST, Grosvenor St, Manchester, M1 7HS, UK

ABSTRACT
A number of thermoplastic polymers are evaluated with respect to their use in 0–3 piezoceramic polymer composites. Materials were conventionally poled in a heated silicone oil bath and characterised by measurements of electrical resistivity, dielectric properties, and piezoelectric coefficients. The composites were prepared using lead titanate and commercial modified lead titanate powders in a number of different thermoplastic polymers. The polymers were found to possess mainly high resistivities ($>10^8$ Ωm) even at the elevated temperatures used for poling. Poling problems were therefore encountered for the majority of the materials investigated due to the unfavourable resistivity mismatch. However, it was found that by using the commercial powder, which possessed a higher resistivity, a more balanced resistivity match was achieved and active composites could be produced with useful piezoelectric coefficients. For example, materials were prepared that exhibited a d_{33} of 26 pCN^{-1} and g_{33} of 88 mV mN^{-1}. Further work on thermoplastic systems is expected to yield more active materials, which also have important processing advantages over conventional thermosetting systems.

1. INTRODUCTION

In 1917, inspired by the submarine detection problem of World War I, Langevin applied the converse and direct piezoelectric effects to the emission and detection of underwater sound waves by means of large size quartz plates[1] and thus opened the field of ultrasonics and hydroacoustics. In 1946 polycrystalline piezoelectric ceramics were developed,[2,3] with the discovery of ferroelectricity in materials such as barium titanate and, later, other perovskites. Recent years have witnessed the growth of materials engineering with the optimisation of specific properties by carefully patterned inhomogeneous solids. Following this trend, the latest step in the evolution of piezoelectric materials is the development of piezoelectric multiphase composites which have some outstanding advantages over conventional ceramics.[4,5] The idea of connectivity proposed by Newnham *et al.*[6] is an important feature of composites since physical properties can change by many orders of magnitude depending on the manner in which each phase is self-connected. For diphasic composites, there are 10 distinct classes. Of these, the 1–3, 3–3 and 2–2 type composites give superior properties when used in the hydrostatic mode, but are of a more complex nature and difficult to manufacture compared with the much simpler 0–3 connectivity.

0–3 composites, comprising piezoelectric ceramic particles dispersed within a polymer matrix, were selected for their ease of processing coupled with respectable hydrostatic sensitivity. To date, most active research in 0–3 composites has been directed towards epoxy-based composites or other common thermosets. The obvious disadvantage of such systems is their inflexibility regarding further shaping after the initial curing stage. Thermoplastics offer the advantage of flexibility in the forming stage, where the material can be shaped upon the application of heat in a mould. Furthermore, any flash or excess from the mould can be reprocessed, saving energy and materials and thus producing less waste for the environment.

The major difficulty in the use of common non-polar thermoplastics (such as polyethylene and polypropylene) in 0–3 piezoceramic polymer composites arises in the poling process, when the relatively high electrical resistivity of the polymer matrix effectively reduces the poling field which can be applied to the ceramic particles. According to the simple two-layer model,[7] the ratio of electric fields within the two phases of the composite is equal to the ratio of their resistivities

$$E_c/E_p = \rho_c/\rho_p$$

where E_c and ρ_c are the electric field and resistivity of the ceramic phase and E_p and ρ_p those of the polymeric phase. If the polymeric phase has a higher resistivity ($\sim 10^{11}\,\Omega$m) than the ceramic phase ($\sim 10^{7}\,\Omega$m)[8] at the poling temperature, then only a small fraction of the applied electric field will develop within the ceramic particles.

The purpose of the present study is to evaluate the feasibility of using various thermoplastic polymers in 0–3 piezocomposite materials selected largely on the basis of electrical resistivity, since this, as demonstrated above, has been identified as one of the most important criteria for successful poling and hence for the production of active composites.

2. EXPERIMENTAL PROCEDURES

Pure lead titanate ($PbTiO_3$) powder was prepared by conventional solid state reaction of the component oxides PbO and TiO_2 (>99%, Fluka Chemicals Ltd). Mixing was carried out using zirconia balls in propan-2-ol for 2 h, followed by calcination at 900°C for 1 h, and finally milling for a further 1 h to break down any large agglomerates. A commercial modified lead titanate powder (PC6, Morgan Matroc Ltd) was also used in the investigation, since this was expected to exhibit a reasonably high resistivity ($\sim 10^{8}\,\Omega$m at a temperature of 80°C, according to the manufacturer) which in principle should facilitate poling of the composite.

Polypropylene and Primacor 3460 (polyethylene–acrylic acid copolymer, DOW-Chemical) composites were prepared by mixing the polymers with the relevant processing aids (stearic acid and microcrystalline wax in the case of polypropylene) and then adding the ceramic powder on a heated two-roll mill. PVC composites were fabricated using a Z blade mixer into which the paste resin, plasticiser and stabiliser were added, followed by the ceramic powder, and these components mixed for 10 min. The paste was deaerated and placed in an oven at 120°C to produce a pre-gel. Modified polyethylene composites were prepared on the two-roll mill but with no external heating as the intensive shearing action of the rollers was found to produce enough thermal energy to plastically deform the material for mixing. The single phase polymers were fabricated according to the same processing routes with the ceramic phase being omitted. After mixing, the polymers and composites were pressed in a heated press at 10 MPa for 2 min to produce flat, square sheets. Test specimens were cut from the pressed sheets and electrodes applied using air-drying silver paint (Acheson Electrodag 915).

Electrical resistivity measurements were carried out as a function of temperature on circular specimens having a diameter ~55 mm, using a three-terminal guarded electrode configuration. The applied potential for these measurements was 6 V, giving an electric field ~5 V mm^{-1}, and using a heating rate of 1 K min^{-1}.

Poling experiments were carried out using a high-voltage amplifier (Chevin Research HVA1B), with specimens being subjected to the appropriate poling field in a heated silicone oil bath at temperatures from 60 to 120°C for the required time (15 min), and then cooled under field to a temperature of 40°C. A Philips PW1380 X-ray diffractometer was used to determine the degree of poling achieved by scanning over the (001) and (100) peaks, which appear over the range 20–23.5° 2θ for lead titanate. Piezoelectric measurements were also used to characterise the composites using a custom-built d_{33} meter. Finally, dielectric properties were measured at a frequency of 1 kHz using a Hewlett Packard 4284A LCR meter.

3. RESULTS AND DISCUSSION

Composites prepared using the Primacor copolymer did not show any indication of falling within the required resistivity range for poling, giving values $>10^{10}$ Ωm over the temperature range investigated. Due to the problems associated with conventional poling above the melting point of the polymer, the Primacor was not studied further.

The PVC used, contrary to quoted figures of up to 10^{18} Ωm,[9] was found to possess a resistivity many orders of magnitude lower at a value of $\sim 3\times10^{5}$ Ωm (Fig. 1) at a temperature of 60°C. Poling studies of the PVC composites, carried out at 60°C, were partially successful; full reversal of the relative intensities of the (001) and (100) peaks did not occur, although the peak intensities did change significantly (Fig. 2). The composites showed moderate piezoelectric and dielectric properties, with the most highly poled specimen having a d_{33} coefficient of 20 pC/N^{-1} but with a significant loss tangent of ~0.14 (Table 1). This high dielectric loss was attributed to the relatively low resistivity of these composites. The fact that full reversal of the peak intensities did not occur suggests that too low a resistivity for the continuous polymeric phase not only increases the dielectric loss of the resulting composites, but also reduces the electric field applied across the ceramic particles. It is clear that the simple two-layer model used to represent the electric field distribution within the composite no longer applies in these circumstances.

The resistivity of the pure polypropylene sample was found to be very high. However, common processing aids such as stearic acid, which is used for

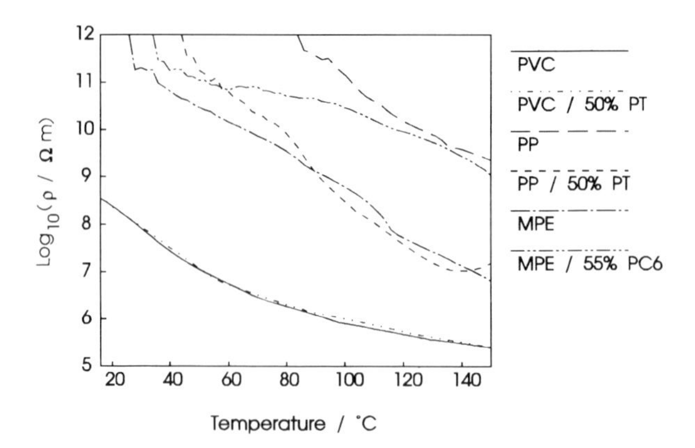

Figure 1 Resistivity results for polymers and composites (PP = polypropylene, MPE = Modified Polyethylene).

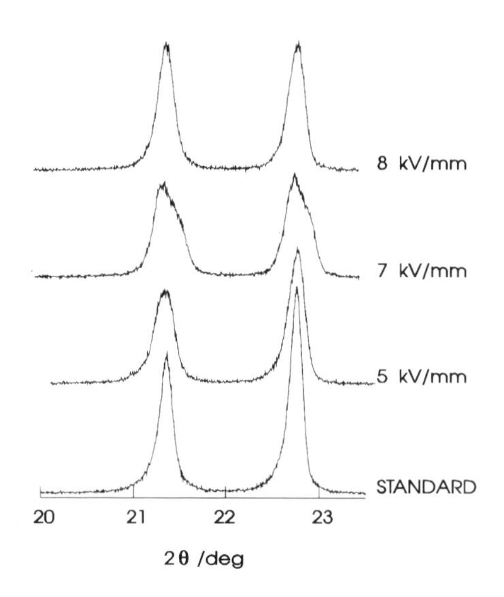

Figure 2 XRD results for poling study of PVC–50 vol.% PT composite.

Table 1
Dielectric and piezoelectric properties of PVC–50 vol.% PT composite

Poling conditions	ε_r (1 kHz)	d_{33} pCN^{-1}	g_{33} $mV\,mN^{-1}$	$g_{33}d_{33}$ $10^{-15}\,m^2\,N^{-1}$	$\tan\delta$
50% VfPT standard	27.1	0	0	0	0.13
50 kV/60°C/15 min	27.2	16	66	1056	0.134
70 kV/60°C/15 min	26.4	18	77	1386	0.149
90 kV/60°C/15 min	29.1	20	78	1560	0.153

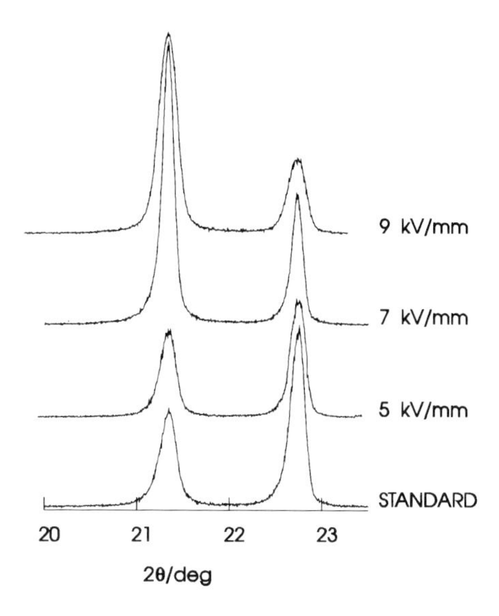

Figure 3 XRD results for poling study of polypropylene–50 vol.% PT composite.

Table 2
Dielectric and piezoelectric properties of polypropylene–50 vol.% PT composite

Poling conditions	ε_r (1 kHz)	d_{33} pCN^{-1}	g_{33} mV mN^{-1}	$g_{33}d_{33}$ 10^{-15} m^2 N^{-1}	tan δ
50% VfPT standard	18.4	0	0	0	0.041
50 kV/120°C/15 min	19.5	6	35	210	0.046
70 kV/120°C/15 min	18.3	10	62	620	0.046
90 kV/120°C/15 min	16.6	12.5	85	1062	0.039

lubrication of the rolls during processing (decreasing tool wear), have ionic character and may contribute to a lowering of resistivity in the presence of water, introduced as moisture adsorbed on the surface of the powder (Fig. 1). The stearic acid content used was 1% by weight and the acid : wax ratio was maintained at 1 : 2. The advantage of using stearic acid to control the resistivity is that it is already used in processing, and therefore the secondary effect of a reduction in resistivity is highly favourable. Unfortunately, in commercial practice, the addition that can be made is limited to less than 5%, as this is the point at which detrimental properties are encountered, such as problems with the extrudate. Even though the relative intensities of the (001) and (100) peaks did reverse (Fig. 3), the resulting piezoelectric properties were poor, with a maximum d_{33} of 12.5 pCN^{-1} (Table 2). This is assumed to be due to the relatively high elastic modulus of the polypropylene compared to that of the other polymers investigated.

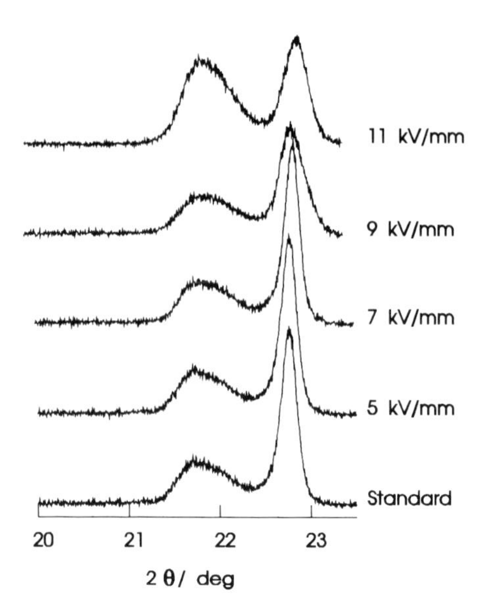

Figure 4 XRD results for poling study of modified polyethylene–55 vol.% PC6 composite.

Table 3
Dielectric and piezoelectric properties of modified polyethylene–55 vol.% PC6 composite

Poling conditions	ε_r (1 kHz)	d_{33} $\mathrm{pCN^{-1}}$	g_{33} $\mathrm{mV\,mN^{-1}}$	$g_{33}d_{33}$ $10^{-15}\,\mathrm{m^2\,N^{-1}}$	tan δ
55% VfPC6 standard	35.1	0	0	0	0.021
50 kV/120°C/15 min	33.1	15	51	765	0.011
70 kV/120°C/15 min	34.2	18	59	1062	0.011
90 kV/120°C/15 min	34.1	23	76	1748	0.011
110 kV/120°C/15 min	33.3	26	88	2288	0.012

The modified polyethylene polymer is considered to be a good candidate for composite fabrication as it was found to be simple to manufacture by cold rolling, and pliable enough to accept loading levels of 55 vol.% of the ceramic powder with no processing problems. Unfortunately, the polymer suffers the usual drawback of non-polar thermoplastics; that of not conforming to a resistivity of the order of $10^7\ \Omega$m at a reasonable poling temperature (Fig. 1).

The XRD results for composites prepared using the PC6 powder were unusual in that the (001) peak showed a substantially greater degree of broadening than the (100) peak (Fig. 4). This seems to indicate internal stress developed preferentially along the *c* axis of the tetragonal particles due to mechanical clamping within the powder agglomerates. The composite prepared with 55 vol.% PC6 powder did, however, show a reasonable degree of piezoelectric activity with a d_{33} value of

26 pCN^{-1} for poling at 120°C for 15 min with an applied field of 11 kV/mm (Table 3). A reasonable degree of 90° domain re-alignment appears to have occurred on poling, as indicated by the growth of the (001) peak. These composites were found to exhibit a relatively low dielectric loss at room temperature ($\tan \delta = 0.011$ at 1 kHz), which may bc of some benefit in hydrophone applications,[5] provided that the degree of poling achieved can be improved.

4. SUMMARY AND CONCLUSIONS

Resistivity measurements were carried out as a function of temperature and were found to be in the range 3×10^5 Ωm to 1×10^{13} Ωm for the polymers and composites investigated over the temperature range 20–150°C. Composites which possessed significant deviation from the apparent optimum of 10^7 Ωm at 90–120°C, were found to be increasingly difficult to pole, and exhibited poor piezoelectric coefficients.

The dielectric loss of composites based on modified polyethylene was found to be of the order of 1% at room temperature, which may be of particular merit for hydrophone applications. The highest piezoelectric coefficients were obtained for the modified polyethylene-based material loaded with 55 vol.% of a commercial modified lead titanate powder (d_{33} = 26 pCN^{-1}, g_{33} = 88 mV mN^{-1}).

Novel thermoplastic polymer-based materials have been developed within the present work which, although exhibiting a reduced performance relative to the epoxy-based materials (d_{33} = 45 pCN^{-1}, g_{33} = 110 mV mN^{-1} for comparable loading levels of powder),[10] are attractive from the point of view of cost of manufacture (being well suited to large-scale production), reduced dielectric loss, and potentially superior moisture resistance.

REFERENCES

1. P. Langevin: French Patent No. 505 703, 1918.
2. A. Von Hippel, R. G. Beckenridge, F. G. Chesley and L. Tisza: *Ind. Eng. Chem.*, 1946, **38**, 1097.
3. B. Wul and I. M. Goldman: *Compt. Rend. Acad. Sci., URSS*, 1946, **51**, 21.
4. H. Banno: *Ferroelectrics*, 1983, **50**, 3–12.
5. R. E. Newnham, *et al.*: *Ferroelectrics*, 1980, **27**, 49–55.
6. R. E. Newnham, *et al.*: *Mat. Res. Bull.*, 1978, **13**, 599–607.
7. K. W. Wagner: *Arch. Electrotech.*, 1914, **2**, 371–374.
8. M. A. Williams, D. A. Hall and A. K. Wood: Proc. ISAF 92, 1992, 508–511.
9. W. J. Roff: *Fibres, Plastics and Rubbers*, 176, Butterworths Scientific Publications, London, 1956.
10. M. A. Williams: unpublished results.

The Morphology of Barium Titanate Powders Produced by the Barium Carbonate–Titanium Dioxide Reaction

I. D. KINNON, L. S. TOVEY and F. L. RILEY*

Wolfson Advanced Ceramics Consortium, School of Materials, University of Leeds, Leeds, LS2 9JT, UK

ABSTRACT
Barium titanate powder of μm dimension is the basis for the production of a large range of high-dielectric electroceramics. The barium titanate is normally formed by high-temperature reaction between barium carbonate and titanium dioxide, and important objectives at this stage are the attainment of maximum completion of reaction, and the formation of a homogeneous powder with a fine, uniform, particle size. This paper examines relevant aspects of the solid state reaction between barium carbonate and titanium dioxide, with particular regard to relationships between reactant and product powder particle sizes and morphologies.

1. INTRODUCTION

Barium titanate ($BaTiO_3$) powder is normally prepared by the solid state reaction between barium carbonate ($BaCO_3$) and titanium dioxide (TiO_2) powders at temperatures in the region of ≈1000°C, according to the overall equation:

$$BaCO_{3(s)} + TiO_{2(s)} = BaTiO_{3(s)}$$

The kinetics of this reaction have been studied by many groups over a wide range of conditions, and models for the reaction mechanism have been constructed.[1–5] In general these models suppose that reaction proceeds by the diffusion of more mobile Ba^{2+} and O^{2-} ions out of $BaCO_3$ particles at their contact points with TiO_2 particles, and into the TiO_2 lattice, within which the Ti and O ionic species are believed to be relatively immobile.[6] With the development of a product layer at the TiO_2/titanate interface the rate-controlling step becomes the similar diffusion of Ba^{2+} and O^{2-} through the $BaTiO_3$ lattice. The development of an intermediate film of Ba_2TiO_4 at the interface between the TiO_2 and the $BaTiO_3$ has been assumed to be possible.

On this basis of understanding it would be expected that the morphology and size of the product $BaTiO_3$ particles should closely approximate to those of the reactant TiO_2 particles, possibly with some size adjustment to allow for the molar volume expansion of approximately 98%, but modified by any sintering of reactants and product taking place during the reaction stage. This is an aspect which does not seem to have received significant attention. The morphology of the $BaCO_3$ particles might be expected to be relatively unimportant, provided that

*To whom all correspondence should be addressed

good physical contact was developed between the two reactant particles to allow transport of Ba^{2+} and O^{2-} ions between the two phases.

$BaTiO_3$ powder is widely used in the electroceramics industry. The degree of completion of this reaction under specific conditions is therefore of considerable interest, as is the size, size distribution, and morphology, of the product powder. Because of the absence of detailed information on these aspects in the literature we have commenced a study of relationships between starting and product powder characteristics, and of the extent to which the course of the reaction might be influenced by the morphology of the reactants, and their state of mixedness.

In the work reported here barium titanate was prepared from a commercial electronic-grade barium carbonate, and two types of titanium dioxide. The morphologies of the reactants and the products after calcination at 1000°C were investigated using transmission electron microscopy (TEM), and particle size analyses were carried out using a laser light-scattering technique. Rates of reaction were measured by isothermal thermogravimetric analysis.

2. EXPERIMENTAL

Barium carbonate ('Electronic Grade', Cookson Ceramics Ltd, Wallsend, UK), titanium dioxide 'T1' (HPT-3 grade, Tioxide (UK) Ltd, Stockton-on-Tees), and a titanium dioxide 'T2' prepared in-house by the controlled hydrolysis of titanium butoxide from isopropanol solution, followed by calcination of the hydrated product, were used in this study. Equivalent spherical particle size distributions of the powders were determined by a standard laser light-scattering technique (Malvern Mastersizer/E, Malvern Instruments, Malvern, UK), over the range 0.1–80 μm. Powders were dispersed in distilled water at pH 7 and ultrasonically treated ($\approx$10 min) until a constant size distribution was obtained at the steady state

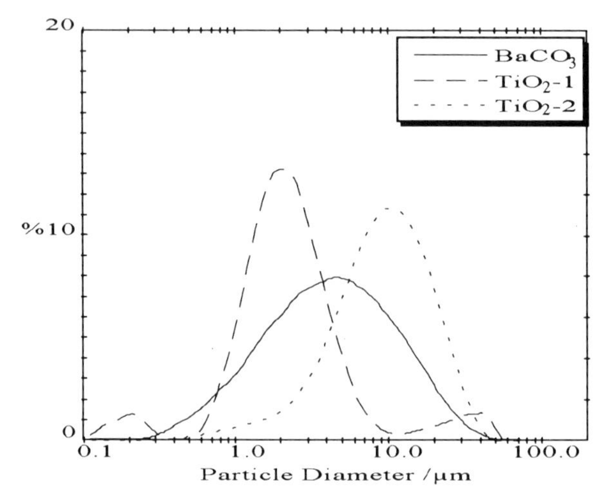

Figure 1 Particle size distribution curves for starting powders.

pH of the suspension. The size distribution curves for the three powders are shown in Fig. 1. Powder morphologies were determined using a transmission electron microscope (Jeol STEM 200CX). The micrographs give, of necessity, an indication of the morphologies of assemblages of particles of less than the mean size, because of the opacity of larger particles.

Equimolar amounts of barium carbonate ($BaCO_3$) and titanium dioxides T1 or T2 were dispersed in propan-2-ol and homogenised with a high shear rate stirrer for 30 min, and then ultrasonically treated for 10 min, using an ice bath to prevent excessive heating. The resulting slurries were dried under an infrared lamp; the dried powders were subsequently calcined without compaction.

Standard thermogravimetric analysis (TGA) of the two powder mixtures, detecting the loss of weight resulting from carbon dioxide evolution, was carried out in flowing laboratory air at a heating rate of 5°C/min^{-1} to determine an optimum isothermal reaction temperature. This showed a conveniently fast rate of weight loss at ≈1000°C. Powder mixtures (≈5 g) contained in a small crucible of high purity alumina were lowered into a furnace preheated to 1000°C, and allowed to react in flowing laboratory air (100 cm^3 min^{-1}) until weight loss had ceased. Powders were re-examined after reaction using TEM.

3. RESULTS and DISCUSSION

The particle size distribution curves (Fig. 1) show that the $BaCO_3$ had a mean dimension of 5 μm, with a fairly broad distribution. The TEM micrograph (Fig. 2) shows the particles to be rod like in morphology with an aspect ratio in the region of 10. The morphology of the $BaCO_3$ particles makes identification with the measured equivalent spherical dimension uncertain, but estimates of mean particle volume correlated reasonably well with the 5 μm dimension obtained. The two

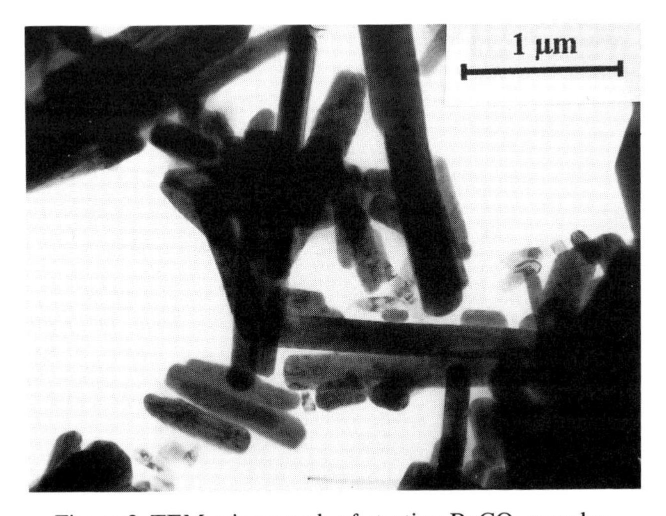

Figure 2 TEM micrograph of starting $BaCO_3$ powder.

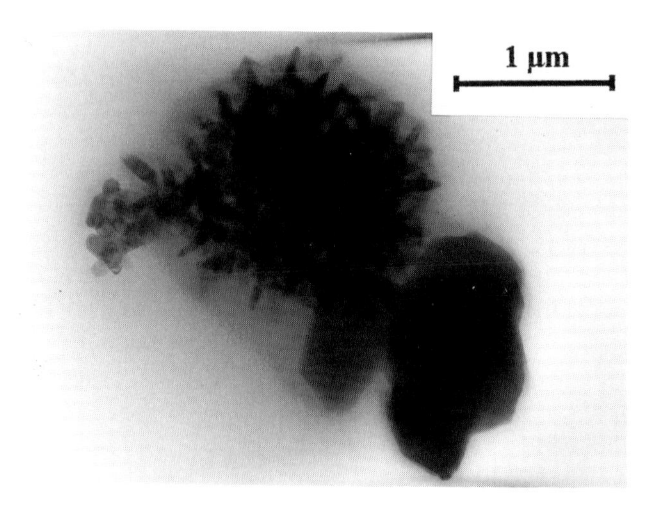

Figure 3 TEM micrograph of starting TiO_2-1 powder.

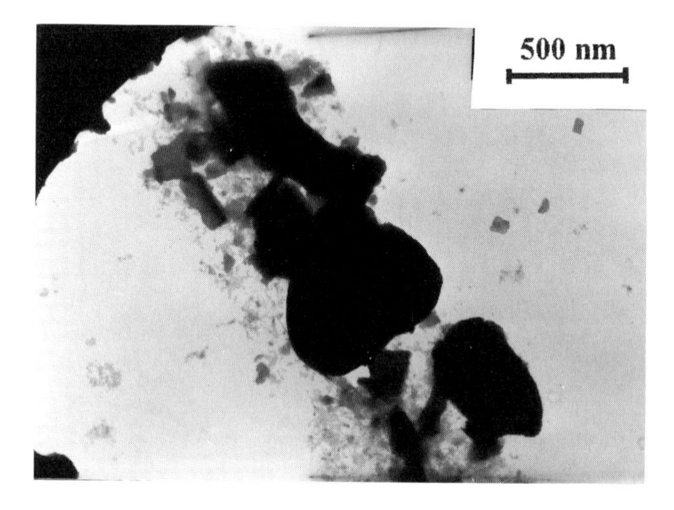

Figure 4 TEM micrograph of starting TiO_2-2 powder.

TiO_2 powders had narrower size distributions, with median values at 1.7 and 6.4 μm; the corresponding TEM micrographs are shown in Figs 3 and 4. The T1 powder consists in the main of strongly agglomerated nano-dimension primary crystals, building up approximately spherical units of 1 to 2 μm dimension, as detected by the particle size analysis. The coarser T2 powder was harder to characterise by TEM; the smallest, 100 to 200 nm, particles seen at the extreme range of the laser analyser trace can, however, be identified with the small, and equiaxed, particles seen in Fig. 4, which is not therefore typical of this powder as a whole. It was clear that the two TiO_2 powders had completely different structures and morphologies.

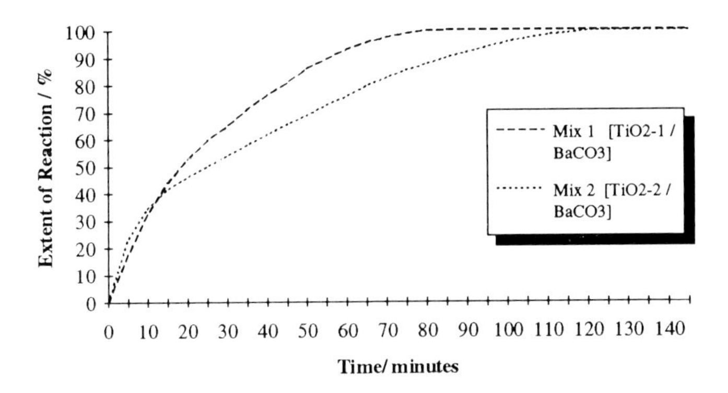

Figure 5 Isothermal reaction curves for the two mixtures.

Thermogravimetric analysis (an example of which is Fig. 5) showed that both mixtures reacted completely in ≈2 h, and that the mixture containing the finer T1 powder reached completion significantly faster. The initially faster reaction rate of the (overall) slower-reacting mixture containing the T2 powder may be explained by the presence of the small proportion of very fine TiO_2 particles (≤200 nm) seen in Fig. 4.

Figures 6 and 7 show the unreacted mixed powders. Again the different morphologies of the two TiO_2 powders are immediately obvious. Some fragmentation of the $BaCO_3$ may have occurred during the mixing stage, as evidenced by the presence of some short, but still non-equiaxed, particles. It is clear that the $BaCO_3$ particles tend to adhere to each other, separating the TiO_2 particles. This indicates some success in the powder homogenisation stage, but a failure to disperse completely the larger, high-aspect-ratio $BaCO_3$ particles. This is one problem which requires more attention. The lack of effective contact between the rod-like $BaCO_3$ and the entrapped, open-structured, TiO_2 particles is striking. This picture provided of the actual configurations of the two reactant powders is thus in striking contrast to the spherical, encapsulating, configuration, assumed in the modelling of reaction mechanisms.[6]

After reaction the powder morphology is markedly different, with complete absence of any relics of the $BaCO_3$ rods (Figs 8 and 9). The $BaTiO_3$ particles are approximately spherical, and show some signs of being agglomerated. Another feature of the reactant powders is illustrated by the particle size distribution curves of Fig. 10, in which the curves of the product powders are shown together with those of the two TiO_2 powders. The similarity is marked, with some shift upscale, possibly the result of the ≈98% molar volume expansion of the TiO_2 crystals (corresponding to a linear dimension expansion of ≈33%) on reaction to form the $BaTiO_3$; it is, however, surprising that an even larger shift in median size was not seen, particularly in view of the likelihood that some sintering of the finer $BaTiO_3$ particles forming from the finer TiO_2 will be starting to occur. This suggests that the mechanistic models proposed do not account for all aspects of the reaction.

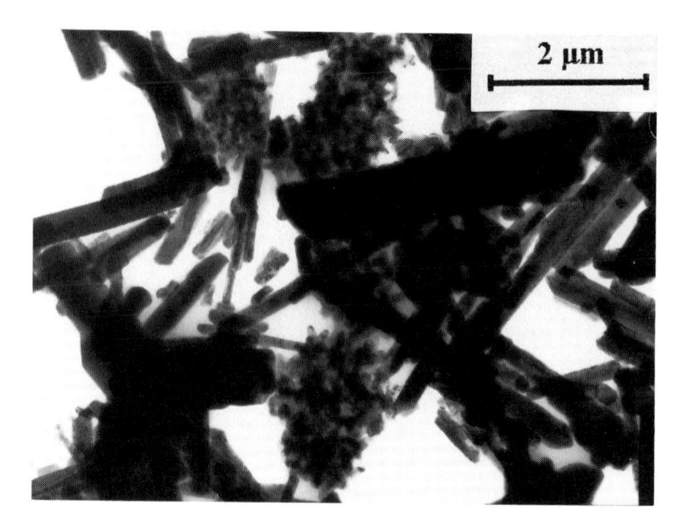

Figure 6 TEM micrograph of mixture 1.

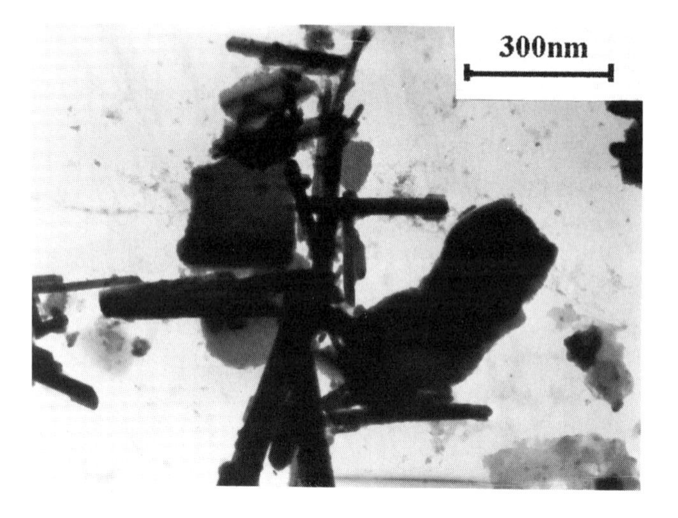

Figure 7 TEM micrograph of mixture 2.

4. CONCLUSIONS

Mixtures of $BaCO_3$ and TiO_2 powders of widely differing morphologies react to form $BaTiO_3$ powder of a median particle size and particle morphology, corresponding most closely to that of the starting TiO_2. This is generally as would be

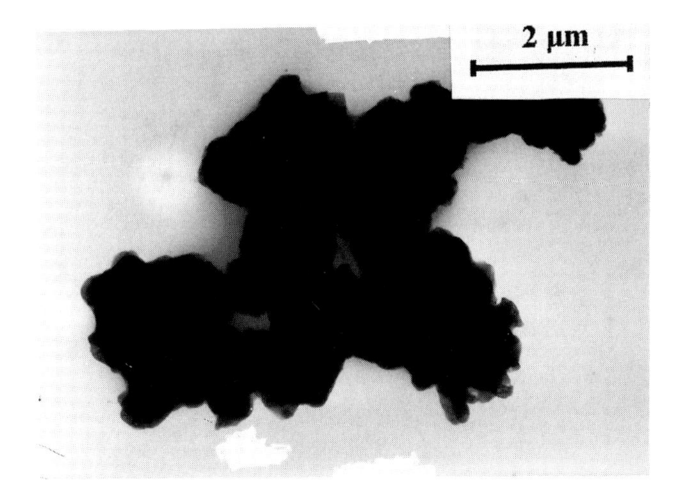

Figure 8 TEM micrograph of reaction products from reaction of mixture 1 at 1000°C for 3 h.

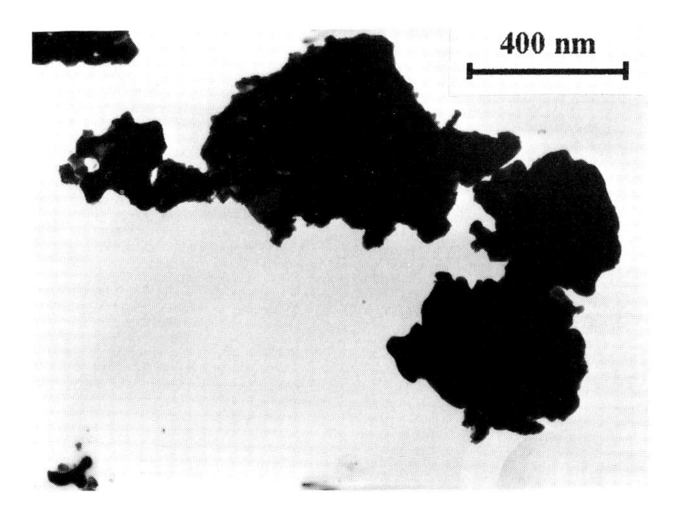

Figure 9 TEM micrograph of reaction products from the reaction of mixture 2 at 1000°C for 3 h.

expected on the basis of a model, suggesting that the reaction mechanism involves the solid state diffusion of Ba^{2+} and O^{2-} ions out of the $BaCO_3$ particle and into the TiO_2. The morphology of the $BaCO_3$ powder particles used here makes good contact between the phases unlikely, and the observed fast rate of reaction is therefore surprising. A considerable improvement in reaction rate would be

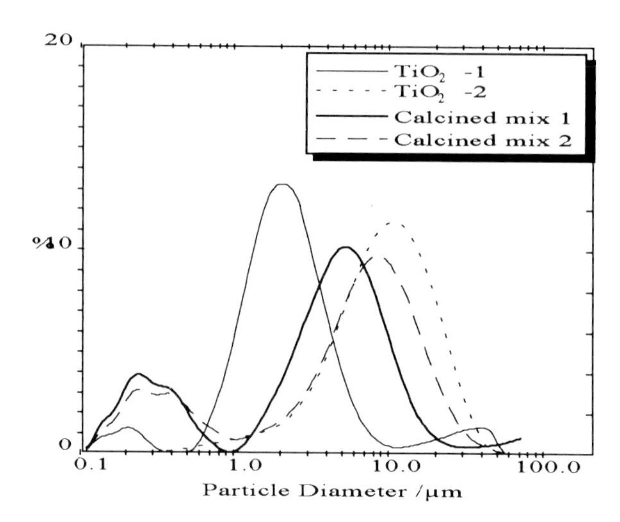

Figure 10 Particle size distribution curves for the two product powders, together with those of reactant TiO_2 powders.

expected if better interdispersion of the two reactant phases could be achieved, and this constitutes a worthwhile goal. Simultaneously, a lesser degree of sintering and agglomeration of the product powder would be permitted, with attendant advantages for succeeding processing stages.

REFERENCES

1. G. Tamman: *Z. Anorg. Allgem. Chem.*, 1923, **149**, 89–98.
2. W. Jander: *Z. Anorg. Allgem. Chem.*, 1927, **163**, 1–30.
3. R. E. Carter: *J. Chem. Phys.*, 1961, **34**, 2010–2105.
4. A. Mangel and J. Doskocil: *Silikaty*, 1969, **13**, 13–28.
5. E. S. Lee, S. Y. Whang and D. Y. Lin: *Youp Hakhoechi*, 1987, **24**, 484–490.
6. J. C. Niepce and G. Thomas: *Solid State Ionics*, 1990, **43**, 69–76.

Aqueous Processing of Barium Titanate Powders

T. J. EADE, I. A. RAHMAN, M. C. BLANCO, L. S. TOVEY and F. L. RILEY*

Wolfson Advanced Ceramics Consortium, School of Materials, University of Leeds, Leeds, LS2 9JT, UK

ABSTRACT

The aqueous phase processing of colloidal barium titanate is currently of considerable interest, in part because of the developing use of aqueous dispersions as a means of tape casting thin films of barium titanate. This paper examines the stability and rheology of aqueous suspensions of barium titanate powder focussing attention on the relative effectiveness as deflocculants of two polymeric surfactants, one anionic, the other cationic.

1. INTRODUCTION

Colloidal dimension ($<2\,\mu m$) barium titanate powder is the basis for a range of high dielectric constant electrolyte materials, which are often required to be produced in thin ($<100\,\mu m$) film form. Thin plastic films of powder are readily obtainable by the tape casting technique, in which barium titanate powder is dispersed with a blend of organo-polymeric plasticiser and binder materials in an organic solvent medium.[1] After formation of the film the solvent is evaporated to leave a thin plasticised sheet of powder ready for further processing stages.

For environmental reasons there is now a move away from the use of non-aqueous solvents towards water as the powder carrier medium. Barium titanate powder is not readily dispersed in water to form a deflocculated suspension,[2–4] unless attention is paid to overcoming the interparticle attractive van der Waals forces, either electrostatically by the development of a sufficiently large surface charge on the particles, or through steric effects resulting from the adsorption of a polymer film on each particle. Attention is thus being directed to controlling the stability and rheology of aqueous barium titanate suspensions through the use of surfactant materials, and recent reports have described the use of polyacrylic acid[5,6] and polyvinyl alcohol[6,7] as adsorbent or deflocculant systems. A feature of barium titanate powders is their tendency, especially if they are slightly stoichiometrically rich with respect to barium oxide, to release Ba^{2+} cations into solution. These cations would be expected to interact with anionic surfactants, of the general type $-[R.COO^-]_n-$, to reduce their activity. For this reason cationic surfactants containing positively charged alkylammonium groups, $-[R' - NR_3^+]_n-$, would as a class seem to be a better alternative. We have therefore carried out a comparative study of the effectiveness at room temperature as a deflocculant of a cationic surfactant, compared with a commonly used ammonium polyacrylate

*To whom correspondence should be addressed

anionic surfactant. Two slightly non-stoichiometric barium titanate powders were used, one BaO-rich, the other TiO_2-rich.

2. EXPERIMENTAL

The barium titanate powders were closely similar: the BaO-rich (Ba:Ti ratio 1.07) powder had a specific surface area of 2.1 $m^2\,g^{-1}$, corresponding to a mean particle dimension of 500 nm; the TiO_2-rich (Ba:Ti ratio 0.99) powder had a specific surface area of 1.7 $m^2\,g^{-1}$, corresponding to a mean particle dimension of 600 nm. Suspensions of powder were prepared in distilled water in polypropene containers, using a standardised ultrasonic treatment (Lucas Dawes 'Soniprobe' on full power for 10 min). Adjustments to pH were made with dilute aqueous hydrochloric acid, or ammonium hydroxide, and pH was measured to ± 0.02 units using a pH meter (Corning 220, Ciba Corning Diagnostics Ltd, Sudbury, UK). The anionic surfactant was an ammonium polyacrylate (Dispex A40, Allied Colloids Ltd, Bradford, UK); the cationic surfactant was an experimental grade (supplied by Allied Colloids Ltd, Bradford, UK) containing quaternary ammonium alkyl groups on a polymer backbone, with charge density closely similar to that of the anionic polymer. All work was carried out at room temperature (21°C).

The extent of powder deflocculation was first assessed semi-quantitatively by dispersing powder in water in standard 25 cm^3 calibrated sedimentation tubes, and measuring sediment density (expressed relative to the true solid density) after 7 days. Particle size distributions were measured with a laser particle size analyser ('Mastersizer', Malvern Instruments, Malvern, UK). To determine the influence of barium ion on the effectiveness of the deflocculant, Ba^{2+} ion was added in the form of a dilute aqueous solution of barium chloride (AnalaR grade, BDH, Poole, UK).

Particle ζ-potential measurements were made over the pH range $\sim$2 to $\sim$10, on dilute ($\sim$0.5 vol.%) suspensions of powder using a laser analyser ('Zetasizer', Malvern Instruments, Malvern, UK), with a 5 $mMol\,l^{-1}$ potassium chloride electrolyte buffer solution. The rheological properties of more concentrated suspensions were measured with a cone-on-plate rheometer (Carri-med SCL500, Rheometer Systems, Dorking, UK), which allowed measurement of shear stress as a function of shear rate, from which apparent viscosity and yield stress values were calculated. Slips containing 30 vol.% of barium titanate powder and varying proportions of surfactant were prepared using a standard 10 min ultrasonic treatment, and cast on to plaster of Paris slabs; cast green densities were measured after drying at 100°C for 2 days.

3. RESULTS AND DISCUSSION

Suspensions of barium titanate powder without deflocculant could not be satisfactorily dispersed in water by adjustment of pH, as judged by the uniformly low sediment relative densities obtained (typically $\sim$20% of theoretical, on the basis of

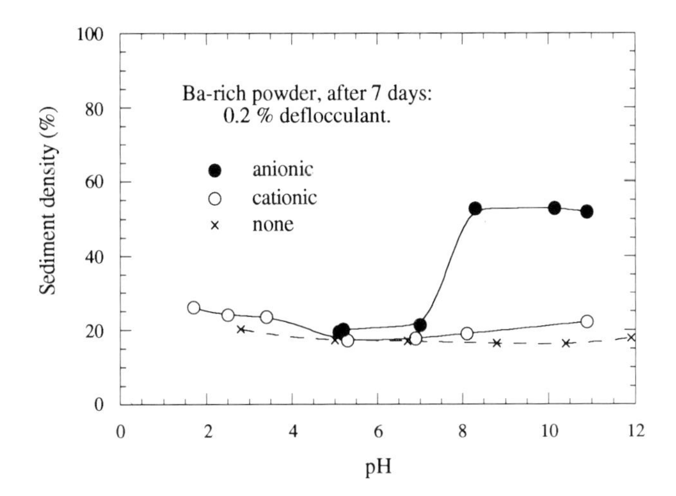

Figure 1 Sediment density as function of pH, without and with addition of (0.2 wt%) deflocculant.

a barium titanate density of 6.02 Mg m^{-3}). The cationic surfactant was almost completely ineffective at deflocculating the powder, at concentrations ranging from 0.1 to 3%, and irrespective of pH. In contrast low concentrations (0.1–0.2 wt%) of the anionic surfactant allowed good deflocculation to be achieved, and consequently high sediment densities of ~60%, provided that the suspension pH was >7. This suggests that the mode of action of the ammonium polyacrylate may in this case be a combination of steric and electrostatic stabilisation (electrosteric stabilisation).

The effectiveness of the two types of surfactant in deflocculating barium titanate is shown in Fig. 1. It seems that the ammonium polyacrylate is probably only effective when the suspension pH is high enough to allow full ionisation of the polyacrylic acid; at pH <7 the concentration of undissociated polyacrylic acid groups is assumed to increase; pH had no influence on the effectiveness of the cationic surfactant. Similar effects were seen with varying surfactant concentration: at pH 9 very low concentrations of anionic surfactant were effective in securing complete deflocculation; the use of large amounts of cationic deflocculant, at concentrations up to 5 times those of the anionic, was without significant effect (Fig. 2).

A more direct measurement of the effectiveness of a surfactant in deflocculating a powder suspension is the apparent particle size: 0.1 wt% of anionic surfactant was effective in breaking down the large proportion of the barium titanate agglomerates to give an apparent mean primary particle size in the region of 0.3 μm. 1 wt% of cationic surfactant achieved some agglomerate breakdown, but a significant proportion of the barium titanate particles was still flocculated, with apparent particle sizes in the range 1–10 μm. A summary of this information, showing the proportion of particles of size <1 μm as derived from the particle size distribution curves for 0.1 wt% of anionic, and 1.0 wt% of cationic surfactant, is given in Fig. 3.

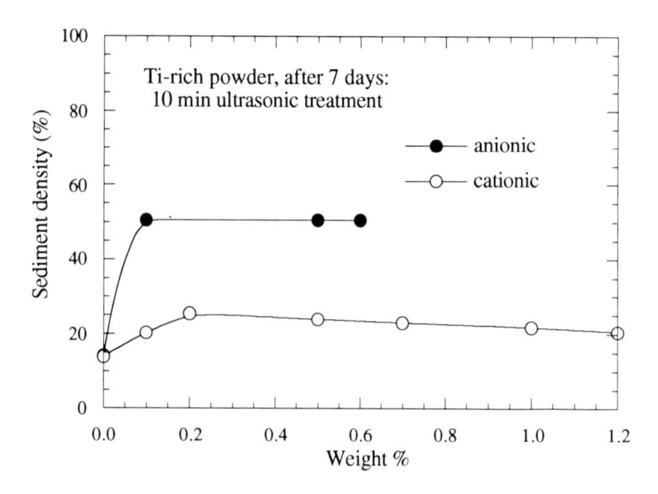

Figure 2 Sediment density as function of amount of deflocculant present, at pH 9.

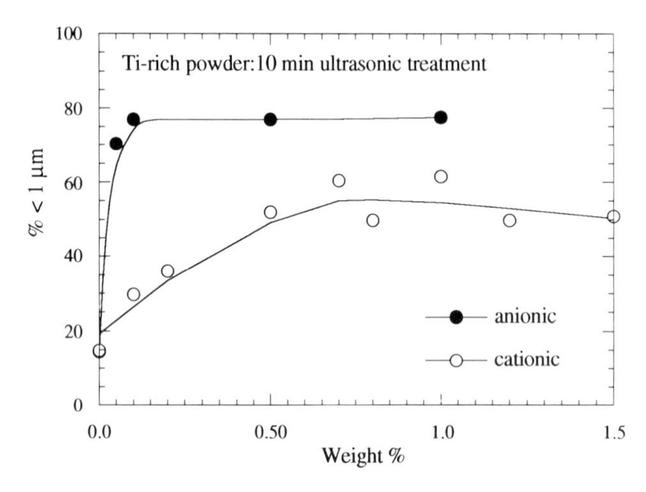

Figure 3 Percentage of particles less than 1 μm in dimension, as function of amount of deflocculant present.

Barium titanate particle ζ-potentials, plotted as a function of pH in Fig. 4, show that there is a slight tendency for the barium-rich powder to develop a higher positive charge at any pH, resulting in a slightly higher isoelectric point (iep). Otherwise the two powders do not appear to differ significantly in behaviour, and for most of the study reported here only the Ti-rich powder was used. Measurements of ζ-potential of suspensions obtained with deflocculant addition show that, in spite of the relative ineffectiveness of the cationic surfactant as a deflocculant, with increasing pH both types of surfactant became strongly adsorbed and, to judge from the surface potentials ($\approx \pm 60$ mV) developed at higher pH values, to approximately the same extent (Fig. 5). This feature was not expected on the basis

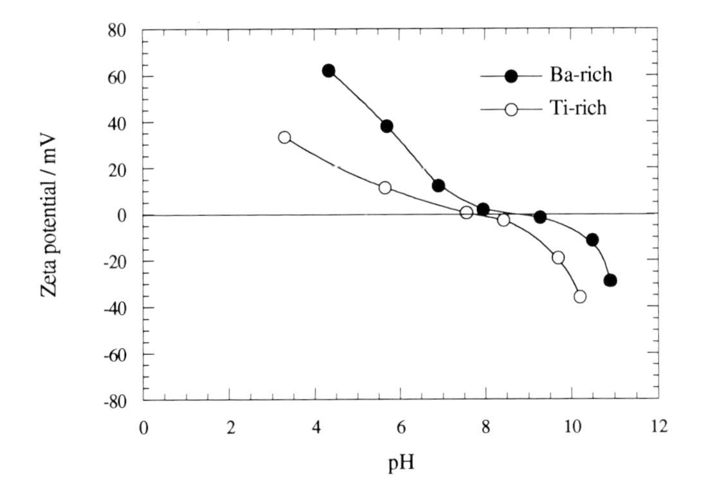

Figure 4 Zeta potential as function of pH for Ba- and Ti-rich powders.

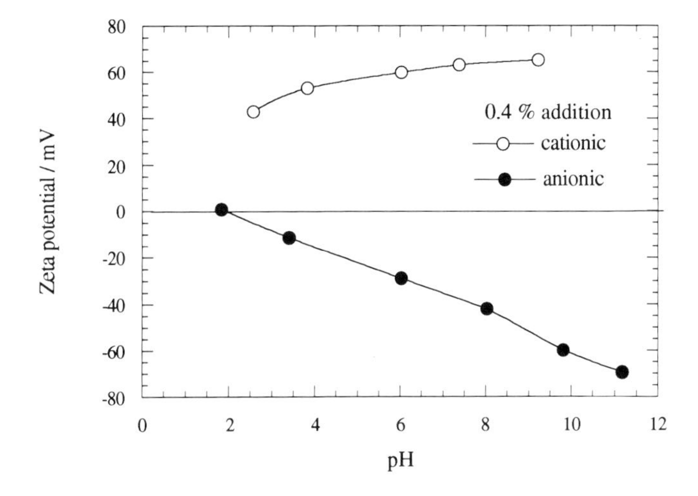

Figure 5 Zeta potential as function of pH with 0.4 wt% deflocculant present.

of the failure of the cationic surfactant to influence the sediment density, and by implication to assist the breakdown of flocs.

The influence of added free Ba^{2+} ion on the ζ-potential developed in the presence of the anionic surfactant is shown in Fig. 6. The anionic surfactant resisted well the presence of small concentrations of Ba^{2+}, and showed itself to be a very effective dispersant for both the Ti-rich and the Ba-rich powders. At very high loadings of Ba^{2+}, as would be expected, the extent of deflocculation increased, as a result of the lowering of the anion activity in solution, and the desorption of the polymeric anions from the barium titanate particle surfaces. This indicates that a considerable excess of cation concentration (with respect to that of

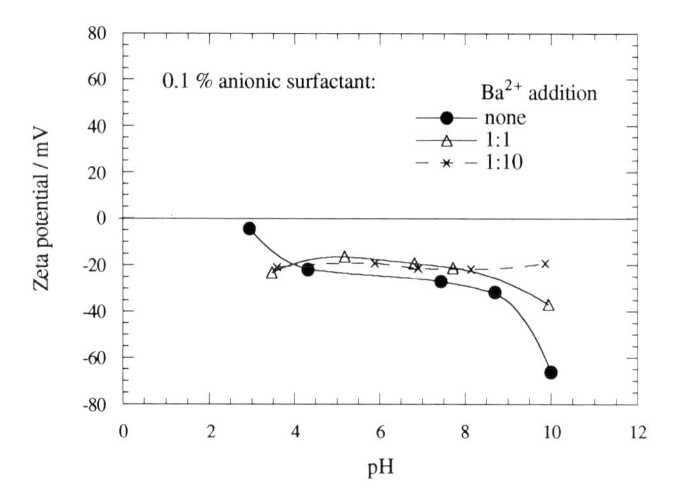

Figure 6 Zeta potential as function of pH with 0.1 wt% deflocculant, and in the presence of free Ba^{2+} ion.

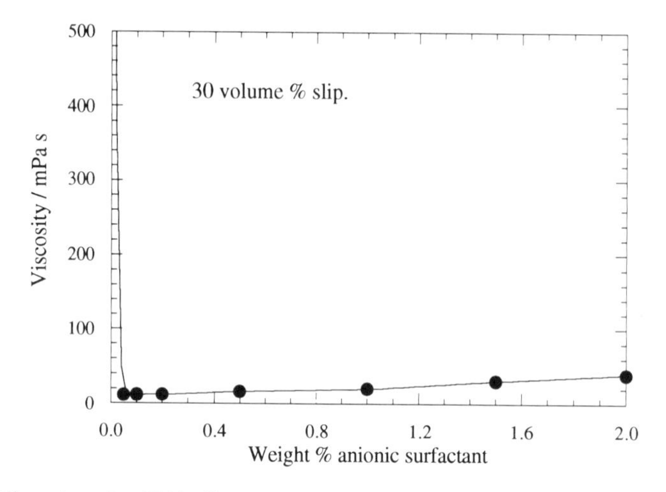

Figure 7 Viscosity of a 30% slip as function of amount of anionic surfactant present.

the anionic charge of the polyacrylic acid chain) is required before surfactant adsorption is affected significantly. The small amounts of Ba^{2+} ion released from the barium titanate particles as a result of natural solubility, or the leaching of uncombined BaO from surfaces, would therefore not be expected to influence significantly the efficiency of the anionic surfactant.

The anionic surfactant was most effective at lowering suspension viscosity and yield stress, with an optimum concentration of ~0.1 wt%. As illustration, the effect of surfactant concentration on the viscosity of a 30 vol.% suspension at pH 9 is shown in Fig. 7. A second practical aspect of suspension processing is the slip-cast green density, which is influenced by the state of flocculation of the suspension. Figure 8 shows the effects of pH on the density of casts prepared from a 30 vol.%

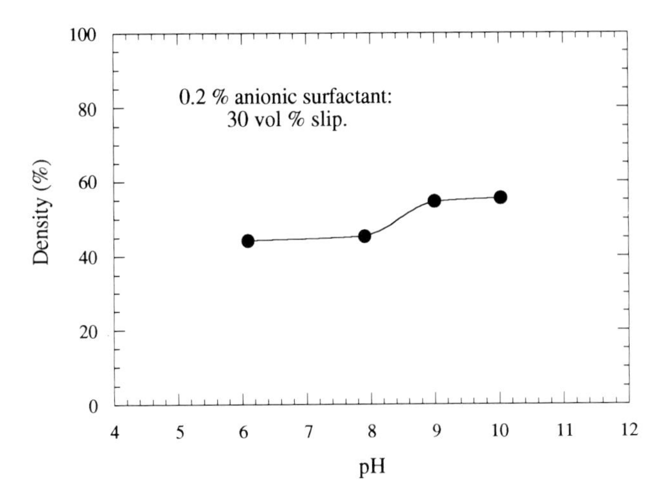

Figure 8 Cast green density as function of pH, obtained from a 30 vol.% suspension containing 0.2 wt% anionic surfactant at pH 9.

suspension of powder at pH 9. This picture essentially mirrors that of Fig. 1, with a marked increase in cast density occurring for pH >8.

The failure of the cationic surfactant to overcome interparticle attractions and to produce good deflocculation even through, on the basis of particle surface charge, it was tending to be strongly adsorbed, was unexpected. At the concentrations of surfactant used (up to 3 wt%) complete coverage of the barium titanate particles would have been expected to be achieved. There are several possible reasons for the failure of the polymer film to maintain adequate interparticle separation distances: as a result of interaction between protruding cationic polymer molecules, particle–particle bridging might occur between contacting barium titanate particles, and lead to floc development; alternatively, adhesion of polymer molecules to the underlying particle surface might locally be weakened through enthalpy and/or entropy effects on the close approach of two coated particles. This would allow movement of polymer away from the approaching surfaces, and a particle–particle distance of approach permitting dominance of the van der Waals attractive component. It is of interest in this context that the input of ultrasonic energy has the effect of increasing the extent of flocculation (Fig. 9), suggesting that an element of thermodynamic stabilisation of the deflocculated suspension may be involved. This behaviour of the surfactant requires a more detailed study, and in particular the use of an adequately wide temperature range to allow identification of a steric stabilisation contribution, and of an entropic component to this effect.[8]

4. CONCLUSIONS

Ammonium polyacrylate at concentrations of ~0.1% on a weight/weight basis, and pH >7, is a very effective dispersant for aqueous suspensions of barium titanate

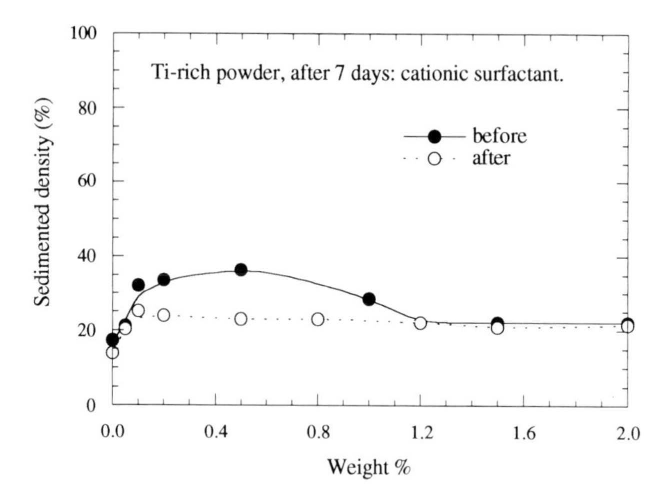

Figure 9 Sediment density as function of amount of cationic surfactant present, before and after ultrasonic treatment.

powder, and allows the formation of stable deflocculated suspensions of ~500 nm particle size powders at their primary particle size level. Concentrations of free Ba^{2+} ions providing a larger equivalent charge than that on the anionic polymer chain considerably reduce the effectiveness of the surfactant. The optimum concentration of the ammonium polyacrylate is determined to be 0.1 wt%, and this level gives, at pH >8, adequately low viscosity suspensions containing up to 40 vol.% of solid, and which yield green cast densities of ~60% of theoretical. A cationic polymer of similar molar mass and charge density to the anionic polymer was relatively ineffective in deflocculating barium titanate suspensions, although ζ-potential measurements showed that strong cation adsorption was occurring. Further work is needed to clarify the behaviour of the cationic polymer and to identify the reasons for its ineffectiveness under the experimental conditions used here.

ACKNOWLEDGEMENTS

Additional financial support in the form of Bursaries and Studentships from the Nuffield Foundation, from the Malaysian Government, and from the Plan Regional de Investigación, Asturias, Spain, is gratefully acknowledged.

REFERENCES

1. Y. Shiraki, N. Otsuke and H. Ninomya: *Yogyo-Kyokai-Shi*, 1974, **82** (9), 470–474.
2. D. A. Bzdawka and D. T. Haworth: *J. Dispersion Sci. Technol.*, 1980, **1** (3), 323–340.

3. S. Mizuta, M. Parish and H. K. Bowen: *Ceram. Int.*, 1984, **10** (2), 43–48.
4. S. Mitzuta, M. Parish and H. K. Bowen: *Ceram. Int.*, 1984, **10** (2), 83–86.
5. Zhien-Chi Chen, T. A. Ring and J. Lemaitre: *J. Am. Ceram. Soc.*, 1992, **75** (12), 3201–3208.
6. A. W. M. de Laat and G. L. T. van den Heuvel: *Colloids and Surfaces A: Physicochem. Eng. Aspects*, 1993, **70**, 179–187.
7. A. W. M. de Laat and W. P. T. Derks: *Colloids and Surfaces A: Physicochem. Eng. Aspects*, 1993, **71**, 147–153.
8. D. H. Napper: *Polymeric Stabilization of Colloidal Dispersions*, Academic Press, London, 1983.

The Effect of ZnO Additions on the Structure and Properties of $Sr_2Nb_2O_7$ Ceramics

F. AZOUGH and R. FREER
Materials Science Centre, University of Manchester/UMIST, Grosvenor Street, Manchester, M1 7HS, UK

ABSTRACT
Ceramics of strontium niobate ($Sr_2Nb_2O_7$) have been prepared by the mixed oxide route. Specimens were sintered at temperatures in the range 1200–1550°C. Fired densities were 92–96% theoretical. Individual grains were needle like or lath like in shape. Hot pressing caused the grains to become oriented. The use of ZnO additions led to a reduction in the sintering temperature and improvement of the microstructure. At 20°C and 1 kHz the relative permittivity was ~57 and $\tan\delta \leqslant 3\times10^{-2}$; the temperature coefficient of capacitance (TCC) was typically −700 ppm/°C.

1. INTRODUCTION

Strontium niobate ($Sr_2Nb_2O_7$) is a member of the family of ferroelectric $A_2B_2O_7$-type compounds having high Curie temperatures. Other important members of this family include $La_2Ti_2O_7$,[1] $Nd_2Ti_2O_7$[2] and $Ca_2Nb_2O_7$.[3] Single crystals of $Sr_2Nb_2O_7$ exhibit strong piezoelectric and electro-optic properties.[4,5] The ferroelectric phase transition is at 1615 K.[5] Ceramics of $Sr_2Nb_2O_7$ exhibit preferential grain growth along the ⟨0k0⟩ direction, causing needle-like or lath-shaped grains. The properties and microstructures of such ceramics depend upon processing conditions. Fukuhara *et al.*[6] investigated the properties of undoped $Sr_2Nb_2O_7$ ceramics and noted the wide variation in dielectric constant and resistivity with frequency and temperature. To increase the density of sintered products they employed hot pressing techniques. In this study, $Sr_2Nb_2O_7$ ceramics have been prepared with additions of ZnO; uniaxial cold pressing and hot pressing methods have been used. The effect of ZnO additions on the microstructure and properties is assessed.

2. EXPERIMENTAL

The starting materials were $SrCO_3$ (Fluka Chemicals, Cat. No. 85893), Nb_2O_5 (99.8% purity, HCST Chemicals), and ZnO (Fluka chemicals, Cat. No. 96484). Powders of $SrCO_3$ and Nb_2O_5 were mixed in a 2:1 molar ratio to yield $Sr_2Nb_2O_7$ after processing. To selected batches, 1 wt% ZnO was added. The mixtures were wet milled for 8 h, dried, calcined at 1100°C for 4 h and pressed uniaxially into discs of 15 mm diameter. Samples were sintered in air in a horizontal tube furnace (Carbolite type CTF-16/75) for 4 h, or hot pressed for 2 h at temperatures in the range 1200–1550°C.

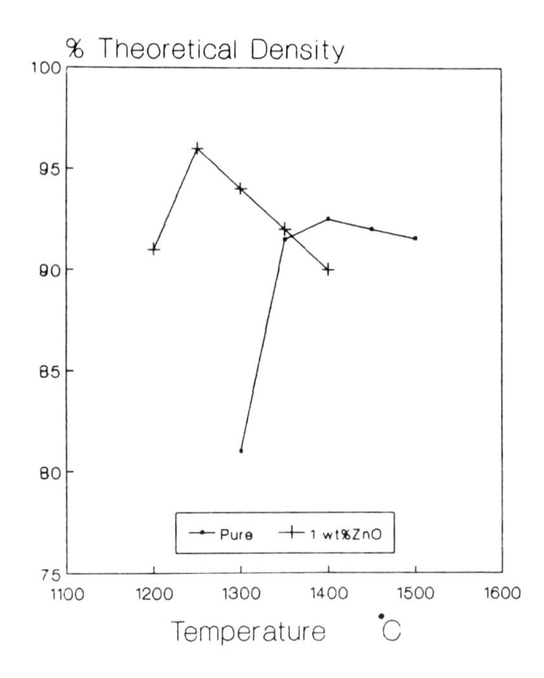

Figure 1 Density of strontium niobate ceramics as a function of sintering temperature: ● undoped, + prepared with ZnO additions.

Products were ground on SiC down to 1200 grade and polished with diamond paste down to 1 μm. X-ray diffraction analysis was performed on a Philips PW 1710 with horizontal goniometer using Cu K_α radiation. Morphologies were studied by optical microscopy and scanning electron microscopy (Philips 505 and 525 instruments), on polished and etched surfaces.

Electrical properties were examined at 1 kHz over the temperature range 20–120°C using a Hewlett Packard HP4284A precision LCR meter in conjunction with a dedicated sample cell.

3. RESULTS AND DISCUSSION

Figure 1 shows the density of strontium niobate ceramics as a function of sintering temperature (at 1 bar total pressure). Undoped specimens had a maximum density of ~92% theoretical for firing temperatures of 1350–1500°C. This is consistent with the findings of Fukuhara *et al.*[6] The addition of 1 wt% ZnO to the starting mixes enabled the sintering temperature to be reduced to 1250°C and increased the maximum fired density to 96% theoretical. Specimens hot pressed at 1300°C had densities of typically 97% theoretical.

The microstructures of undoped strontium niobate ceramics were characterised by wide variation in the shape and size of grains, plus a significant amount of

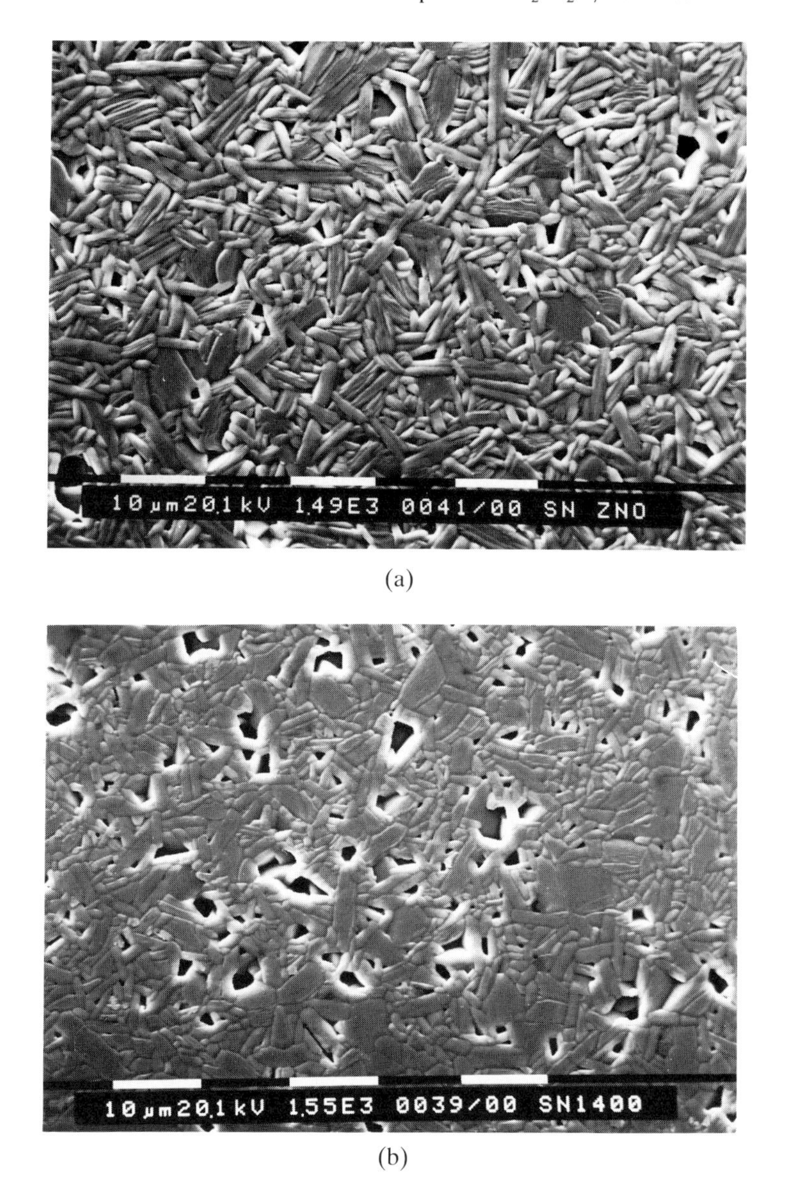

Figure 2 SEM micrographs of uniaxially cold-pressed strontium niobate ceramics (a) undoped, sintered at 1400°C (b) prepared with ZnO additions, sintered at 1250°C.

porosity (Fig. 2a). In contrast, strontium niobate ceramics prepared with additions of ZnO exhibited a more uniform microstructure, less porosity and a smaller range of grain sizes. Individual grains were needle-like or lath-like in shape (Fig. 2b). Specimens sintered at 1250°C had grains $\leqslant 10\ \mu m$ in size (Fig. 2b).

The use of hot pressing enabled high-density specimens, with a high degree of grain orientation, to be obtained. Figure 3a shows the typical microstructure of

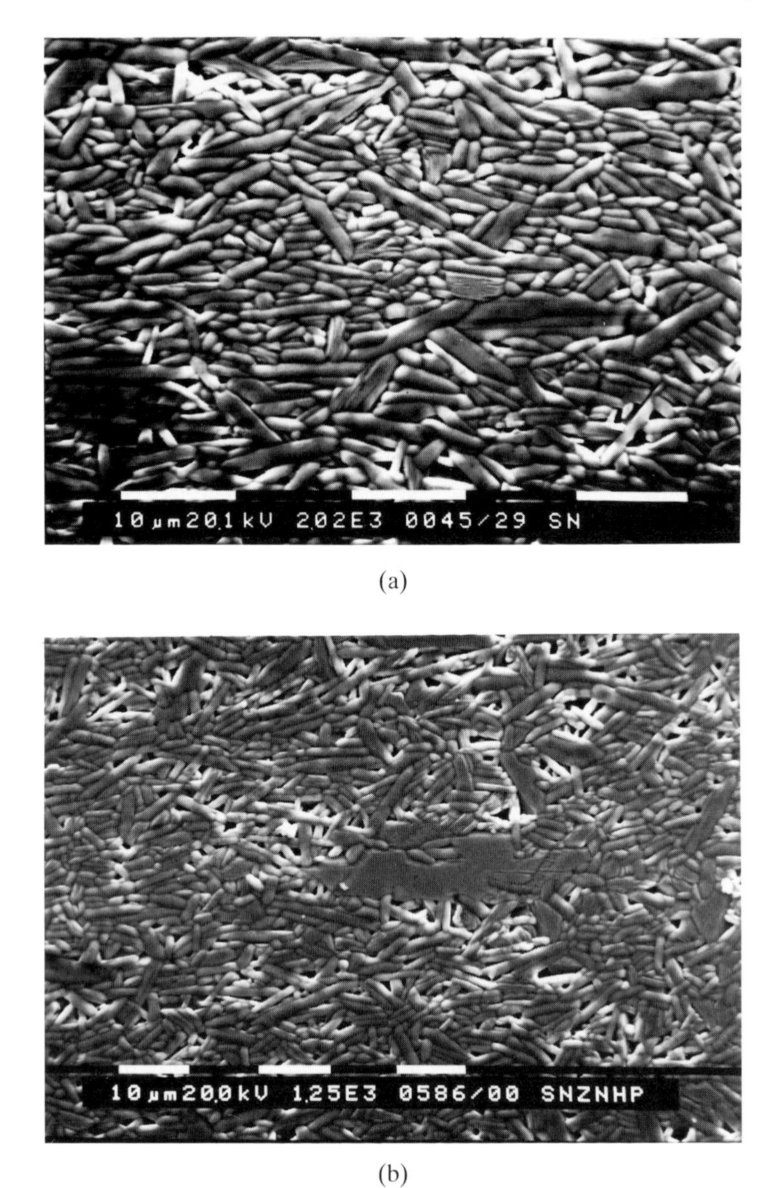

Figure 3 SEM micrographs of hot-pressed strontium niobate ceramics sintered at 1300°C (a) undoped (b) prepared with ZnO additions.

undoped, hot-pressed strontium niobate ceramics. Grain sizes were controlled within narrow bands, and the needle-like morphology was enhanced when ZnO additives were used (Fig. 3b).

X-ray diffraction patterns for uniaxially, cold-pressed specimens of strontium niobate (e.g. Fig. 4a) can be indexed as single-phase products (in accordance with JCPDS data)[7] having randomly oriented grains. As the sintering temperature was

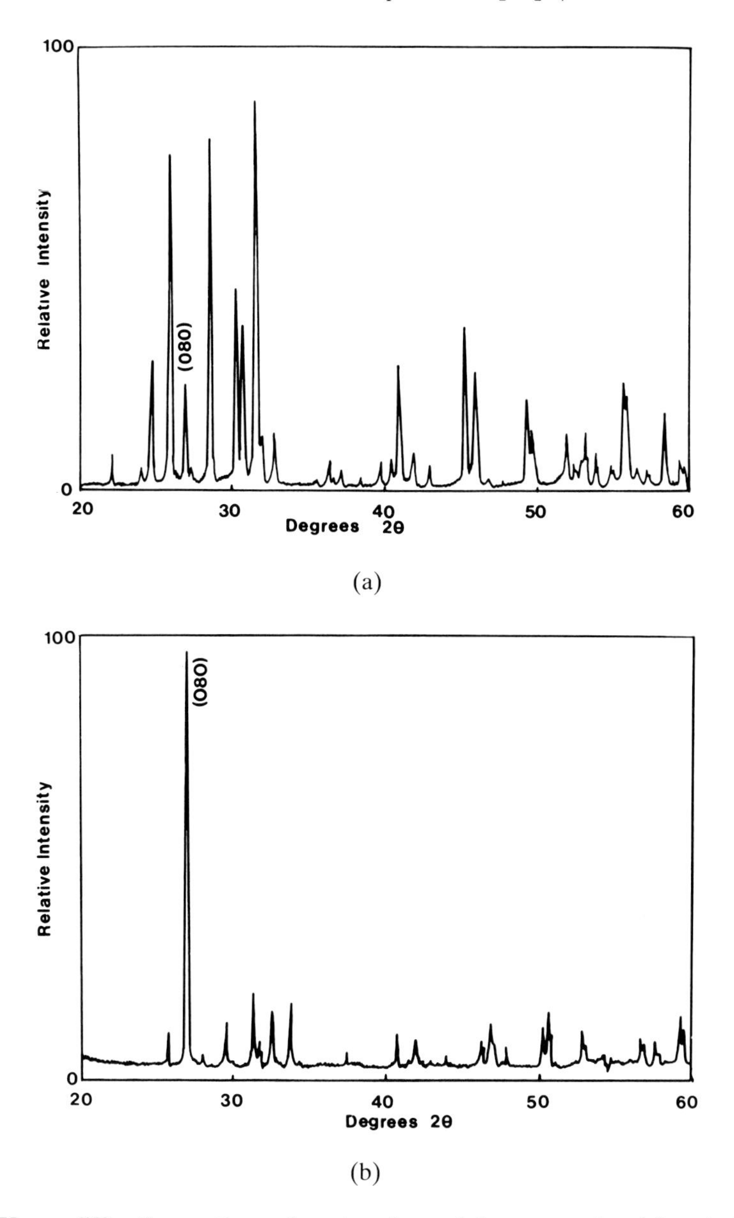

Figure 4 X-ray diffraction patterns for strontium niobate ceramics (a) uniaxially cold pressed and sintered at 1400°C (b) hot pressed at 1300°C.

increased, there was evidence of increasing grain orientation, with reflections of (0k0) increasing in intensity.[6]

For the hot-pressed specimens, X-ray analysis performed on surfaces perpendicular to the forging direction (FD) yielded even higher intensities for (0k0) reflections (Fig. 4b), especially (080) confirming an enhancement of grain orientation. The narrow dimension of the needles/plates (Fig. 3b) therefore corresponds

Table 1
Dielectric properties of strontium niobate (SN) ceramics (at 1 kHz and 20°C)

Specimen	ε_r	$\tan\delta$	TCC (20–120°C) ppm/°C
SN (undoped)	53	3×10^{-2}	−690
SN + 1% ZnO	57	3×10^{-4}	−690
SN hot pressed			
(i) perpendicular to FD*	52	1×10^{-2}	−200
(ii) parallel to FD*	58.6	1×10^{-2}	−720

*FD = Forging direction.

to the *b* axis (i.e. perpendicular to the perovskite slabs in the crystal structure). Fukuhara *et al.*[6] obtained grain orientation factors of approximately 0.38–0.46 for hot-pressed strontium niobate ceramics which had been annealed at a range of temperatures after sintering.

The dielectric properties of strontium niobate ceramics (measured at 20°C and 1 kHz) are summarised in Table 1. Relative permittivity was typically ≥57 for high-density samples and $\tan\delta$ values down to 5×10^{-4} were obtained. These data are broadly comparable with the findings of previous workers,[5,6] although the relative permittivity is sensitive to measurement temperature and frequency.

From Table 1 it may be noted that the temperature coefficient of capacitance (TCC), based on the measurement range 20–120°C, is negative in all cases, and grossly different for measurements made parallel and perpendicular to the hot-pressing direction. Since the structural analogue calcium niobate has a positive TCC,[3] an appropriate composition along the $Sr_2Nb_2O_7$–$Ca_2Nb_2O_7$ solid solution join yields material with zero TCC.

4. CONCLUSIONS

The use of ZnO additions enabled the sintering temperature of $Sr_2Nb_2O_7$ ceramics to be reduced to 1250°C and increased the fired density from 92% to 96% theoretical. Individual grains were needle-like or lath-like in shape, and generally ≤10 μm in size. The presence of ZnO led to a marked improvement in the microstructure, and the use of hot pressing caused an increase in the orientation of the grains. Relative permittivities were typically ≥57 and $\tan\delta$ values 10^{-2}–10^{-4} (at 20°C and 1 kHz). Electrical properties were enhanced in specimens prepared with ZnO additions. The temperature coefficient of capacitance, over the range 20–120°C, was approximately −700 ppm/°C.

REFERENCES

1. P. A. Fuierer and R. E. Newnham: *J. Am. Ceram. Soc.*, 1991, **74**, 2876–2881.
2. G. Winfield, F. Azough and R. Freer: *Ferroelectrics*, 1992, **133**, 181–186.
3. F. Azough: unpublished work.
4. S. Nanamatsu, M. Kimura, K. Doi and M. Takahashi: *J. Phys. Soc. Jpn*, 1971, **30**, 300.
5. S. Nanamatsu, M. Kimura and T. Kawamura: *J. Phys. Soc. Jpn*, 1975, **38**, 817.
6. M. Fukuhara, C.-H. Huang, A. S. Balla and R. E. Newnham: *J. Mater. Sci.*, 1991, **26**, 61–66.
7. JCPDS card number: 28-1247.

Structural and Electrical Characterisation of a New Bismuth Vanadium Oxide

A. K. BHATTACHARYA,* R. G. BISWAS,* K. K. MALLICK* and P. A. THOMAS†

**Centre for Catalytic Systems and Materials Engineering, Department of Engineering, University of Warwick, Coventry, CV4 7AL, UK*

†Department of Physics, University of Warwick, Coventry, CV4 7AL, UK

ABSTRACT

Recently, the family of bismuth vanadium compounds has generated considerable interest due to their high oxide ion conductivity at low temperatures. Using a novel preparation technique a new compound has been synthesised at room temperature. The material is found to be amorphous at room temperature and becomes fully crystalline at 593 K. Preliminary X-ray fluorescence and thermogravimetric studies showed that the compound has a chemical composition of $Bi_4V_{2.1}O_{11\pm\delta}$. X-ray powder diffraction measurements showed that the material has a body centred tetragonal structure at room temperature which is analogous to the well-known high-temperature γ-$Bi_4V_2O_{11}$ phase. Electrical characterisation has been performed using impedance spectroscopy in the temperature range 300–573 K.

1. INTRODUCTION

Recently, a new compound $Bi_4V_2O_{11}$ has been synthesised.[1] This has generated considerable interest due to its high oxide ion conductivity. This compound has a face-centered orthorhombic structure (α phase) at room temperature. On heating, it transforms to tetragonal (β phase) at 720 K and subsequently to a body-centered tetragonal structure (γ phase) at 840 K. On further heating, the γ phase melts congruently at 1160 K via the intermediate formation of a γ' phase. The phases formed on cooling the melt are identical to those observed when heating except that an additional α' phase appears at 720 K. The phase transition sequence can be described as

$$\alpha \xrightarrow{720\,\mathrm{K}} \beta \xrightarrow{840\,\mathrm{K}} \gamma \xrightarrow{1150\,\mathrm{K}} \gamma' \xrightarrow{1160\,\mathrm{K}} \text{liquid}$$

$$\alpha \xleftarrow[680\,\mathrm{K}]{} \alpha' \xleftarrow[720\,\mathrm{K}]{} \beta \longleftarrow \gamma \longleftarrow \gamma' \longleftarrow$$

Although the α' and γ' phases have been observed by thermal studies, these were not detected by X-ray crystallography. It is believed that this is due to either the narrow stability domain of these phases or the imperceptible evolution of cell parameters.[1]

The $Bi_4V_2O_{11}$ structure has been described by several authors[1,2] as the vanadium analogue of the Bi_2WO_6 structure which belongs to the Aurivillius family of phases. This structure consists of alternating layers of corner-sharing M–O octahedra and $Bi_2O_2^{2+}$ sheets made up from edge-sharing square pyramidal BiO_4

groups. In a recent single crystal structural study of α-$Bi_4V_2O_{11}$, it has been shown[3] that the orthorhombic face-centred phase (space group *Aba2*) is better described as alternating $Bi_2O_{2.75}^{0.5+}$ layers of highly irregular corner-sharing BiO_4 tetrahedra and $VO_{2.75}^{0.5-}$ layers of distorted corner-sharing octahedra.

It is believed that disorder in the high-temperature phase (γ form) leads to increased anionic conductivity compared with the ordered α and β phases. It was found that partial substitution of V^{5+} by transition metal ions such as Cu^{2+} and Ni^{2+} stabilised the disordered tetragonal phase γ at room temperature.[4]

A new bismuth vanadium oxide has been synthesised in this work using a novel room temperature preparation technique and its initial electrical property measurements have been carried out.

2. EXPERIMENTAL PROCEDURE

Hydrated bismuth nitrate and ammonium metavanadate, mixed in a molar ratio of 2:1, were treated in a manner described in earlier work by the present authors.[5] Bulk compositional analysis was carried out on a Philips Sequential X-ray fluorescence spectrometer (Model PW2400). LiB_4 was used as a standard fluxing agent to fuse the samples. Due to possible volatilisation loss of Bi_2O_3 at the temperature of fusion, compositional accuracy of the analysis was assessed by using a series of accurately weighed mixtures of Bi_2O_3 and V_2O_5.

Thermogravimetric analysis of the samples was carried out to determine the oxygen content in the compound using a Polymer Laboratories thermal analyser (Model STA1500); 5% H_2/N_2 was used as the reducing atmosphere and the temperature was ramped at a rate of 5°C/min.

X-ray powder diffraction patterns for samples treated at various temperatures were recorded in the region of 2θ = 5–120° with a step scan of 0.1°/min on a Philips diffractometer (Model PW1710) using CuK_α radiation. Cell parameters were calculated and further refined using linear regression procedures (Philips APD 1700 software) applied to the measured peak positions of all major reflections up to 2θ = 90°. This software was also used to calculate the average crystallite size using the well-known Scherrer equation.

Electrical conductivity measurements were performed on pressed discs of 13 mm diameter. Both faces were coated with Pt paste and the sample placed between platinum discs in a temperature-controlled furnace. Complex impedance spectra were obtained from a Schlumberger 1260 frequency response analyser. Measurements were performed in dry air over the frequency range 0.5 Hz–1 MHz, using a signal voltage of 0.1 V. A typical result is shown in Fig. 1.

3. RESULTS AND DISCUSSION

The XRD pattern of the freshly prepared powder, shown in Fig. 2, is broad and diffuse, indicating that the material is amorphous at room temperature and long-

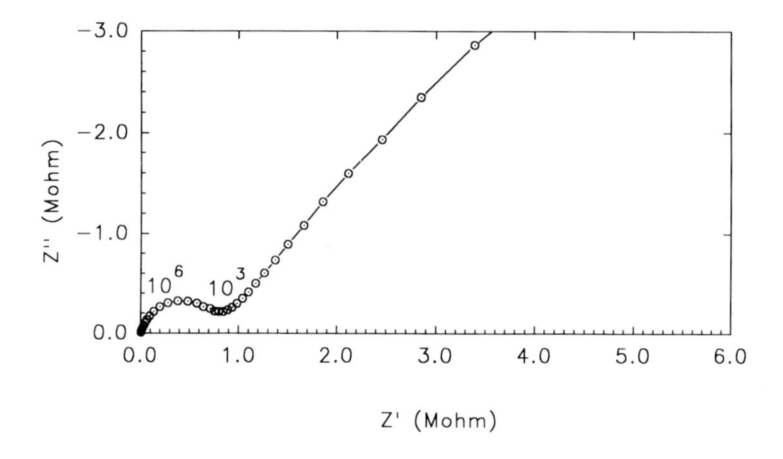

Figure 1 Typical impedance plot obtained for $Bi_4V_{2.1}O_{11\pm\delta}$ in dry air.

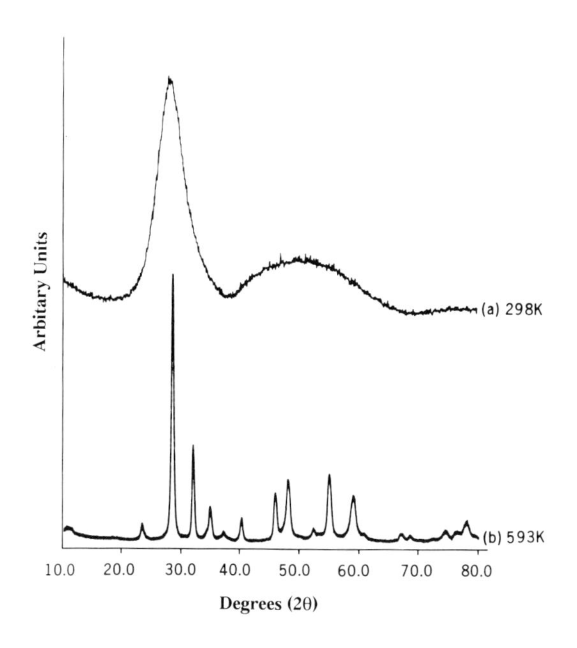

Figure 2 X-ray diffraction pattern for $Bi_4V_{2.1}O_{11\pm\delta}$.

range order in the structure is absent. At 593 K the material is transformed to a fully crystalline state. The unit cell is body-centred tetragonal with a = 3.934(2) Å and c = 15.37(1) Å. The composition of this compound was found to be $Bi_4V_{2.1}O_{11\pm\delta}$, from X-ray fluorescence and thermogravimetric analysis. A detailed analysis has been given in work.[5]

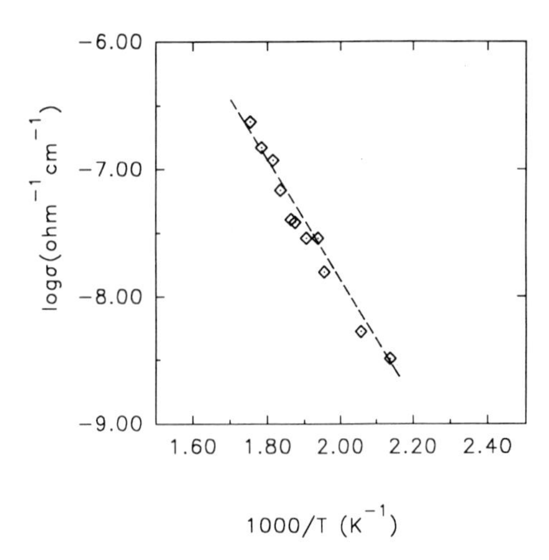

Figure 3 Temperature dependence of conductivity for $Bi_4V_{2.1}O_{11\pm\delta}$.

The average crystallite size of the sample heat treated at 593 K was determined from the line broadening of the strongest diffraction peak; this was estimated to be 10 nm.

A typical impedance plot for the new bismuth vanadium compound shows a well-resolved semicircle and a low-frequency 'spike'. The intersection of the two is a measure of the true bulk conductivity of the compound. This well-defined bulk characteristic occurs with a relaxation frequency ranging $10^{2.7}$–10^6 Hz. The low-frequency spike on the other hand is assumed to be due to electrode polarisation effects. Figure 3 shows the preliminary conductivity results obtained. This material exhibited a conductivity of $10^{-7}\ \Omega^{-1}\ cm^{-1}$ at 593 K, as compared to $10^{-4}\ \Omega^{-1}\ cm^{-1}$ reported for γ-$Bi_4V_2O_{11}$.[1]

4. CONCLUSION

A new bismuth vanadium oxide phase, isomorphous with γ-$Bi_4V_2O_{11}$, has been identified. Initial conductivity studies show that the material has a lower conductivity over a similar temperature range compared with the data for $Bi_4V_2O_{11}$. Further work is in progress to correlate the structure of the new compound to its electrical properties.

REFERENCES

1. F. Abraham, M. F. Debreuille-Gresse, G. Mairesse and G. Nowogrocki: *Solid State Ionics*, 1988, **28–30**, 529.
2. K. B. R. Varma, G. N. Subbana, T. N. Guru Row and C. N. R. Rao: *J. Mater. Research*, 1990, **5** (11), 2718.

3. M. Touboul, J. Lokaj, L. Tessier, V. Kettman and V. Vrabel: *Acta Crystall. C*, 1992, **48**, 1176.
4. F. Abraham, J. C. Boivin, G. Mairesse and D. G. Nowogrocki: *Solid State Ionics*, 1990, **40/41**, 934.
5. A. K. Bhattacharya, K. K. Mallick and P. A. Thomas: submitted for publication.

Structure and Electrical Properties of Ceria Based Oxide Ion Conductors Prepared at Low Temperatures

A. K. BHATTACHARYA, R. G. BISWAS, A. HARTRIDGE, K. K. MALLICK and J. L. WOODHEAD

Centre for Catalytic Systems and Materials Engineering, Department of Engineering, University of Warwick, Coventry CV4 7AL, UK

ABSTRACT

Solid solutions of the general formula $Ce_{1-x}Ln_xO_{2-x/2}\ \square_{x/2}$ (Ln = lanthanide (III) and $\square$ = anion vacancy), were prepared by a novel sol-gel route. These materials were characterised by powder diffraction and scanning electron microscopy. The gels formed on sol evaporation at 100°C were found to be fully crystalline solid solutions with fluorite structure and crystallite size of approximately 5 nm. This is the lowest temperature of formation to date. The lattice parameter of these materials was found to be directly proportional to the ionic radius of the dopant. Electrical properties of these materials were compared with similar materials made by a ceramic preparative method.

1. INTRODUCTION

Lanthanide doped CeO_2 of the general formula $Ce_{1-x}Ln_xO_{2-x/2}\ \square_{x/2}$ (Ln = lanthanide (III) and $\square$ = anion vacancy) possesses fluorite structure over a range of dopant concentrations. The substitution range of fluorite structure stability depends on the individual dopant. Partial replacement of Ce^{IV} ion with Ln^{III} ions in the lattice creates a corresponding number of anion vacancies. The resulting solid solutions have high ionic conductivity and have therefore been widely studied for possible applications as oxygen sensors[1] and in electrochemical fuel cells.[2,3]

The magnitude of conductivity does not increase linearly with increasing dopant concentration, but increases to a maximum between 10 and 20 mol.% before falling off sharply.[4] The magnitude of this maximum in turn depends on the ionic radius of the dopant. Several authors have attempted to explain these complex phenomena in terms of oxygen vacancy–dopant cation pairs and clustering of oxygen ion vacancies.[5–8]

Material of this type is conventionally prepared by high-temperature (1200–1600°C) ceramic powder processing techniques. Recently, solid solutions have been prepared at a relatively low temperature of 100°C using a novel sol-gel preparation technique developed in this laboratory. Advantages of this method over high-temperature methods include:

1. Controlled crystallite size
2. Less point defects

3. Almost complete theoretical densification at low temperatures
4. Low temperature of production

2. EXPERIMENTAL PROCEDURE

Solid solutions of general formula $Ce_{1-x}Ln_xO_{2-x/2}\,\square_{x/2}$, where Ln = La, Gd, Sm, Nd, Yb, and □ = anion vacancy, were prepared using a novel method described in earlier work.[9]

Morphology and elemental analysis of the samples were carried out on a JEOL scanning electron microscope (JEM 6100) operating at 20 kV and equipped with an energy-dispersive X-ray analyser. Conducting samples were prepared by either carbon coating or gold sputtering finely ground powder specimens. Both broad beam and point energy-dispersive analysis of the powder particles was performed using pure manganese as a reference standard.

X-ray powder diffraction patterns of the samples were recorded in the region of $2\theta = 10$–$80°$ with a scanning speed of $\frac{1}{4}°$/min on a Philips diffractometer (Model PW1710) using CuK_α radiation. The Philips APD 1700 software was used to calculate the average crystallite size of each composition from the broadening of a specific diffraction peak, using the well-known Scherrer equation.

Electrical conductivity measurements were performed on pressed discs of 13–16 mm diameter and 1–2 mm thick. The discs were sintered at various temperatures. Prior to measurement, both faces of the disc were coated with platinum paste and the sample placed between platinum discs in a temperature-controlled furnace. Complex impedance spectra were obtained from a Schlumberger 1260 frequency response analyser. Measurements were performed in dry air over the frequency range 0.5 Hz–1 MHz, using a signal voltage of 0.1 V.

3. RESULTS AND DISCUSSION

Materials produced at 100°C were all fully crystalline and their analysed compositions by X-ray fluorescence were, within experimental error, nearly identical to their nominal values. A doping level of $x = 0.2$ has been considered in the present investigation.

The XRD pattern of a samarium doped ceria gel dried at 100°C with a nominal composition of $x = 0.2$ is shown in Fig. 1. For comparison, the XRD pattern of Sm_2O_3 obtained by nitrate decomposition, and that of CeO_2 reference standard (JCPDS No: 34-394) are also included. It can be seen that there are no reactant lines present indicating the formation of a solid solution.

The XRD patterns of the other lanthanide doped materials with $x = 0.2$ calcined to 100°C are shown in Fig. 2. Similar peak widths at half height indicate similar crystallite sizes but the position of the (220) diffraction peak decreases linearly with increasing ionic radius of the dopant and is shown in Fig. 3. The

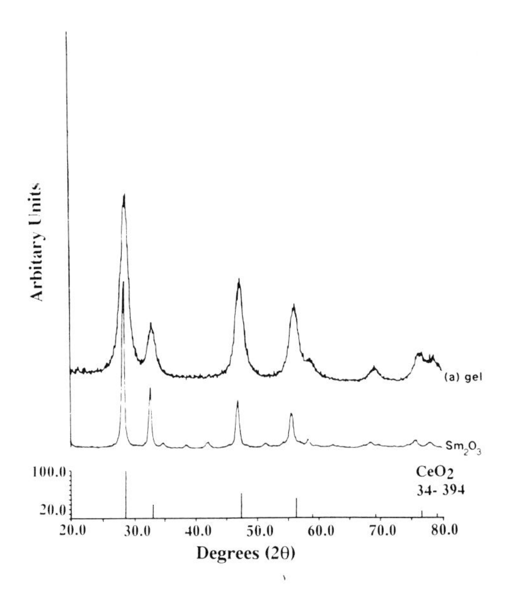

Figure 1 X-ray diffraction pattern of $Ce_{0.8}Sm_{0.2}O_{1.9}\,\square_{0.1}$ at 100°C.

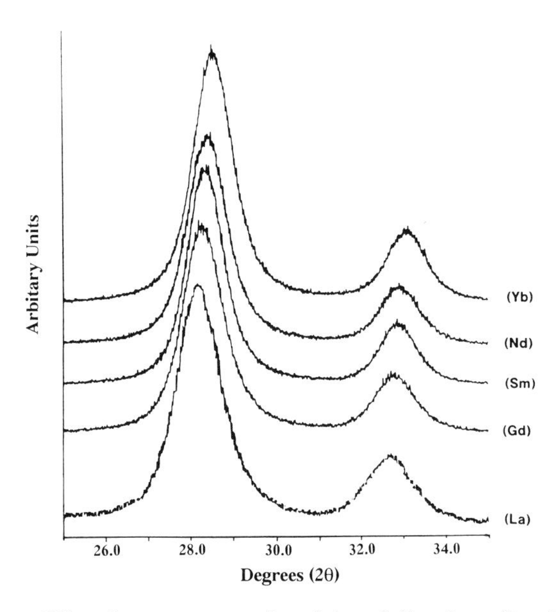

Figure 2 X-ray diffraction pattern peaks of doped $Ce_{0.8}Ln_{0.2}O_{1.9}\,\square_{0.1}$ at 100°C.

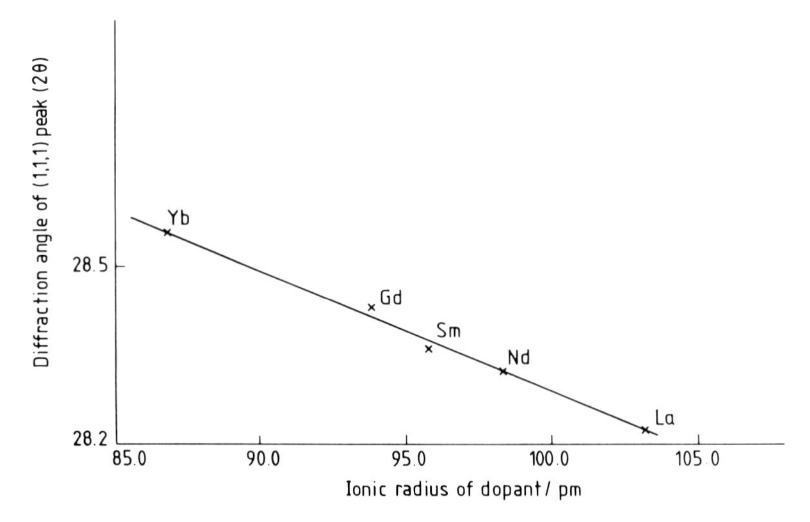

Figure 3 Peak shift versus ionic radii of doped $Ce_{0.8}Ln_{0.2}O_{1.9}\,\square_{0.1}$ calcined at 500°C.

Table 1
Average crystallite size (nm) of $Ce_{0.8}Ln_{0.2}O_{1.9}\,\square_{0.1}$ solid solution at 100°C

Sample	100°C
CeO_2	5.03
$Ce_{0.8}Yb_{0.2}O_{1.9}$	4.82
$Ce_{0.8}Gd_{0.2}O_{1.9}$	4.78
$Ce_{0.8}Nd_{0.2}O_{1.9}$	4.66
$Ce_{0.8}La_{0.2}O_{1.9}$	5.10

average size of crystallites in the lanthanide doped samples, with $x = 0.2$, is given in Table 1.

A typical impedance plot for Sm doped $Ce_{0.8}Ln_{0.2}O_{1.9}\,\square_{0.1}$ compound is shown in Fig. 4. The plot shows a well-resolved semicircle and a low-frequency 'spike'. The intersection of the two is a measure of the true bulk conductivity of the compound. This well-defined bulk characteristic occurs with a relaxation frequency ranging $10^{3.5}$–10^6 Hz. The low frequency spike on the other hand is assumed to be due to electrode polarisation effects. Figure 5 shows the conductivity data obtained from samples doped with 20 mol.% lanthanides. For comparison, the results also include a single compound of $Ce_{0.8}Sm_{0.2}O_{1.9}\,\square_{0.1}$ prepared by conventional ceramic method at 1550°C for 12 h. No discernible difference in the conductivity values was observed between the sample prepared by sol-gel method and subsequently sintered at 1200°C and the sample prepared by ceramic method. A detailed analysis of the conductivity results is in progress.[10]

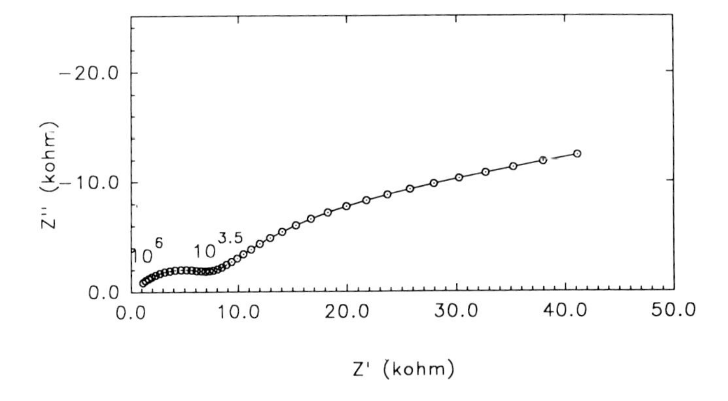

Figure 4 Typical complex impedance plot for $Ce_{0.8}La_{0.2}O_{1.9}\,\square_{0.1}$.

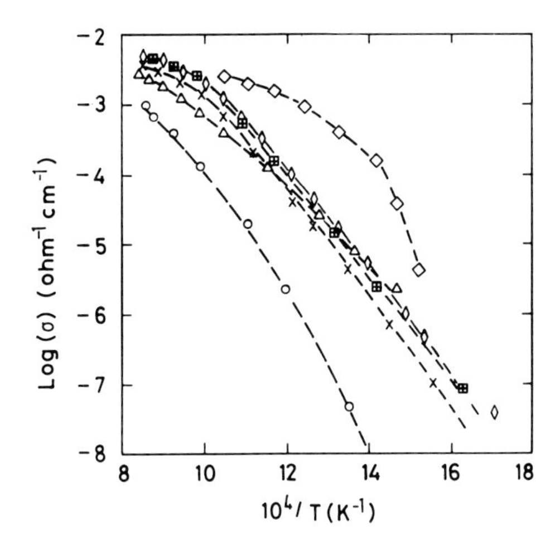

Figure 5 Arrhenius plot of $Ce_{0.8}Ln_{0.2}O_{1.9}\,\square_{0.1}$ calcined at 1200°C for 2 h: Ln = Sm(◊) sol-gel, Sm(⊞) ceramic, Nd(◇), Yb(×), La(△), CeO_2(○).

4. CONCLUSIONS

The novel preparative technique used to prepare lanthanide doped ceria with fluorite structure has enabled these materials to be synthesised at the lowest temperature reported to date. These solid solutions were homogeneous on a nanoscale and were found to have crystallite sizes approaching 5 nm. This small crystallite size enables these materials to be densified at relatively lower temperatures. Equivalent conductivity values were obtained for sol-gel-derived lanthanide doped materials sintered at a temperature significantly lower than that employed for the ceramic method.

REFERENCES

1. T. Takahashi, K. Ito and H. Iwahara: *Review energy primaire*, 1966, **2**, 42.
2. M. J. Verkerk, M. W. J. Hammink and A. J. Burggraaf: *J. Electrochem. Soc.*, 1983, 130.
3. H. Yahiro, K. Eguchi and H. Arai: *Solid State Ionics*, 1989, **36**, 71.
4. C. R. A. Catlow: *Solid State Ionics*, 1984, **12**, 67.
5. V. Butler, C. R. A. Catlow, B. E. F. Fender and J. H. Harding: *Solid State Ionics*, 1983, **8**, 109.
6. J. A. Kilner: *Solid State Ionics*, 1983, **8**, 201.
7. J. A. Kilner and R. J. Brook: *Solid State Ionics*, 1982, **6**, 237.
8. T. H. Etsell and S. N. Flengas: *Chem. Rev.*, 1970, **70**, 339.
9. A. K. Bhattacharya, A. Hartridge, K. K. Mallick and J. L. Woodhead: submitted for publication.
10. A. K. Bhattacharya, R. G. Biswas, A. Hartridge, K. K. Mallick and J. L. Woodhead: in preparation.

Development and Evaluation of Oxide Cathodes for Ceramic Fuel Cell Operation at Intermediate Temperatures

J. A. LANE, H. FOX, B. C. H. STEELE and J. A. KILNER

Centre for Technical Ceramics, Department of Materials, Imperial College of Science, Technology and Medicine, Prince Consort Road, London SW7 2BP

ABSTRACT

Ceramic materials of the perovskite family $La_{1-x}Ca_xFe_{1-y}Co_yO_3$ show promise as cathode materials in solid oxide fuel cell (SOFC) systems operating at intermediate temperatures (450–850°C). It is necessary that they be able to sustain large oxygen fluxes while also possessing high electronic conductivity ($>10^2$ S cm^{-1}) and mechanical and chemical stability with the electrolyte substrate. The electrocatalytic behaviour is monitored by $^{18}O/^{16}O$ isotopic exchange followed by dynamic secondary ion mass spectrometry (SIMS) investigations to determine the diffusion profile and surface exchange kinetics. The polarisation due to charge transfer at the electrode/electrolyte interface and diffusion–adsorption in the bulk electrode is examined by AC impedance spectroscopy of symmetrical electrodes applied to dense pellets of the electrolyte substrate, $Ce_{0.9}Sm_{0.1}O_{2-\delta}$. Electrical conductivity is measured using a four-point DC technique between 25 and 800°C.

1. INTRODUCTION

The cost of ceramic fuel cell systems can be reduced by operating at intermediate temperatures (450–850°C). This enables conventional materials to be used to supply the heated fuel and oxidant gases to the fuel cell stack and for sealing the stack itself. Developments in processing thin ceramic sheets now allow existing oxide electrolytes (such as doped ceria, $Ce_{1-x}A_xO_{2-\delta}$, $x = 0.1$–0.2, A = Gd, Sm) to be used over this temperature range, and so further development activities are principally focused on the fabrication and evaluation of alternative cathode materials. These cathode materials have to be compatible with the electrolyte substrate (thermal expansion match, no chemical reactions during co-firing procedures and fuel cell operation), and must also be able to sustain large oxygen fluxes (~0.5 A cm^{-2}). Ceramic materials of the perovskite, ABO_3, family are extensively used as cathodes in present-day solid oxide fuel cell (SOFC) technology. The doped lanthanum manganite, $La_{1-x}Sr_xMnO_{3\pm\delta}$ ($x = 0.15$–0.5), is the chosen material because,[1,2]

(i) it has a close thermal expansion match to yttria-stabilised zirconia (YSZ) the established electrolyte ($\sim 12 \times 10^{-6}$ K^{-1} for $La_{0.85}Sr_{0.15}MnO_{3\pm\delta}$ compared with $\sim 10 \times 10^{-6}$ K^{-1} for YSZ),

(ii) it is stable under cathodic conditions and undergoes no chemical reaction with the electrolyte substrate at typical SOFC operating temperatures (1000°C) and

(iii) it has acceptable levels of electronic conductivity for SOFC operation ($\sigma > 10^2$ S cm^{-1}).

However, $La_{1-x}Sr_xMnO_{3\pm\delta}$ perovskites possess very low ionic conductivity even at high temperatures, as demonstrated by their low oxygen self-diffusion coefficients ($D^* \approx 10^{-13}$ cm^2 s^{-1} at 900°C)[3] and so are unable to sustain high oxygen fluxes through the bulk electrode material.

Doped lanthanum cobaltite perovskites of the form $La_{1-x}Sr_xCoO_{3-\delta}$ exhibit much higher ionic and electronic conductivity ($D^* \approx 10^{-8}$ cm^2 s^{-1} at 700°C, $\sigma > 10^3$ S cm^{-1}) and also possess high activity for the exchange of oxygen between the gas and solid phases (surface exchange coefficient, $k \approx 10^{-6}$ cm s^{-1} at 700°C).[3] However, these materials show long-term instability under fuel cell operating conditions owing to a large thermal expansion mismatch with YSZ ($\sim 20 \times 10^{-6}$ K^{-1})[4] and chemical reaction with the electrolyte at temperatures greater than 800°C producing insulating phases ($La_2Zr_2O_7$, $SrZrO_3$) at the electrode/electrolyte interface.[5]

By replacing most of the cobalt on the *B* site sub lattice of the perovskite structure with manganese or iron the thermal expansion of the material is reduced significantly to levels approaching those of suitable electrolytes, the chemical stability between cathode and electrolyte is increased while a significant amount of ionic and electronic conductivity is maintained.[6]

In this work the O^{2-} mass transport properties of the perovskite material $La_{0.6}Ca_{0.4}Fe_{0.8}Co_{0.2}O_3$ have been examined using $^{18}O/^{16}O$ isotopic exchange and SIMS. Self-diffusion coefficients (D^*) and surface exchange coefficients (k) have been determined at several temperatures giving an activation energy for self diffusion (E_a). The electrical conductivity of this material has been established as a function of temperature using a four-point contact technique. AC impedance spectroscopy has been used to ascertain the polarisation associated with the electrode/electrolyte interface as a function of temperature.

2. EXPERIMENTAL

Dense ceramic samples of the perovskite $La_{0.6}Ca_{0.4}Fe_{0.8}Co_{0.2}O_3$ were prepared by the amorphous citrate route.[7] The required proportions of the metal nitrates (Aldrich Chemical Co., 99.9% pure) were dissolved in distilled water. An amount of citric acid equal to the molar quantity of metal cations was added to this aqueous solution together with $\frac{2}{3}$ molar quantity of ethylene glycol (Aldrich Chemical Co., 99.9% pure) to aid the polyesterification reaction. This solution was heated at 100°C until all the water had been removed and the polymerisation reaction had taken place. The oxide was obtained by drying the polymerised gel at 400°C followed by calcination in air at 1000°C for 10 h. The oxide powder obtained was milled for 48 h and isostatically pressed at 300 MPa into circular pellets and rectangular bars. These were sintered in air for 5 h at 1250°C and in all cases density was greater than 95% theoretical.

Ceramic pellets were prepared for $^{18}O/^{16}O$ isotopic exchange by polishing to 0.25 μm surface finish with diamond spray (Hyprez) and cleaned in an ultrasonic bath. They were first pre-annealed in an atmosphere of dry ^{16}O at a pressure of 1 atm. at the same temperature as the subsequent ^{18}O anneal but for a longer time, to ensure equilibrium of the sample. The ^{18}O anneals were carried out over several temperatures for 500 s each in an ^{18}O enriched atmosphere (97% ^{18}O) at 1 atm. Secondary ion mass spectrometry (SIMS Atomika 6500) was used to obtain the depth concentration profile of ^{18}O in the solid ceramic using a line scanning technique[8] developed for fast ion diffusers. This requires the exchanged specimens to be sliced perpendicular to the first polished surface and repolished before the SIMS primary beam is traversed across the required length obtaining an $^{18}O/(^{16}O + ^{18}O)$ depth concentration profile over several hundred microns. The flux of exchanging species across the solid surface, $x = 0$, is directly proportional to the difference in concentration in the gas and the solid surface such that:

$$-D^* \frac{dC}{dx} = k(C_g - C_s), \; x = 0 \tag{1}$$

where $D^* = O^{2-}$ self-diffusion coefficient (cm^2 s^{-1}), k = surface exchange coefficient (cm^{-1}), C_g = the ^{18}O isotopic concentration of the anneal gas, C_s = the ^{18}O isotopic concentration at the surface, $x = 0$.

The profile is fitted using non-linear least-squares fitting to a solution for the semi-infinite medium conditions of the exchange anneal.[8]

$$\frac{C_x - C_{bg}}{C_g - C_{bg}} = erfc\left[\frac{x}{2\sqrt{D^*t}}\right] - [\exp(hx + h^2 D^* t)] \times erfc\left[\frac{x}{2\sqrt{D^*t}} + h\sqrt{D^*t}\right] \tag{2}$$

where C_x = the ^{18}O isotopic concentration at depth x, C_{bg} = the natural ^{18}O isotopic background concentration, $h = k/D^*$, t = time of ^{18}O anneal(s).

The activation energy for O^{2-} self-diffusion was obtained from the slope of the Arrhenius plot of $\log D^*$ versus $1/T$.

For analysis by AC impedance spectroscopy symmetrical electrodes were applied to both sides of a thin, dense pellet of an electrolyte substrate, $Ce_{1-x}Sm_xO_{2-\delta}$, $x = 0.1$ or 0.2. This was achieved by making a slurry of the oxide powder in a solution of ethanol, binder (polyvinyl butyral) and plasticiser (polyethylene glycol 200) in the ratio 20:12:4:3 by weight. This slurry was then painted on to the electrolyte pellet, dried under a heat lamp and then isopressed at 300 MPa to assure maximum contact between electrode and electrolyte. The pellet was sintered at 1150°C for 5 h with heating and cooling rates of 2°C/min. Platinum gauze contacts were attached to either side of the pellet and connected by platinum wire to a frequency response analyser (FRA Sclumberger 1260). AC impedance spectra at frequencies 0.01–1 × 10^6 Hz were taken at temperatures between 100°C and 930°C. From these spectra, values for the polarisation associated with the electrode/electrolyte interface are obtained, in addition to the grain interior and grain boundary conductivity of the ceria electrolyte. An activation energy for the electrode polarisation resistance (R_{el}) is obtained from an Arrhenius plot of $\log R_{el}$ versus $1/T$.

The sintered bar specimens were used to make DC four-point conductivity measurements over the temperature range 25–800°C. Four platinum wire contacts were made to the bar, those at either end providing a constant current supply of 100 mA and the two inner contacts measuring the potential drop, ΔV, along a known length of the sample with known cross-sectional area. The sample was heated in a furnace at 1°C/minute and the temperature and ΔV (mV) were continuously recorded on a chart. The total electrical conductivity, σ_T (S cm^{-1}), was calculated according to:

$$\sigma_T = \frac{I}{\Delta V} \times \frac{L}{A} \tag{3}$$

where I = applied current, 100 mA, L = separation of inner contacts (cm), A = cross-sectional area of the sample (cm^2).

3. RESULTS AND DISCUSSION

A typical diffusion profile for $La_{0.6}Ca_{0.4}Fe_{0.8}Co_{0.2}O_3$ (Fig. 1), is shown together with the fit to the diffusion equation (1) and the corresponding values for oxygen self-diffusion, D^* (cm^2 s^{-1}), and surface exchange coefficient, k (cm s^{-1}). D^* and k

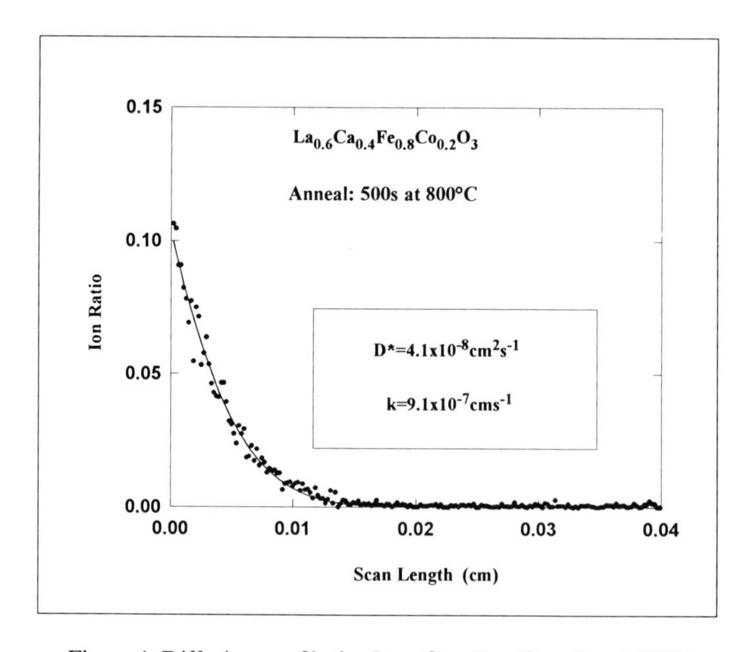

Figure 1 Diffusion profile for $La_{0.6}Ca_{0.4}Fe_{0.8}Co_{0.2}O_3$ at 800°C.

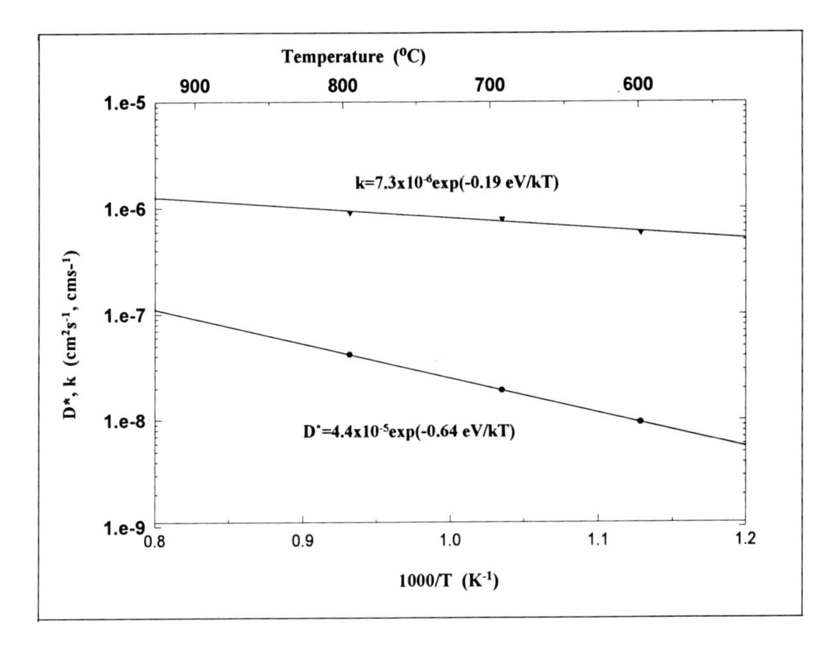

Figure 2 Arrhenius plot of self-diffusion coefficient, D^*, and surface exchange coefficient, k, showing activation energies (eV).

are also displayed on an Arrhenius plot for several temperatures and an activation energy calculated for each slope (Fig. 2).

A spectrum obtained from AC impedance spectroscopy for electrodes on a $Ce_{0.9}Sm_{0.1}O_{2-\delta}$ is shown (Fig. 3). The electrode polarisation resistance is calculated from the difference between the low- and high-frequency intercepts with the real axis and is displayed on an Arrhenius plot of $\log R_{el}$ versus $1/T$ (Fig. 4). Data from the literature[9] for $La_{0.6}Sr_{0.4}Fe_{0.8}Co_{0.2}O_3$ on YSZ electrolyte are included for comparison.

Electrical conductivity versus temperature from four-point DC conductivity is shown (Fig. 5).

3.1 Oxygen Self-Diffusion

$La_{0.6}Ca_{0.4}Fe_{0.8}Co_{0.2}O_3$ exhibits lower oxygen self-diffusion coefficient than perovskites which contain more cobalt on the B site cation sub lattice and similar A site cation doping by less than an order of magnitude (Table 1).[3] However, D^* is much greater than manganese containing perovskites, even those containing significant B site substitution by cobalt.

This is attributed to the relative oxygen vacancy concentrations in each material, as oxygen diffusion is facilitated via a vacancy mechanism. Teraoka *et al.*[10] suggested that Fe^{3+} tends to bind the O^{2-} more strongly than Co^{3+} and this seems to decrease the concentration of the oxygen vacancies and hence the oxygen

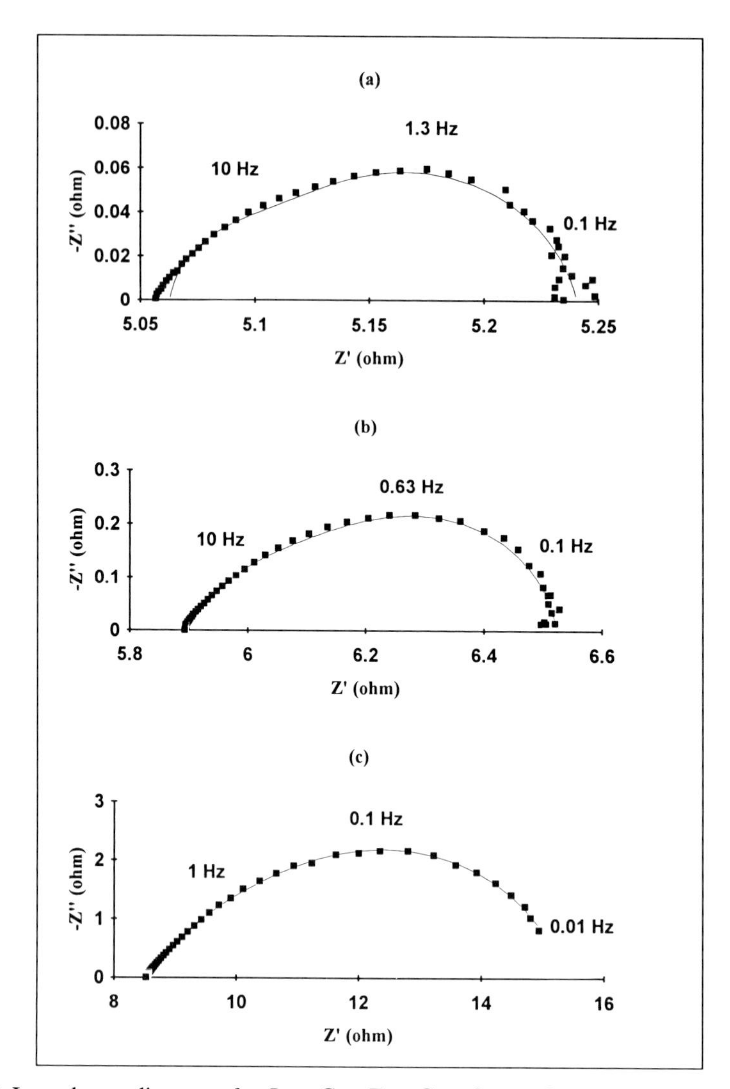

Figure 3 Impedance diagrams for $La_{0.6}Ca_{0.4}Fe_{0.8}Co_{0.2}O_3$ on $Ce_{0.9}Sm_{0.1}O_{2-\delta}$ electrolyte at (a) 800°C, (b) 700°C and (c) 600°C. Numbers indicate frequency in Hertz. High-frequency intercept does not approach zero owing to resistance of electrolyte and measuring leads.

self-diffusion coefficient is decreased as the Fe:Co ratio is increased. Manganite perovskites tend to have a much lower oxygen vacancy concentration.

By doping the perovskite lanthanum sub lattice with a lower valence cation, oxygen vacancies are introduced into the crystal lattice to maintain charge neutrality according to the equation (in Kröger–Vink notation):[11]

$$2\text{CaO} \xrightarrow{\text{La}B\text{O}_3} 2\text{Ca}'_{\text{La}} + 2\text{O}_{\text{O}}^{\times} + \text{V}_{\text{O}}^{\cdot\cdot} \quad (4)$$

where B is Fe and/or Co and so the oxygen self-diffusion coefficient tends to increase as the amount of lanthanum substitution increases.

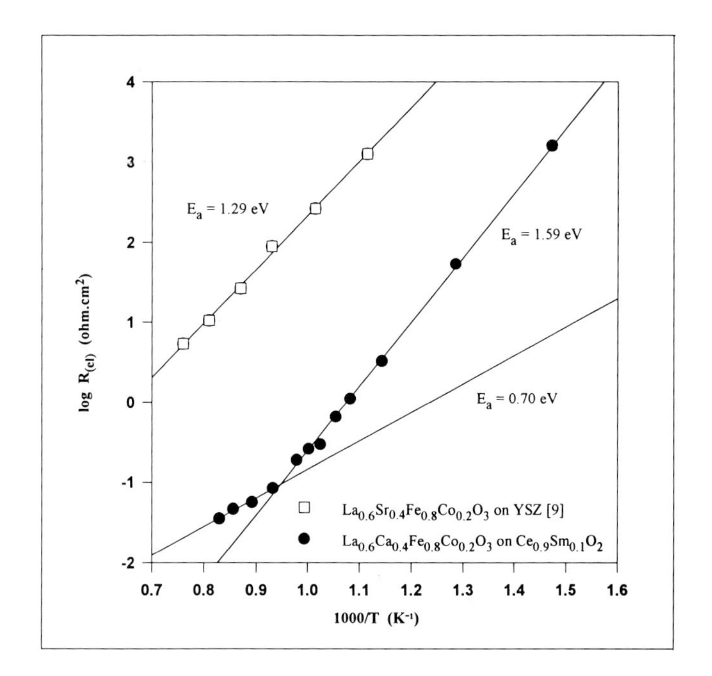

Figure 4 Arrhenius plot of electrode interfacial polarisation for $La_{0.6}A_{0.4}Fe_{0.8}Co_{0.2}O_3$ electrodes on $Ce_{0.9}Sm_{0.1}O_{2-\delta}$ and YSZ electrolytes.

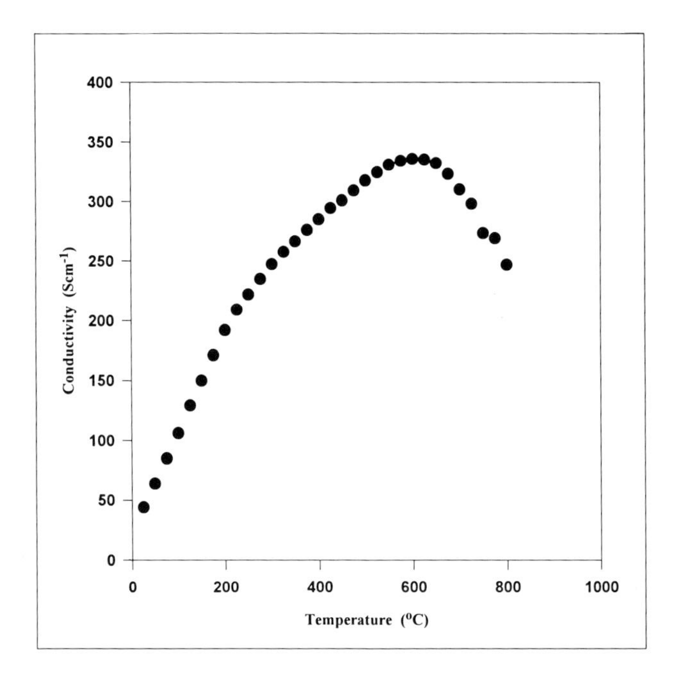

Figure 5 Temperature dependence of electrical conductivity of $La_{0.6}Ca_{0.4}Fe_{0.8}Co_{0.2}O_3$.

Table 1
Comparison of D^*, k and E_a for selected perovskite materials

Sample	T (°C)	D^* ($cm^2 s^{-1}$)	E_a (eV)	k (cms^{-1})
$La_{0.6}Ca_{0.4}Fe_{0.8}Co_{0.2}O_3$	800	4.1×10^{-8}	0.64 ± 0.10	9.1×10^{-7}
	693	1.9×10^{-8}		7.9×10^{-7}
	612	9.5×10^{-9}		5.9×10^{-7}
$La_{0.5}Sr_{0.5}MnO_3$[3]	900	3×10^{-12}	3.63 ± 0.20	9×10^{-8}
	800	8×10^{-14}		1×10^{-7}
	700	2×10^{-15}		1×10^{-8}
$La_{0.6}Ca_{0.4}Fe_{0.2}Co_{0.8}O_3$[3]	900	3×10^{-7}	1.45 ± 0.20	4×10^{-5}
	800	1×10^{-7}		2×10^{-5}
	700	2×10^{-8}		4×10^{-6}

3.2 Surface Exchange Coefficient

The surface exchange coefficients exhibited by $La_{0.6}Ca_{0.4}Fe_{0.8}Co_{0.2}O_3$ (10^{-6}–10^{-7} cms^{-1}) are lower than those of materials containing a greater proportion of cobalt (10^{-5}–10^{-6} cms^{-1}).[3] In this case, as for many materials with high diffusivity, the value of k is often the limiting factor in maintaining high oxygen fluxes through the material. Steele[12] has examined the requirements for D^* and k in technological applications and found the oxygen transport across the gas/solid interface to be the limiting process.

3.3 AC Impedance Spectroscopy

Results from AC impedance spectroscopy are represented in the form of complex impedance diagrams, imaginary versus real components. The response due to the electrolyte is observed as two semicircles at the high frequency end of the spectrum due to parallel combination of the electrolyte grain interior and grain boundary resistances with their respective capacitances. These responses are not observed above 500°C in the frequency range used in this study, 0.01–1×10^6 Hz, however, their total resistance and that due to the measuring leads is indicated by the resistance at the high-frequency intercept of the electrode polarisation arc.

The complex impedance diagrams for $La_{0.6}Ca_{0.4}Fe_{0.8}Co_{0.2}O_3$ electrodes on $Ce_{1-x}Sm_xO_{2-\delta}$ electrolyte consisted of two overlapping arcs between 500°C and 930°C corresponding to the interfacial resistance. These arcs are a result of two processes occurring at the electrode/electrolyte interface. The arc at high frequency is attributed to charge transfer of O^{2-} ions between electrode and electrolyte and the low-frequency arc is due to a concentration polarisation caused by the diffusion of oxygen species to the electrode/electrolyte interface being a limiting process.[9] The latter process dominates at lower temperatures and decreases as the temperature increases and so the two arcs are more clearly seen at higher temperatures. The change in activation energy of the total electrode interfacial impedance over this temperature range supports this hypothesis and is in close agreement with results using a different technique.[13] Modelling of this behaviour using equivalent circuits is in progress and will be reported in the future.

3.4 Electrical Conductivity

Electrical conductivity increases with temperature to 600°C where it reaches a maximum of 340 $S\,cm^{-1}$ and then decreases with further increases in temperature. This semiconductor-type conduction behaviour is attributed to a localised hopping of charge carriers between B^{3+} and B^{4+} sites (B = Fe or Co) which is thermally activated. This mixture of B^{3+} and B^{4+} sites is created by the A site substitution, where as well as ionic charge compensation as in equation (4) there is also an electronic charge compensation mechanism:

$$2CaO + \tfrac{1}{2}O_2\ (g) \xrightarrow{LaBO_3} 2Ca'_{La} + 3O_O^{\times} + 2\,h^{\cdot} \tag{5}$$

Above 600°C conductivity decreases, oxygen vacancies are formed according to equation (4) and in order to maintain charge neutrality, B^{4+} sites are reduced to B^{3+} sites so the number of charge carriers decreases.

4. CONCLUSIONS

$La_{0.6}Ca_{0.4}Fe_{0.8}Co_{0.2}O_3$ has a sufficiently large diffusion coefficient to be used at the higher temperatures of the intermediate range, 750–850°C. However, oxygen flux through the material is likely to be limited by the surface exchange process. Replacement of most of the cobalt on the B site by iron has not reduced oxygen diffusion significantly but has had a greater effect on the surface exchange kinetics.

AC impedance spectroscopy shows that the interfacial polarisation at the electrode/electrolyte interface between $La_{0.6}Ca_{0.4}Fe_{0.8}Co_{0.2}O_3$ and $Ce_{0.9}Sm_{0.1}O_{2-\delta}$ is low compared to similar materials on YSZ electrolyte. The interfacial polarisation is attributed to two distinct processes occurring at the interface.

Electrical conductivity of $La_{0.6}Ca_{0.4}Fe_{0.8}Co_{0.2}O_3$ is less than that of materials containing a greater proportion of cobalt but is greater than similar manganite perovskites and is well within the limits defined for useful SOFC operation.

ACKNOWLEDGEMENTS

I would like to acknowledge the help of Dr H. Middleton at Imperial College for help with the electrical conductivity measurements, and British Nuclear Fuels plc for financial aid.

REFERENCES

1. E. Ivers-Tiffee, W. Wersing and M. Schiessl: *Ber. Bunsenges. Phys. Chem.*, 1990, **94**, 978.
2. S. Otoshi, H. Sasaki, H. Ohnishi, M. Hase, K. Ishimaru, M. Ippommatsu, T. Higuchi, M. Miyayama and H. Yanagida: *J. Electrochem. Soc.*, 1991, **138**, 1519–1523.
3. S. Carter, A. Selcuk, R. J. Chater, J. Kajda, J. A. Kilner and B. C. H. Steele: *Solid State Ionics*, 1992, **53–56**, 597.
4. Y. Ohno, S. Nagata and H. Sato: *Solid State Ionics*, 1983, **9 & 10**, 1001–1008.
5. T. Kawada, N. Sakai, H. Yokokawa, M. Dokiya and I. Azai: *Solid State Ionics*, 1992, **50**, 189–196.
6. L.-W. Tai, M. M. Nasrallah and H. U. Anderson: in Proc. 3rd Int. Symp. on SOFC: 183rd Electrochemical Society Meeting, S. C. Singhal and H. Iwahara eds, 242–251, Electrochem. Soc. Inc., New Jersey, 1993.
7. M. S. G. Baythoun and F. R. Sale: *J. Mater. Sci.*, 1982, **17**, 2757–2769.
8. R. J. Chater, S. Carter, J. A. Kilner and B. C. H. Steele: *Solid State Ionics*, 1992, **53–56**, 859–867.
9. C. C. Chen, M. M. Nasrallah and H. U. Anderson: in Proc. 3rd Int. Symp. on SOFC: 183rd Electrochemical Society Meeting, S. C. Singhal and H. Iwahara eds, 598–612, Electrochem. Soc. Inc., New Jersey, 1993.
10. Y. Teraoka, H. M. Zhang, S. Furukawa and N. Yamazoe: *Chem. Lett.*, 1985, 1743–1746.
11. F. A. Kröger and H. J. Vink: *Solid State Physics Vol. 3*, F. Seitz and D. Turnbull eds, Academic Press, New York, 1965.
12. B. C. H. Steele: *Mater. Sci. Eng.*, 1992, **B13**, 25.
13. Y. Takeda, R. Kanno, M. Noda, Y. Tomida and O. Yamamoto: *J. Electrochem. Soc.*, 1987, **134**, 2656–2661.

Hydrothermal Synthesis of Strontium Hexaferrite: Powder Composition, Morphology and Magnetic Properties

A. ATAIE,* I. R. HARRIS* and C. B. PONTON*†
*School of Metallurgy and Materials, †IRC in Materials for High Performance Applications, The University of Birmingham, Edgbaston, Birmingham, B15 2TT, UK

ABSTRACT
Fine particles of strontium hexaferrite, $SrFe_{12}O_{19}$, have been synthesised hydrothermally from mixed aqueous solutions of iron and strontium nitrates at a relatively low temperature of 200°C. The effect of the synthesis conditions (temperature, time, alkali molar ratio and solution stirring) on the sol particle morphology, crystal structure and magnetic properties has been investigated. The results have shown that, as the synthesis temperature increases, the particle size and measured saturation magnetisation at 14 kOe* increase and coercivity decreases. The higher magnetisation of 54 emu/g was obtained using a synthesis temperature of 200°C, alkali molar ratio of 2, time of 1 h and stirring speed of 1000 rpm.

1. INTRODUCTION

The principal characteristics of and preparation methods for strontium hexaferrite have been reviewed by Kojima[1] and Stäblein[2] respectively. Strontium hexaferrite, $SrFe_{12}O_{19}$, possesses a high coercivity due to its relatively high magnetocrystalline anisotropy factor ($K_1 = 3.5 \times 10^6$ erg/cm^3) and this property makes it an attractive material for use in permanent magnet applications.[1] Strontium hexaferrite is produced mainly by a conventional mixed oxide ceramic method which involves calcining of $SrCO_3$ and Fe_2O_3 at around 1200°C. However, in order to improve the material properties, non-conventional routes such as the coprecipitation method,[3–5] organometallic precursor method,[6] glass crystallisation method,[7–8] salt bath method[9] and the hydrothermal method[10] have also been employed to synthesise strontium hexaferrite. According to a recent investigation,[10] only 39 wt% $SrFe_{12}O_{19}$ was obtained when strontium and iron nitrates in Fe : Sr molar ratio of 12 were processed hydrothermally at 250°C for 9 h; the coercivity, remanence and saturation magnetisation were also reported as 1300 Oe, 13.85 emu/g and 21.84 emu/g respectively.

In the present work, the hydrothermal synthesis of strontium hexaferrite has been investigated systematically and uniform size high-purity strontium hexaferrite particles of platelet morphology were obtained at a relatively low temperature and in a short time.

*Relationship between the c.g.s. and SI units which are used in this paper is as follows: 1 erg = 10^{-7} J, 1 emu/cm^3 = 12.57×10^{-7} weber/metre^2 (tesla), 1 oersted (Oe) = 79.6 amperes/metre, 1 gauss = 10^{-4} tesla (T).

2. EXPERIMENTAL PROCEDURE

A 250 ml capacity PTFE lined Berghof laboratory autoclave was used to synthesise strontium hexaferrite from mixed aqueous solutions of iron nitrate ($Fe(NO_3)_3 . 9H_2O$) and strontium nitrate ($Sr(NO_3)_2$), which were obtained from Aldrich Ltd. The synthesis procedure was as follows:

1. Dissolving the strontium nitrate into 4 moles of distilled water.
2. Adding the iron nitrate into the strontium nitrate aqueous solution with an Fe : Sr molar ratio 8 and dissolving completely.
3. Coprecipitation of the resultant solution by the addition of NaOH in different molar ratios of $OH^- : NO_3^-$ (this ratio denoted by R) and agitating for 0.5 h.
4. Processing of 100 ml volumes of the precipitated solution in the autoclave at temperatures in the range of 160 to 220°C for times of 0.5 to 5 h under autogenous pressure.
5. Washing of the synthesised material twice with distilled water, then drying at about 60°C using a hotplate.

The crystal structure and morphology of the synthesised materials were analysed using X-ray powder diffraction (Co radiation) and scanning electron microscopy (SEM), respectively. The isotropic magnetic values were measured using a vibrating sample magnetometer (VSM) at a maximum field of 14 kOe.

3. RESULTS AND DISCUSSION

3.1 Effect of Synthesis Time

Figure 1 shows the X-ray powder diffraction patterns for strontium hexaferrite synthesised hydrothermally at 220°C with a $OH^- : NO^-$ molar ratio of 2 for synthesis times of 0.5, 1, 2 and 5 h without stirring. Analysis of these patterns confirms the formation of pure strontium hexaferrite under the given experimental conditions, even after only 0.5 h synthesis time.

Hexagonal and platelike particles of the $SrFe_{12}O_{19}$ synthesised at 220°C with a $OH^- : NO_3^-$ molar ratio of 2 for 1 h (with a narrow size distribution) can be seen in the SEM micrograph shown in Fig. 2. As the synthesis time decreased, the intrinsic coercivity increased due to the decrease in particle size, except for the 1 h process time sample. According to the Wohlfarth and Stoner expression,[11] the lower coercivity value of the 1 h process sample may be caused by the high particle aspect ratio as well as the higher value of the M_s.

The average particle diameter (d), thickness (t) and aspect ratio (d/t) and sample intrinsic coercivity (H_{ci}), remanence (σ_r) and saturation magnetisation (σ_s) are given in Table 1 as a function of process time. The lower saturation magnetisation value for the 0.5 h process time sample is probably due to the existence of some precipitates such as Fe_2O_3 (less than 10 wt%) which cannot be identified by the X-ray diffraction powder method.

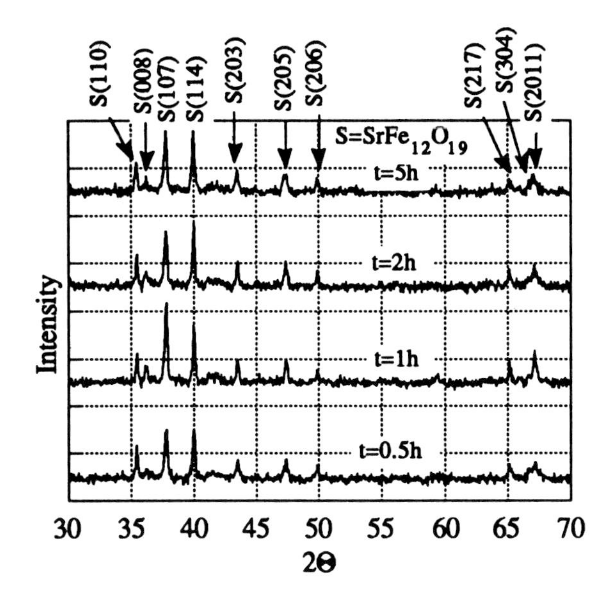

Figure 1 X-ray powder diffraction patterns for strontium hexaferrite synthesised hydrothermally for 0.5, 1, 2 and 5 h at a temperature of 220°C with a $OH^-:NO_3^-$ molar ratio of 2, without stirring.

Figure 2 SEM micrograph of strontium hexaferrite synthesised hydrothermally for 1 h at a temperature of 220°C with a $OH^-:NO_3^-$ molar ratio of 2, without stirring.

Table 1
Average particle diameter d, thickness t and aspect ratio d/t and the intrinsic coercivity H_{ci}, remanence σ_r and saturation magnetisation σ_s of strontium hexaferrite synthesised hydrothermally for 0.5, 1, 2 and 5 h at a temperature of 220°C with a $OH^-:NO_3^-$ molar ratio of 2, without stirring

Time h	H_{ci} Oe	σ_r emu/g	σ_s emu/g	d μm	t μm	d/t
0.5	1275	16.76	46.62	1.56	0.3	5.2
1.0	1055	16.53	49.7	1.65	0.24	6.87
2.0	1185	16.83	49.54	1.68	0.35	4.8
5.0	1153	16.79	49.46	1.68	0.24	7

The formation of platelike particles of strontium hexaferrite in these samples with roughly the same particle size confirms the hypothesis that the nucleation and growth of $SrFe_{12}O_{19}$ particles is completed within 0.5 h.

3.2 Effect of $OH^-:NO_3^-$ Molar Ratio

Figure 3 shows the X-ray powder diffraction patterns for strontium hexaferrite synthesised hydrothermally at 220°C for 5 h with $OH^-:NO_3^-$ molar ratios of 2, 3 and 5, without stirring.

The characteristics of the prepared particles are shown in Table 2. Both the X-ray powder diffraction patterns and the magnetic measurements (which give roughly the same saturation magnetisation values for samples synthesised with

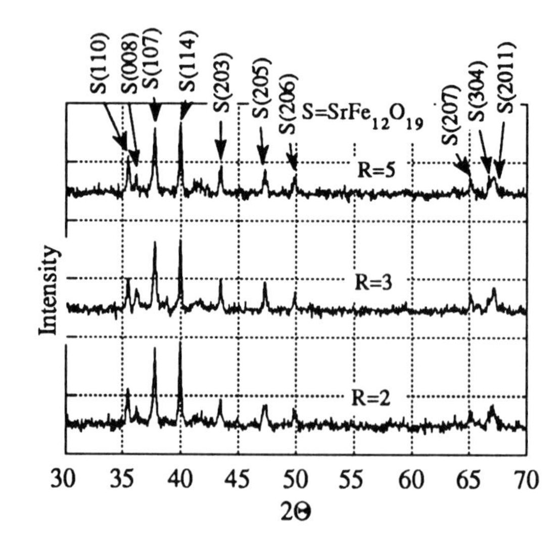

Figure 3 X-ray powder diffraction patterns of strontium hexaferrite synthesised hydrothermally at 220°C for 5 h with $OH^-:NO_3^-$ molar ratios (R) of 2, 3 and 5, without stirring.

Table 2
Average particle diameter d, thickness t and aspect ratio d/t and the intrinsic coercivity H_{ci}, remanence σ_r and saturation magnetisation σ_s of strontium hexaferrite synthesised hydrothermally at 220°C for 5 h with $OH^-:NO_3^-$ molar ratios of 2, 3 and 5, without stirring

$OH^-:NO_3^-$ molar	H_{ci} Oe	σ_r emu/g	σ_s emu/g	d μm	t μm	d/t
2	1153	16.79	49.46	1.68	0.24	7
3	1070	15.57	48.85	1.80	0.32	5.6
5	1061	16.13	50.64	1.83	0.32	5.7

different $OH^-:NO_3^-$ molar ratios), show that the powder composition is not influenced significantly by the $OH^-:NO_3^-$ molar ratio. However, in contrast to the behaviour of hydrothermally synthesised barium hexaferrite,[12] increasing the $OH^-:NO_3^-$ molar ratio increases the particle size slightly and as a result the coercivity decreases. This result shows that the reaction rate decreases with decreasing $OH^-:NO_3^-$ molar ratio, i.e. alkaline concentration, which is in agreement with the kinetics of barium hexaferrite formation in an alkaline medium proposed by Ling Wang.[13] The SEM micrograph of sample synthesised at 220°C for 5 h with a $OH^-:NO_3^-$ molar ratio of 3 is shown in Fig. 4.

3.3 *Effect of Synthesis Temperature*

As the synthesis temperature decreases the particles size decreases and the coercivity increases; however, the saturation magnetisation of samples prepared at

Figure 4 SEM micrograph of Sr-hexaferrite synthesised hydrothermally at 220°C for 5 h with $OH^-:NO_3^-$ molar ratio of 3, without stirring.

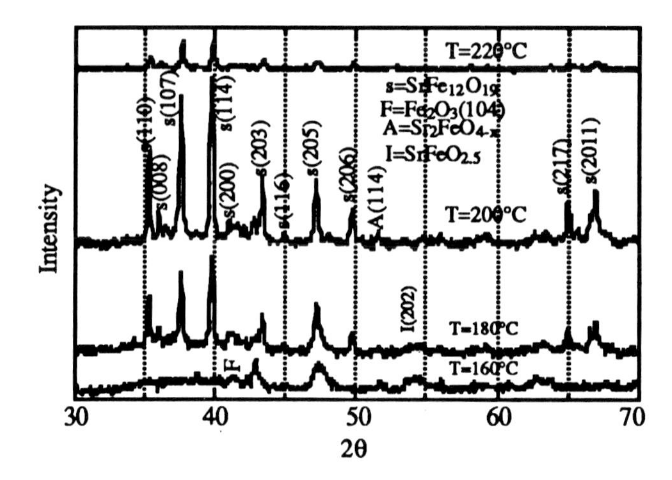

Figure 5 X-ray powder diffraction patterns of particles synthesised hydrothermally at 160, 180, 200 and 220°C for 1 h with $OH^-:NO_3^-$ molar ratio of 2, without stirring.

160°C and 180°C drops markedly owing to the presence of precipitates which were identified as αFe_2O_3, $SrFeO_{2.5}$ and Sr_2FeO_{4-x} from the X-ray diffraction analysis, as shown in Fig. 5.

Differences between the relative line intensities in the X-ray powder diffraction patterns can be caused by preferred orientation of the sample.[14] The particle characteristics (diameter, thickness and aspect ratio) and magnetic properties (coercivity, remanence and saturation magnetisation) of particles synthesised at 160, 180, 200 and 220°C for 1 h with a $OH^-:NO_3^-$ molar ratio of 2, without stirring, are shown in Table 3. Figure 6 shows the coercivity and magnetisation values of particles as a function of the synthesis temperature, while demagnetisation curves of the synthesised materials under different temperatures are also illustrated in Fig. 7.

The SEM micrograph of a sample prepared at 180°C is shown in Fig. 8 which confirms the partial formation of the submicron and platelet particles of strontium hexaferrite.

Table 3

Average particle diameter d, thickness t and aspect ratio d/t and the intrinsic coercivity H_{ci}, remanence σ_r and saturation magnetisation σ_s of particles synthesised hydrothermally for 1 h with a $OH^-:NO_3^-$ molar ratio of 2, without stirring, as a function of process temperature

Temp. °C	H_{ci} Oe	σ_r emu/g	σ_s emu/g	d μm	t μm	d/t
160	1783	0.9	2.47	0.7	0.3	2.3
180	1643	10.88	24.68	0.9	0.2	4.5
200	1453	20.97	48.53	1.4	0.3	4.66
220	1055	16.79	49.7	1.68	0.24	7.0

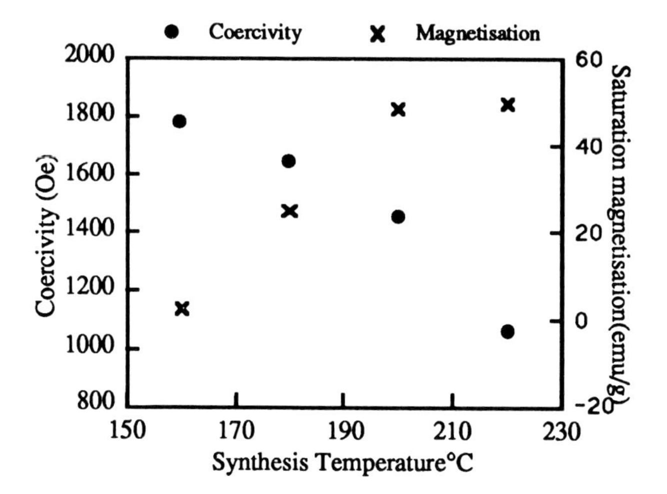

Figure 6 Coercivity H_{ci} and saturation magnetisation σ_s of particles synthesised for 1 h with a $OH^-:NO_3^-$ molar ratio of 2, without stirring, as a function of temperature.

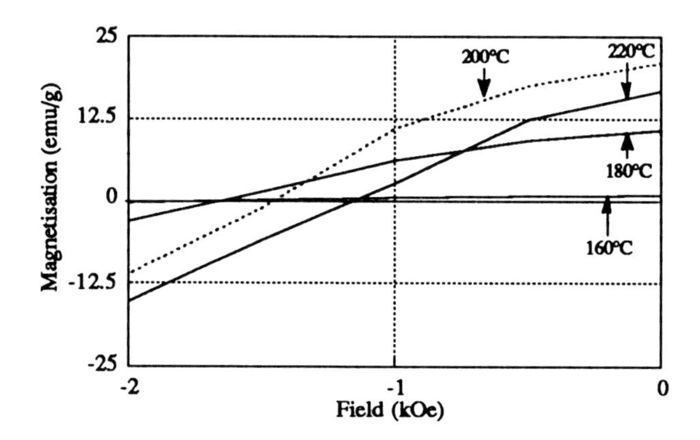

Figure 7 Demagnetisation curves of particles synthesised for 1 h with a $OH^-:NO_3^-$ molar ratio of 2, without stirring, as a function of temperature.

3.4 Effect of Stirring

From the literature it has been reported that the nucleation and growth of particles and their composition can be affected by stirring in the autoclave.[15] Figure 9 shows the magnetisation curves versus applied field for strontium hexaferrite produced by hydrothermal synthesis at a temperature, time and alkali molar ratio of 200°C, 1 h and 2, respectively, with stirring speeds of 0 and 1000 rpm. The saturation magnetisation increased from 48.53 to 54 emu/g for the stirred sample; this increase may be caused by changes in the chemical composition and morphology of particles, as mentioned above.

Figure 8 SEM micrograph of strontium hexaferrite synthesised hydrothermally at 180°C for 1 h with a $OH^- : NO_3^-$ molar ratio of 2.

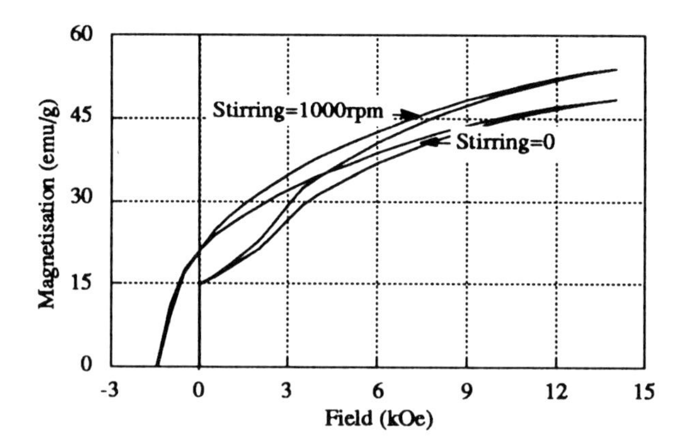

Figure 9 Magnetisation curves of strontium hexaferrite synthesised hydrothermally at a temperature of 200°C, time of 1 h, and in $OH^- : NO_3^-$ molar ratio of 2, with stirring speeds of 0 and 1000 rpm.

4. CONCLUSION

Strontium hexaferrite particles were synthesised at different synthesis temperatures, times, alkali molar ratios and stirring speeds. The results have shown that the solid-solution particles of strontium hexaferrite can be formed at relatively low temperatures, about 200°C (comparing with 1250°C for the conventional ceramic method and about 800°C for the coprecipitation method); and that the morphology of the particles, and hence, the magnetic parameters, are altered as a result of the change in the synthesis conditions.

ACKNOWLEDGEMENTS

The authors wish to thank the members of the Applied Alloy Chemistry Group (School of Metallurgy and Materials) and the Ceramic Group (IRC/School of Metallurgy and Materials) for their help and cooperation. The provision of facilities by Prof. J. F. Knott, FRS FEng (Metallurgy and Materials) and Prof. M. H. Loretto (IRC) is acknowledged. The financial support of A. Ataie by the Iranian government during his PhD studentship is also gratefully acknowledged.

REFERENCES

1. H. Kojima: in *Ferromagnetic Materials: A Handbook on the Properties of Magnetically Ordered Substances*, E. P. Wohlfarth ed., Vol. 3, 305–391. North-Holland, Amsterdam, 1982.
2. H. Stäblein: in *Ferromagnetic Materials: A Handbook on the Properties of Magnetically Ordered Substances*, E. P. Wohlfarth ed., Vol. 3, 441–602. North-Holland, Amsterdam, 1982.
3. W. S. Sutarno Bowman and G. E. Alexander: *Canad. Ceram. Soc. J.*, 1970, **39**, 33–41.
4. S. K. Date, C. E. Deshpande, S. D. Kulkarni and J. J. Shrotri: in Proc. Intern. Conf. on Ferrites, India, 1989, 55–60.
5. V. V. Pankov, M. Pernet, P. Germi and P. Mollard: *Mag. Mag. Mater. J.*, 1993, **120**, 69–72.
6. K. Haneda, C. Miyakawa and K. Goto: *IEEE Trans. Mag. MAG.*, 1987, **23** (5), 3134–3136.
7. D. Bahadur and D. Chakravorty: in Proc. Intern. Conf. on Ferrites, India, 1989, 189–193.
8. H. Sato and T. Umeda: *Mater. Trans. J.*, 1993, **34** (1), 76–81.
9. R. J. Routil and D. Barham: *Canad. J. Chem.*, 1974, **52**, 3235–3246.
10. C. H. Lin, Z. W. Shih, T. S. Chin, M. L. Wang and Y. C. Yu: *IEEE Trans. Mag. MAG.*, 1990, **26** (1), 15–17.
11. E. P. Wohlfarth and E. C. Stoner: *Philos. Trans.*, 1948, **A240**, 599.
12. E. Sada, H. Kumazawa and H. M. Cho: *Ind. Eng. Chem. Res. J.*, 1991, **30**, 1319–1323.
13. M. Ling Wang and Z. Whie Shih: *J. Crystal Growth*, 1992, **116**, 483–494.
14. B. D. Cullity: *Elements of X-ray Diffraction*, 397–417, Addison-Wesley Publishing Co., USA, 1978.
15. M. Yoshimura, N. Kubodera, T. Noma and S. Somiya: *Ceram. Soc. Jap. J.*, 1989, **97**, 16–21.

Properties of Reaction Sintered Manganese–Zinc Ferrites

I. P. KILBRIDE and A. J. BARKER
IRC in Materials for High Performance Application, and School of Chemical Engineering, The University of Birmingham, Edgbaston, Birmingham, B15 2TT, UK

ABSTRACT
Processing parameters and properties of a reaction sintered manganese–zinc ferrite were determined. Metal–metal oxide powder mixes were prepared by attrition milling. Compacts were produced by die pressing at pressures between 8 and 80 MPa. Light milling treatments led to high green densities (maximum of 88% of the theoretical density of the sintered ferrite) but did not realise dense sintered ceramics. More vigorous milling treatments led to sintered ceramics of up to 96% density. Acceptable magnetic permeabilities (1180 μ_0) were achieved in samples with 0.5% linear shrinkage. Green density and metal particle size play key roles in the oxidation and sintering behaviour of the material.

1. INTRODUCTION

Reaction sintering and reaction bonding of ceramics have several advantages over conventional processing methods. Typically, the volume expansion of the reactive phase during reaction, partially or totally compensates for the normal sintering shrinkage. This obviously has benefits in terms of near net shape forming. However, most research in this field has concentrated on non-oxide ceramic materials,[1] such as reaction bonded silicon nitride or structural oxide ceramics such as reaction bonded aluminium oxide.[2] Little attention has been focused on the reaction sintering of functional ceramics where near net shape forming would be desirable in the manufacture of the large numbers of small complicated shape components. This is particularly true of soft ferrite manufacture, with a wide variety of shapes used for transformer cores. The most complicated core shapes tend to be made in materials for power or telecommunication applications with a relative initial magnetic permeability in the range of 1000–2000 μ_0.[3] Control or reduction of shrinkage during the sintering is important in order to optimise the final properties and dimensions of the ceramic component.

2. EXPERIMENTAL

Powder compositions were prepared to give a final composition of Fe_2O_3 70.2, MnO 16.5 and ZnO, 13.3 wt%. Powder compositions are shown in Table 1. The powders were prepared in the following manner. Manganese and zinc oxide were pre-milled in isopropanol for 1 h in the attrition mill (water-cooled stainless steel liner with 1.5 kg of 6.3 mm steel ball bearings for a 0.1 kg charge), then dried

Table 1
Powder compositions

Mix 1	MnO–ZnO pre-mill and iron powder mixed in isopropanol 8 h
Mix 2	MnO–ZnO pre-mill, iron powder and 2 wt% oleic acid dry milled 8 h
Mix 3	MnO–ZnO pre-mill and iron powder dry milled 8 h
Mix 4	MnO 1 h pre-mill, iron and zinc powder dry milled 8 h

before use. Elemental iron and, where appropriate, elemental zinc, were added and either mixed or attrition milled for a further 8 h. The particle size distributions of the raw materials (measured by laser diffraction particle size analysis) are shown in Fig. 1. Test pieces were pressed in a 12.6 mm diameter die at various loads between 1 and 10 kN on a load frame. Samples then underwent various treatments; heat treatment at 800°C in air for 4 h then cooled; heat treatment at 800°C for 4 h then sintering at 1400°C for 2 h followed by cooling in air. Finally, optimised samples were treated as above but were cooled in a protective argon atmosphere after treatment to prevent oxidation of the ferrite.

Density (green and treated) and shrinkage were measured from the dimensions and mass of the samples. Photomicrographs were taken from polished sections or polished and etched sections (conc. HCl for 2 min). Magnetic permeability and dielectric loss tangent versus frequency were measured on samples cooled in argon using an impedance analyser and a test flux <1 mT.

3. RESULTS AND DISCUSSION

3.1 Powder Preparation and Compaction

As shown in Fig. 2, the mixed powder 1 had a broad particle size distribution similar to that of the raw materials. Powder 2, with an oleic acid process control additive, had a mean particle size of 21 μm. This contrasts with the two dry milled samples which were considerably finer (5 μm and 8 μm for powders 3 and 4 respectively). Figure 3 (effect of milling time on the particle size distribution of powder 2) displays evidence that while the mean particle size reduces, relatively, with time the number of fine particles below 4 μm has also reduced. This may suggest that a type of mechanical alloying mechanism may operate in the system; fine material is being incorporated in the coarser metal particles. Photomicrographs of the milled powders, Fig. 4, appear to confirm the particle size analysis. The more aggressive dry milling produces fine irregular-shaped particles.

The effect of applied pressure during die pressing on the density of the green compacts is shown in Fig. 5. High green densities, about 80% of the theoretical density of the ceramic, were achieved in the mixed powder. The milling treatments achieved lower green densities for similar pressures, but were generally higher than the equivalent conventionally processed ceramic. In the case of powder 2 this was because its intermediate particle size distribution between a conventional ceramic

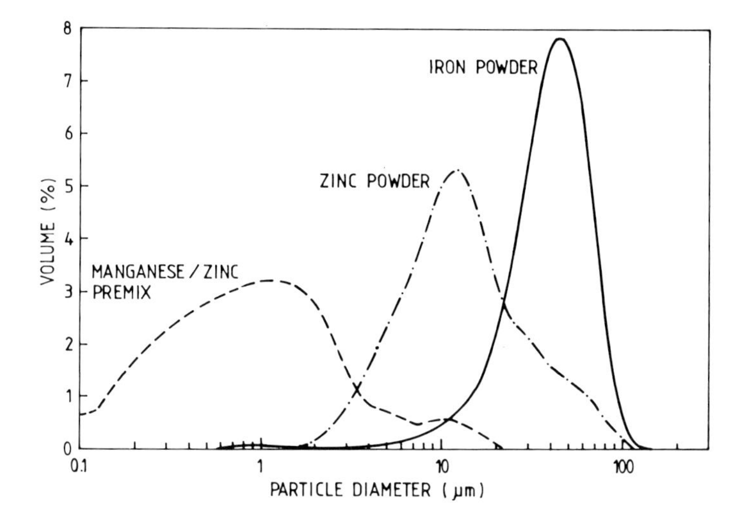

Figure 1 Particle size distributions of raw materials.

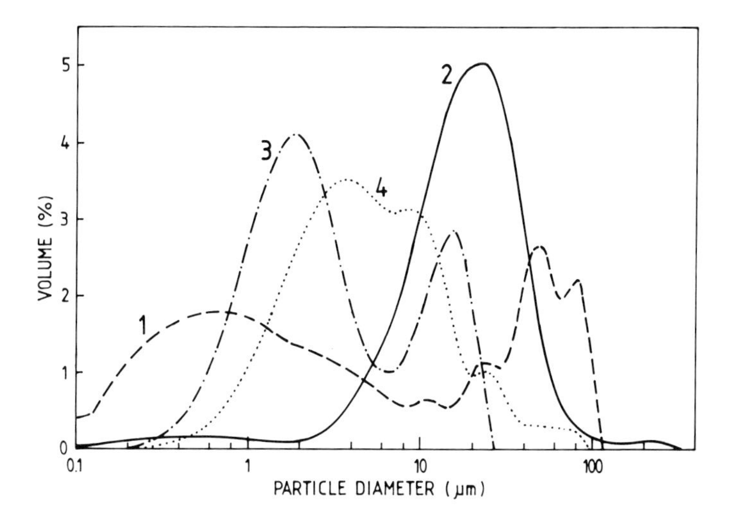

Figure 2 Particle size distributions of powders 1 to 4 after milling treatments.

powder and the mixed powder allows an intermediate packing density. For powders 3 and 4, with a similar particle size to a conventional ceramic, the maximum green density is greater because of the high density of the pure iron component.

3.2 Effect of Green Density on Oxidation Behaviour at 800°C

The behaviour of samples, made from powders 1 and 2, after oxidation at 800°C for 4 h is shown in Figs 6 and 7. Figure 6 is a graph of sintered density versus green

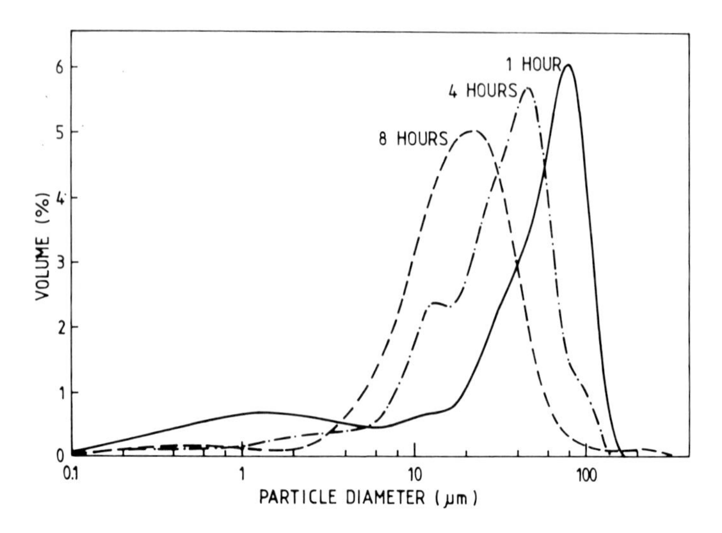

Figure 3 Effect of milling time on the particle size distribution of powder 2.

density after heat treatment at 800°C of powders 1 and 2. The mechanically alloyed powder 2 achieved a higher density for any given green density relative to the mixed powder. Figure 7 shows that while the volume expansion (2% to 8%) due to oxidation is constant (powder 2), or increases (powder 1) with increasing green density, in both samples, the weight gain falls from 23% (maximum) to about 10%. This leaves perhaps 50% residual iron in the samples pressed at 80 MPa after oxidation at 800°C. For any given green density powder 1 has a larger weight increase. This is likely to be due to some minor oxidation of the iron powder during milling and is the probable cause of the increased volume expansion relative to the mechanically alloyed powder.

Photomicrographs of powders 1 and 2 pressed at 40 MPa after heat treatment are shown in Fig. 8. The pores are larger in the mixed sample than in the milled sample. Figure 8(c) shows residual iron in the mixed sample occupying what appear to be pores in Fig. 8(a). The mechanism of oxidation seems to occur by the diffusion of iron atoms into the surrounding structure, thereby leaving pores. This is in contrast to massive iron where oxidation usually occurs by the diffusion of oxygen though a thick $FeO/Fe_3O_4/Fe_2O_3$ layer.[4] However, the finely divided nature of the iron powder, and the presence of zinc and manganese oxides, may alter the oxidation mechanism or products. The oxidation of iron to FeO is thought to occur by the diffusion of iron metal through the oxide layer.

3.3 Effect of Green Density on the Sintering Behaviour at 1400°C

Figure 9 is a graph of sintered density at 1400°C versus green density. A lower green density paradoxically leads to a higher sintered density. At a lower green density, full oxidation of iron is possible at 800°C so that sintering can occur

(a)

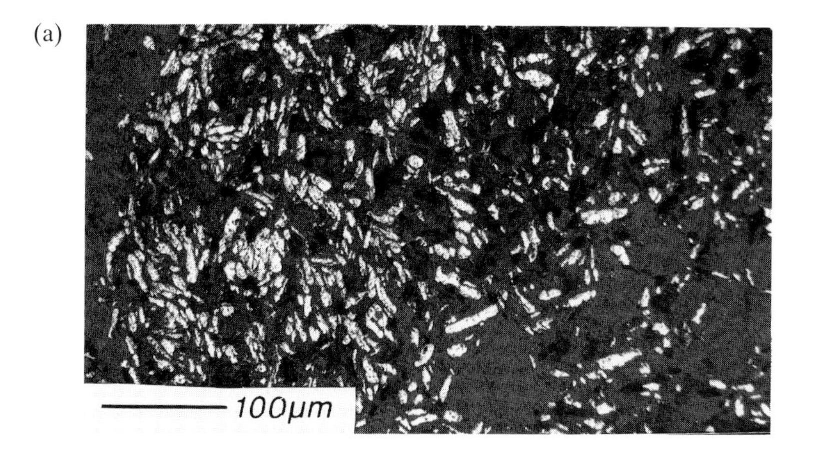

(b)

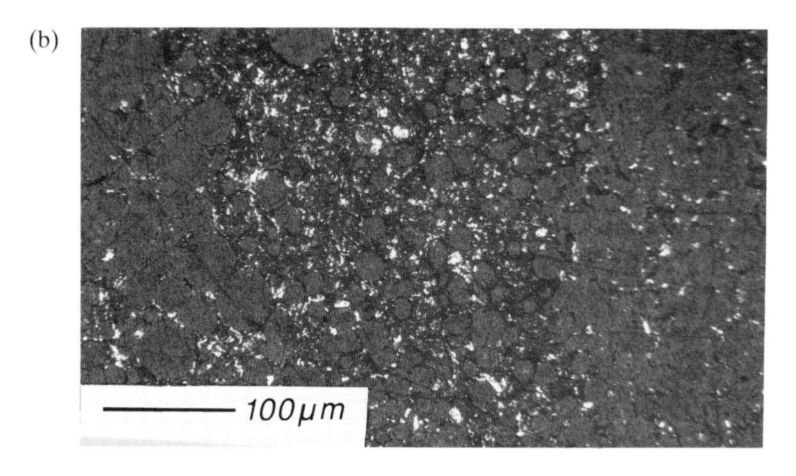

(c)

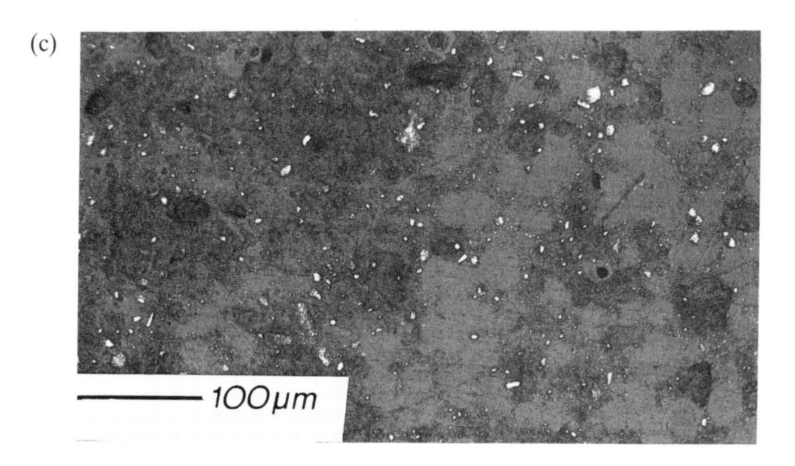

Figure 4 Photomicrographs of (a) powder 1 (b) powder 2 and (c) powder 3 after milling treatment.

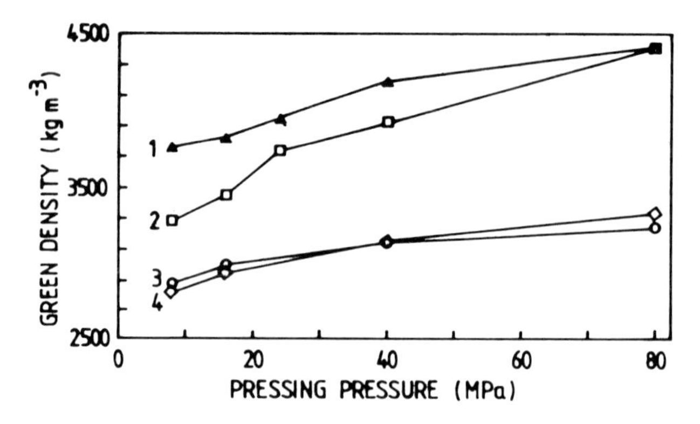

Figure 5 Green density versus die pressing pressure for powders 1 to 4.

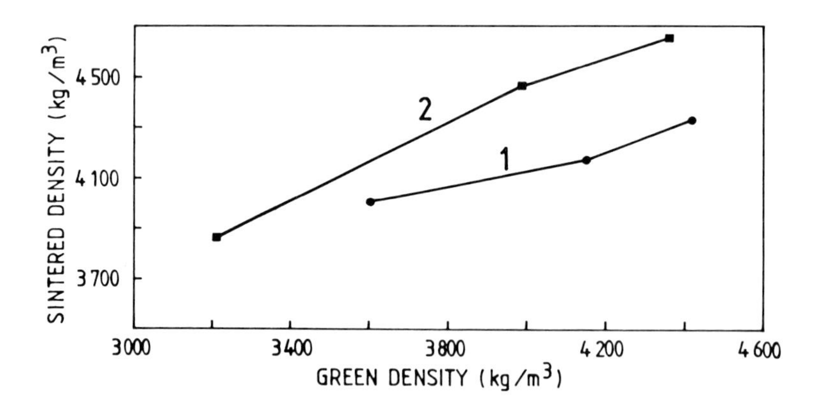

Figure 6 Graph of sintered density versus green density after heat treatment at 800°C.

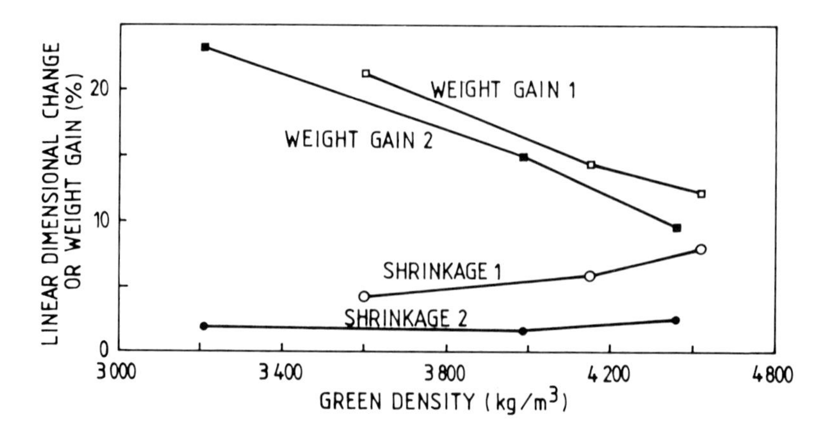

Figure 7 Linear dimensional change and weight gain versus green density for powders 1 and 2 after heat treatment at 800°C.

(a)

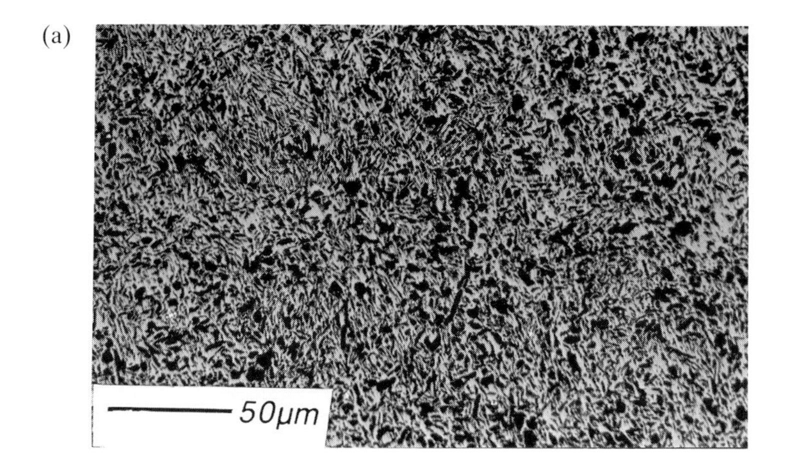

(b)

(c)

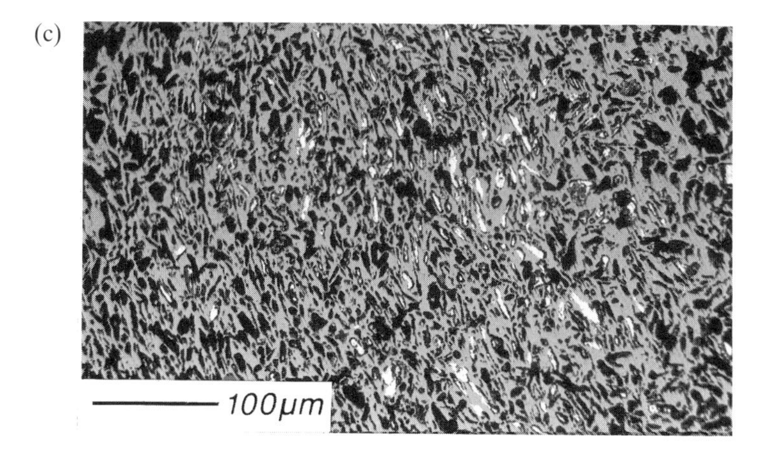

Figure 8 Photomicrographs of (a) powder 1 (b) powder 2 after heat treatment at 800°C and (c) powder 1 at higher magnification, showing residual metal content.

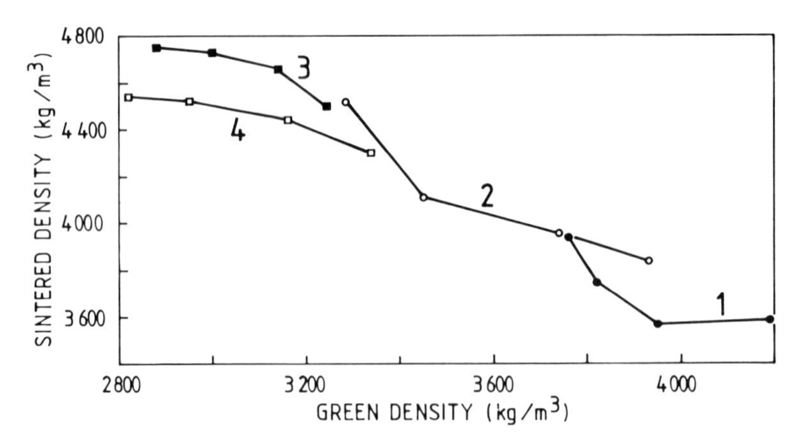

Figure 9 Graph of sintered density versus green density for powders 1 to 4 after sintering at 1400°C.

independently of the volume expansion associated with this oxidation. The residual metal, as a second phase, may also inhibit densification while samples with a higher green density tend to retain more iron after heat treatment. The particle size of the metal phase may also have an effect. The pores left after oxidation will be larger in the less aggressively milled powders, and will therefore require higher sintering temperatures to remove.

3.4 Properties of Samples Cooled in Protective Argon Atmosphere

Samples pressed at 8 MPa were heat treated, sintered at 1400°C, and cooled in an argon atmosphere. The physical and magnetic properties are shown in Table 2. The highest permeability was found in powder 2, while the lowest was found in the

(a)

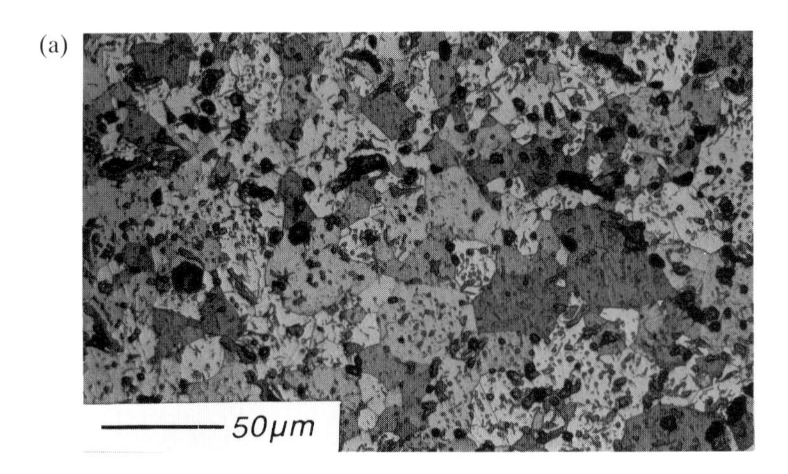

Figure 10 (a) to (d) Polished and etched samples of powders 1 to 4 respectively after cooling in protective argon atmosphere.

(b)

(c)

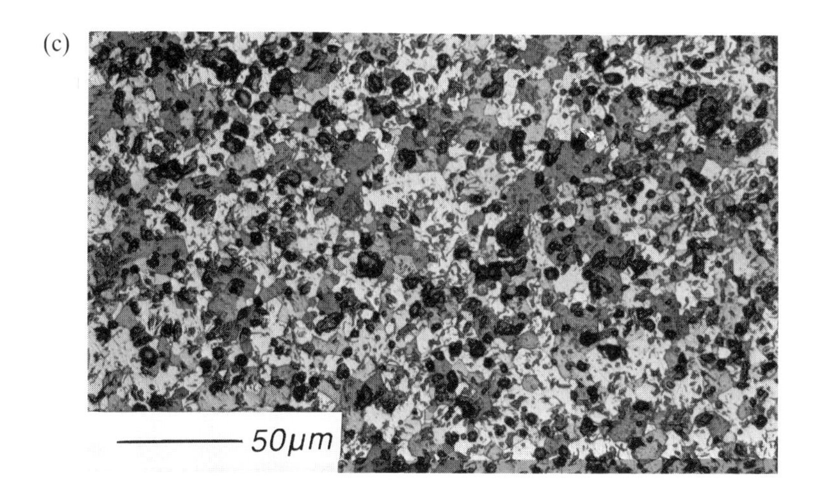

(d)

Table 2
Physical and magnetic properties of sintered ferrite

Sample	Density kg m^{-3}	Linear dimensional change %	Relative initial permeability μ_i 1 kHz	Loss factor 100 kHz $\times 10^6$
1	3.91	7.0	336	77.8
2	4.38	−0.5	1180	13.5
3	4.52	−11.2	503	24.1
4	4.32	−8.5	603	18.4

mixed sample. The low magnitude of the permeability in the mixed powder sample is likely to be a function of its low density relative to the maximum theoretical achievable, and it may also be a function of chemical homogeneity. The lower permeability found in powders 3 and 4, despite their having higher or equivalent density to powder 2, appears to be a function of grain size and the dimensions and quantity of internal defects. This is shown in Fig. 10 (photomicrographs of polished and etched surfaces) where powder 2 has a continuous microstructure with dispersed porosity at the grain boundaries. Samples 3 and 4 appear to be more seriously affected by microcracking.

4. CONCLUSIONS

Attrition milling was used to significantly reduce the particle size of a soft metal powder, and some evidence for incorporation of oxide particles within the metal matrix was found. The production of a high green density compact relative to the final ceramic product was made possible by the incorporation of the metal powder. However, a high green density did not improve final sintered properties, as full oxidation at 800°C of the metal component was only achievable in the lower green density samples. Sintered densities of 95% of theoretical were achieved with reduced shrinkage (11.8%) over conventional ceramics processing. Acceptable magnetic properties (1180 μ_0) were achieved in samples with low linear shrinkage (0.5%).

REFERENCES

1. S. J. Scheider: *Engineered Materials Handbook Vol. 4,* Ceramics and Glasses, ASM International, 1991.
2. S. Wu, D. Holz and N. Clausen: *J. Am. Ceram. Soc. 76*, 1993, **4**, 970–980.
3. E. C. Snelling: *Soft ferrites and their applications*, Butterworth, London, 1988.
4. A. Cottrell: *An Introduction to Metallurgy*. Edward Arnold, London, 1961.

Electrically Conducting Composite Ceramics Produced by Hydrothermal Synthesis

R. D. B. NORFOLK, M. H. L. WISE and C. B. PONTON

IRC in Materials for High Performance Applications and School of Metallurgy and Materials, The University of Birmingham, Edgbaston, Birmingham, B15 2TT, UK

ABSTRACT

Mullite-type material has been produced by hydrothermal synthesis as the basis of a multiphase ceramic system, with TiN, to render an electrical conducting sintered body. Its conducting nature will enable sintered material to be electrical discharge (ED) machined, provided that the material bulk resistivity is <200 Ωcm.[1] This can be used to circumvent the distortions occurring during the sintering of ceramics, which are a major problem for accurate component production. It does so by permitting the shaping process to be performed after sintering. Intimate powder mixing and composite particle formation have been achieved by hydrothermal synthesis of mullite precursor acetates in the presence of TiN and by ball-mill mixing of dry powders. Characterisation of the material in both the green and sintered states has been performed using SEM, XRD and laser diffraction particle size analysis, together with preliminary mechanical and electrical property evaluation by micro-hardness and electrical resistance measurements, respectively.

The two routes investigated have indicated that the electrical and mechanical properties obtainable are significantly affected by the processing route. For the material mixed *in situ* during hydrothermal synthesis the final sintered body comprised a multicomponent ceramic alloy with similar microhardness to that of 3:2 mullite but with a significantly decreased sample electrical resistance. The structure of the ball-mill mixed specimens differed greatly from that seen in the autoclave mixed material but they also showed increased electrical conduction, implying a greater potential for ED machining.

1. INTRODUCTION

The ability to ED machine structural and engineering ceramics into complex shapes is desirable, since the production of dimensionally accurate components after sintering is not readily achievable by abrasive diamond machining. Therefore, the aim of the present work was to use 3:2-type mullite ($3Al_2O_3.2SiO_2$), which has potentially useful high-temperature/structural properties, as the basis of an ED machinable ceramic material. This initially precluded the use of an intrinsically conducting ceramic although the possible influence of hydrothermal processing on the TiN and mullite precursors, and the product material's electrical/electronic properties, has yet to be investigated.

Electrodischarge machining of ceramic materials can generally be achieved in any of three ways, all of which rely on reducing the effective workpiece resistance to <200 Ωcm. Firstly, the use of an intrinsically electrically conducting ceramic, either naturally occurring or manufactured to the appropriate specification, e.g. valency control and/or doping.[2] Secondly, the formation of a well-distributed, fully interpenetrating, three-dimensional second-phase network with a suitably low resistivity. Thirdly, a combination of these two methods, such that during processing the conducting network renders some of the surrounding material bulk electrically conducting by interdiffusion of the conducting species.

The minimum volume for the formation of a conducting network in a discrete two-phase system is generally taken as 16 vol.%.[3] The degree to which such a conducting network has formed can be assessed by simple electrical resistance measurements, and the connectivity modelled (after McLachlan *et al.*).[4]

TiN was added as a conducting second phase to produce a machinable ceramic workpiece with sufficient conduction after sintering, while retaining the high temperature properties of mullite. The hydrothermal processing of aluminium and silicon acetate salts, producing mullite composition material permits the very intimate mixing of an insoluble powder such as TiN with the mullite material. This dispersion of TiN within the ceramic bulk is expected to enable ED machining only after the formation of a three-dimensional conducting network by diffusion of TiN during sintering. The isolation or connectivity of conducting material within the final composite ceramic is dependent upon the mixing characteristics of the material during the process. Thus, to investigate the effect of mixing, another mullite–TiN composite material with a coarser distribution of TiN was produced by rotary ball-milling. This enabled a comparison with the *in situ* hydrothermally produced material.

2. EXPERIMENTAL PROCEDURE

Material of mullite composition was produced by hydrothermal synthesis, in an autoclave heated to 220°C for 3 h under autogenous pressure, using aluminium and silicon acetates as precursors. Material referred to as autoclave-mixed in this paper contained TiN powder which was mixed in during this process. The other material, referred to as ball-mill mixed, was produced by drying autoclave-derived mullite to which TiN powder was added; the resulting material was mixed in a rotary ball-mill using a polypropylene container and ceria-stabilised zirconia milling balls. The two routes were used to produce mixed powders with TiN contents ranging from 20 to 60 wt%. Both types of mixed powder were uniaxially pressed into compacts which were sintered in a vacuum furnace at temperatures in the range 1200°C to 1500°C for periods of 1 to 12 h.

Measurement of the electrical resistance of both green and sintered compacts was made both by point-probe and cross-sectional resistance tests. Electrodeposition of copper from a 0.05 M $CuSO_4$ solution, onto the surface of the sintered discs was performed using a low-voltage power supply. Both optical and scanning electron microscopy were used to characterise the microstructure, the latter method in both secondary electron imaging (SEI) and backscattered electron imaging (BEI) modes.

Laser diffraction particle size analysis was employed for all stages of processing to evaluate the extent of the mullite sol particle deposition on the TiN particles during hydrothermal synthesis, and to determine the particle agglomeration characteristics during all modes of mixing and drying. Microhardness measurements were conducted to provide mechanical property trend data.

3. RESULTS

All green compacts (of both autoclave and ball-mill derived material) exhibited a very high electrical resistance ($>10^{12}\,\Omega$) over the whole composition range investigated. In contrast, sintered discs of both types exhibited a low, measurable resistance ($10^3\,\Omega$), with a 60 wt% TiN content. However, as expected, decreasing the TiN content increased the electrical resistance of both types of disc as shown in Fig. 1. The influence of process variation on the sintered (at 1500°C for 3 h) sample electrical resistance was investigated using material containing 20 wt% TiN, produced by both routes. It was found that the dry ball-mill mixed material exhibited a lower electrical resistance than autoclave-mixed material (i.e. $10^3\,\Omega$ as against $10^7\,\Omega$ for ball-mill derived and autoclave-derived material respectively). Increasing the sintering temperature caused a continuous decrease in the measured resistance of the autoclave-derived discs and an increase in the electrical resistance of the ball-mill derived discs (Fig. 2). Whereas an increase in the sintering time caused a sudden decrease in the measured resistance of the autoclave-derived discs at 3 h followed by a roughly linear increase and no significant change in the electrical resistance of the ball-mill derived discs (Fig. 3).

Particle size analysis of the as-produced autoclave-derived material showed that the particle size distribution widened significantly with a decrease in the amount of TiN present. This was also the case for the dried powders of the same material. Analysis of the ball-mill derived powders showed that increased TiN content

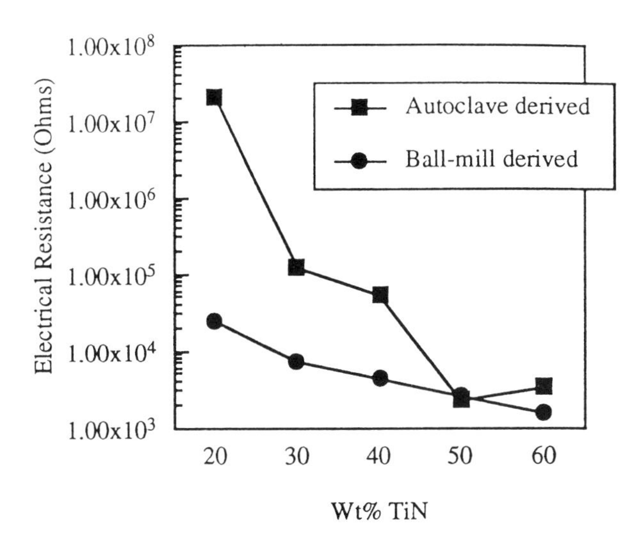

Figure 1 Electrical resistance as a function of TiN wt% for autoclave and ball-mill derived sintered discs.

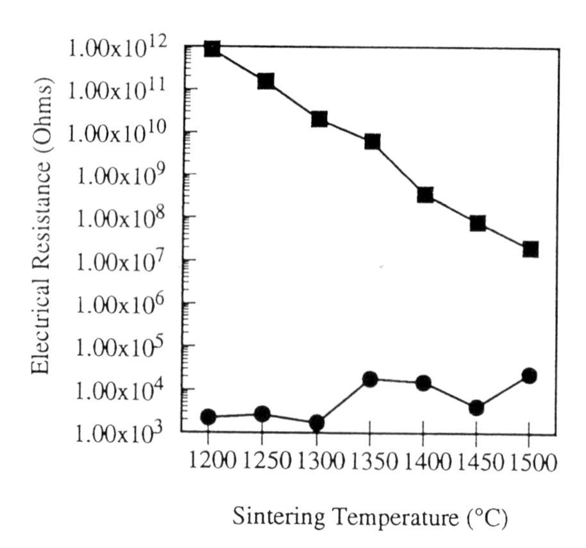

Figure 2 Electrical resistance as a function of sintering temperature for autoclave and ball-mill derived materials, sintered for 3 h.

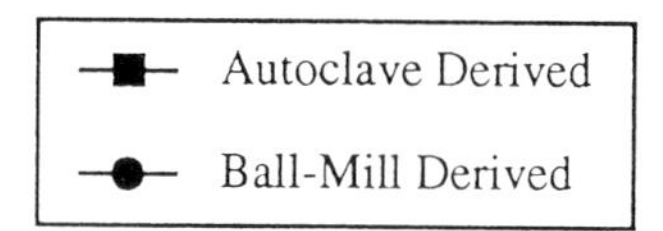

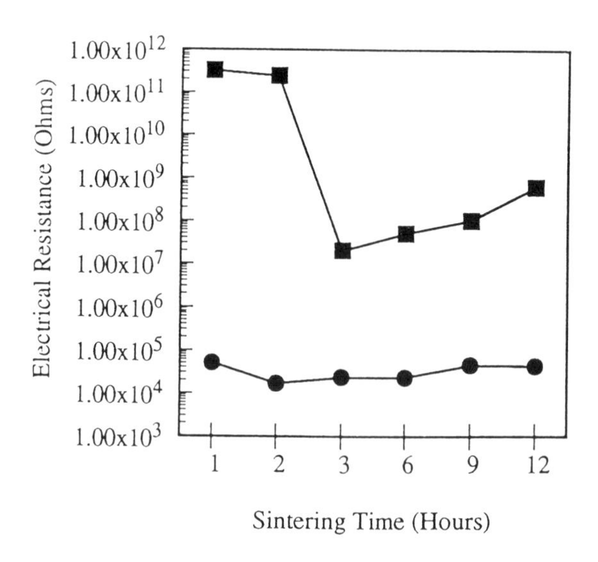

Figure 3 Electrical resistance as a function of sintering dwell time for autoclave and ball-mill derived materials, sintered at 1500°C.

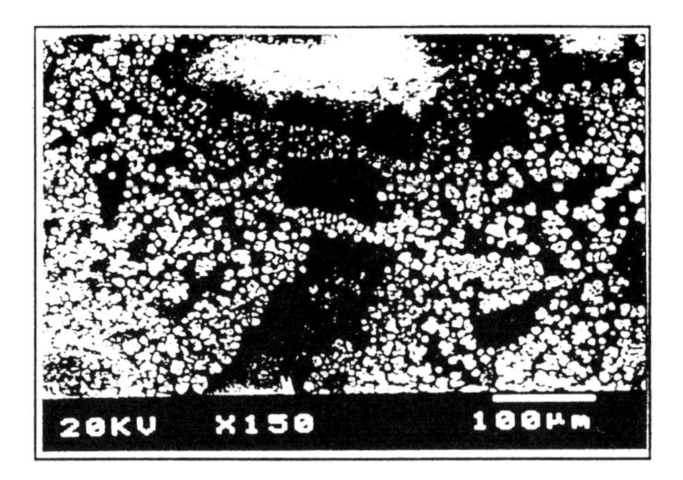

Figure 4 SEI micrograph showing electrodeposit of metallic copper on electrically conducting ball-mill derived mullite–TiN composite ceramic.

narrowed the particle size distribution slightly, whereas increasing the milling time had the reverse effect, as did increasing the powder/milling-ball mass ratio.

Vickers indentation microhardness measurements showed that sintered discs derived from ball-mill mixed material had a comparable microhardness to those produced from autoclave-mixed material. The technique also indicated a decrease in microhardness with increasing TiN content (from 10 to 16 GPa at 20 wt% TiN to ≈0.5 GPa at 60 wt% TiN). The incorporation of 20 wt% TiN into the mullite composition material resulted in an increase in hardness over that predicted for a mullite–TiN composite by the rule of mixtures, for material from both processing routes used (16 GPa cf. 13.5 GPa respectively). This occurred at longer sintering dwell times in both materials. A great increase in the microhardness of the autoclave-derived discs was also evident on using increased sintering temperatures.

The decoration technique was only successful with sintered discs derived from ball-mill mixed material; the deposition of copper onto suitably conducting material within a ball-mill derived disc is shown in Fig. 4.

Scanning electron microscopy (SEM) showed very similar fracture surfaces and polished surfaces in both types of compact with SEI. In the backscattered imaging mode the prepared specimens showed a network of high atomic number material surrounding isolated regions of low atomic number material in the ball-mill derived discs (Fig. 5). However, BEI of the autoclave-derived discs showed much less contrast with no discernible network of either high or low atomic number material (Fig. 6).

4. DISCUSSION OF RESULTS

The two differing production routes were selected because they were expected to result in final ceramic bodies with significant microstructural differences, leading to

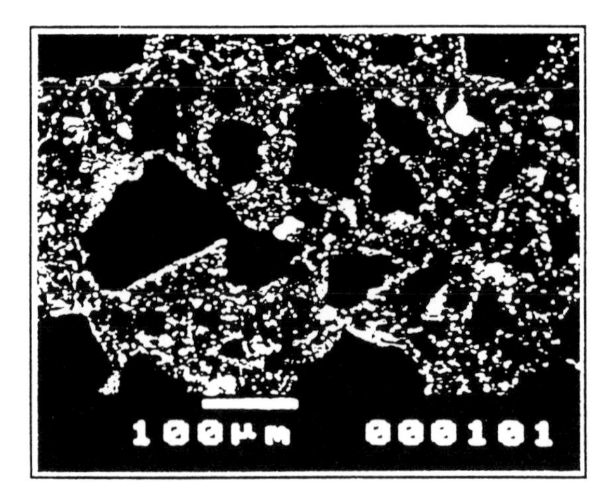

Figure 5 BEI ball-mill derived composite ceramic showing interconnected network of high atomic number material.

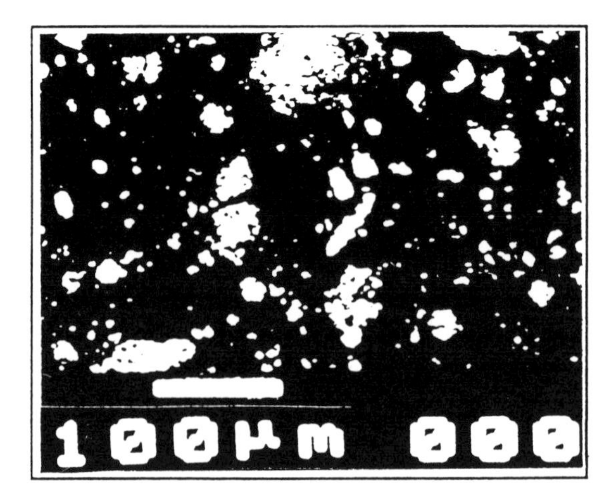

Figure 6 BEI autoclave derived composite ceramic showing isolated regions of high atomic number material.

dissimilar physical properties. This is indeed the case, since the ball-mill derived discs showed a much lower electrical resistance than the autoclave-derived discs. This is a result of the coarser distribution of TiN within the ball-milled powder, leading to the formation of a conducting phase network, by diffusion, throughout the ceramic body during sintering. The high resistance of the autoclave-derived discs is a result of the near-uniform distribution of TiN within the mullite composition powder after hydrothermal synthesis. The isolation of the conducting component occurred because of deposition, during synthesis, of mullite composition material onto the TiN particles which acted as nucleation sites. This supposition is supported by particle size measurement, which shows that when lower amounts of TiN were added, fewer nucleation sites were present, leading to greater deposition on those TiN particles remaining, which resulted in an increasing particle size distribution with decreasing TiN content.

The BEI micrographs show that ball-mill derived discs contain a network of high atomic number material, this being discrete TiN, and material into which TiN has diffused. This arises because the harder TiN particles agglomerate during ball-milling, and thus areas are formed within the green compact which are TiN rich. Thus, during sintering, the diffusion of TiN into the surrounding material is inhomogeneous leaving some regions TiN rich and others TiN depleted. The TiN network is responsible for electrical conduction within the compact; the low electrical resistance network is only formed during sintering when diffusion of TiN into the mullite composition material reaches a threshold point beyond which electronic conduction is possible. The BEI micrographs of autoclave-derived material show no high atomic number material network; instead, the near-homogeneous discrete distribution of TiN and the hindered diffusion of TiN through the mullite material coating on the TiN particles have resulted in the threshold level for conduction not being achieved in any region of the disc. At higher TiN contents this threshold is reached, as demonstrated by the lower electrical resistance measurements for the autoclave-derived discs with TiN contents of 50 and 60 wt%.

The sintering regime controlled the degree of TiN diffusion in both types of disc. In ball-mill derived discs the lowest electrical resistance was achieved after sintering at 1450°C for 3 h. Increasing the sintering temperature or time beyond this resulted in increased electrical resistance, due to the gradual depletion of the TiN-rich areas below the threshold point for conduction.

The copper decoration technique showed the presence of conducting paths both on the surface and within the bulk of the ball-mill derived discs, since the deposition of copper was uniform and independent of electrical contact placement. Overgrowth of the copper deposit has prevented the further use of this technique for quantitative assessment of the conducting network volume.

The material formed during sintering is not a pure mullite–TiN composite; XRD has shown that pure 3:2 mullite is not present. Therefore it is assumed that the incorporation of TiN with the mullite composition powders has prevented mullite formation during sintering. The ceramic alloy formed during sintering does, superficially, show superior mechanical properties to those of pure mullite. The composition and structure of this new material will be investigated in future work by TEM and other analytical techniques.

However, the limited machining trials indicate that the composite material's resistance is still too great for ED machining. Thus, lowering of the material resistance by changing the nature of the conducting phase is a priority.

5. CONCLUSIONS

The production of an engineering ceramic composite with greatly reduced electrical resistance has been achieved by ball-mill mixing of TiN powder and hydrothermally synthesised material of mullite composition. An attempt to combine the synthesis and mixing stages of production in autoclave mixing did not produce material with greatly lowered resistance. However, the near-homogeneous distribution of TiN in the autoclave-derived material indicates that the uniform discrete dispersion of an insoluble second phase within hydrothermally synthesised material is possible. Following on from this, the ability to selectively diffuse the TiN second phase into the hydrothermal synthesised mullite during sintering, and thereby change the material's microstructure and properties, has been demonstrated.

Superior electrical and mechanical properties were exhibited by the ball-mill mixed powder-derived composites which contained regions of alumino-silicate surrounded by a network of TiN-rich material.

ACKNOWLEDGEMENTS

The authors acknowledge the financial support of the SERC via the IRC; RDBN would also like to thank Prof. J. F. Knott, FRS, FEng (Head of School) for the provision of a School of Metallurgy and Materials funded studentship.

REFERENCES

1. R. F. Firestone: *Abrasionless machining methods for ceramics; The science of machining and surface finishing*, Natl. Bur. Stand. Spec. Publ., **562**, 261–281, 1979.
2. W. Konig, D. F. Dauw and G. Levy: EDM – *Future steps towards the machining of ceramics*, Ann. CIRP, **37** (2), 1988.
3. D. J. Phelps and C. P. Flynn: *Phys. Rev. B.*, 1979, **20**, 3653–3659.
4. D. S. McLachlan, M. Blaszkiewiecz and R. E. Newnham: *J. Am. Ceram. Soc.*, 1990, **73** (8), 2187–2203.

Glass-Ceramic Coatings for Stainless Steel

A. ÇAPOǦLU and P. F. MESSER
Department of Engineering Materials, University of Sheffield, Sheffield, UK

ABSTRACT
Glass ceramic coatings for stainless steel have been developed for use as substrates for hybrid circuit use. Bloating occurred during the sintering of powder layers on steel and pellets of the powder. The tendency to bloat was found to be highly dependent on the ZnO content of the glass and could be significantly reduced by filtering the suspensions of milled glass powders. This enabled a composition that crystallised in a short time to be successfully densified.

1. INTRODUCTION AND AIM OF STUDY

A glass-ceramic based on barium disilicate has been developed to coat stainless steel to produce substrates suitable for hybrid circuitry. Barium disilicate has a high coefficient of thermal expansion ($\cong 13.6 \times 10^{-6}/°C$), which enables the expansion coefficient of the glass-ceramic to be matched to that of the steel. This study continues work of J. Collins[1] and Wu Tie[2] in this Department who showed that coatings based on $BaO.2SiO_2$ could be applied to Fecralloy[1] and stainless steel.[2]

In order to reduce processing temperatures, zinc oxide and boron oxide are included in the composition of the glass-ceramic along with the major constituents, to form $BaO.2SiO_2$. The coating is applied to the steel as a glass powder and sintered to form a dense layer. During the heat treatment the coating crystallises. A highly crystallised coating is required both to prevent the thick-film resistors and conductor tracks sinking into the surface, and for matching the thermal expansion coefficients.

To obtain quantitative data on sintering behaviour of various compositions, compacted pellets of glass powder were employed. The sintering behaviour of pellets is different from that of coatings which are bonded to a non-shrinking material. Pellets can shrink freely in all directions, whereas coatings are constrained and can shrink freely only in the direction perpendicular to the steel/coating interface. Consequently, a coating requires a somewhat more severe heat treatment than that given to a pellet of the same composition to achieve the same density.

The pellets and coatings were found to bloat, i.e. pore growth occurred with consequent reduction in bulk density and blistering of their surfaces. The severity of the bloating was very dependent on the zinc oxide content of the glass-ceramic. Although pellets and coatings could attain a high bulk density in a short sintering time, this time was too short for the layer to crystallise sufficiently. Therefore, it was important to minimise bloating.

The aim of this study was to find the reason for the bloating behaviour.

2. SINTERING TRIALS

2.1 Preparation of Glass Powders

A number of compositions, some of which are given in Table 1, were melted in a platinum crucible at 1450°C. The starting materials used are listed in Table 2. The glasses were poured into water so that they became stressed and cracked to facilitate crushing and milling.

The glasses were crushed to pass through a 63 mesh (approx. 250 μm) sieve by placing the frit in steel pots containing a few steel balls. The pots were shaken on an OP–PO vibratory mill (Podmore & Sons Ltd, Stoke-on-Trent).

The crushed powders were milled wet in a small vibroenergy mill (W. Bolton, Stoke-on-Trent), which had a stainless steel milling chamber, using zirconia cylpebs (12.8 mm diameter by 13 mm long). The initial millings were performed with a solids content of 30 vol.%. Samples of powder were withdrawn periodically and the size distribution determined using a Coulter LS 130 laser diffraction and PIDS system. Milling was continued in all cases until a median size of $\cong$15 μm was obtained. A typical distribution is shown in Fig. 1.

Table 1
Compositions of some of the glasses studied

Class code	BaO wt%	SiO_2 wt%	ZnO wt%	B_2O_3 wt%	Al_2O_3 wt%
CAM-6	50.28	40.06	2.10	4.06	3.50
CAM-4	49.29	39.27	4.03	3.98	3.43
CAM-7	47.25	37.64	8.01	3.82	3.28

Table 2
Suppliers and purity of glass components

Chemical	Purity	Supplier
Barium Carbonate ($BaCO_3$)	Lab. reagent	B.D.H. Chemicals
Quartz (SiO_2)	99%	Colin McNeal Ltd
Zinc Oxide (ZnO)	Lab. reagent	B.D.H. Chemicals
Aluminium Hydroxide ($Al(OH)_3$)	General-purpose reagent (99%)	B.D.H. Chemicals
Boric Acid ($H_3(BO)_3$)	Lab. reagent	B.D.H. Chemicals

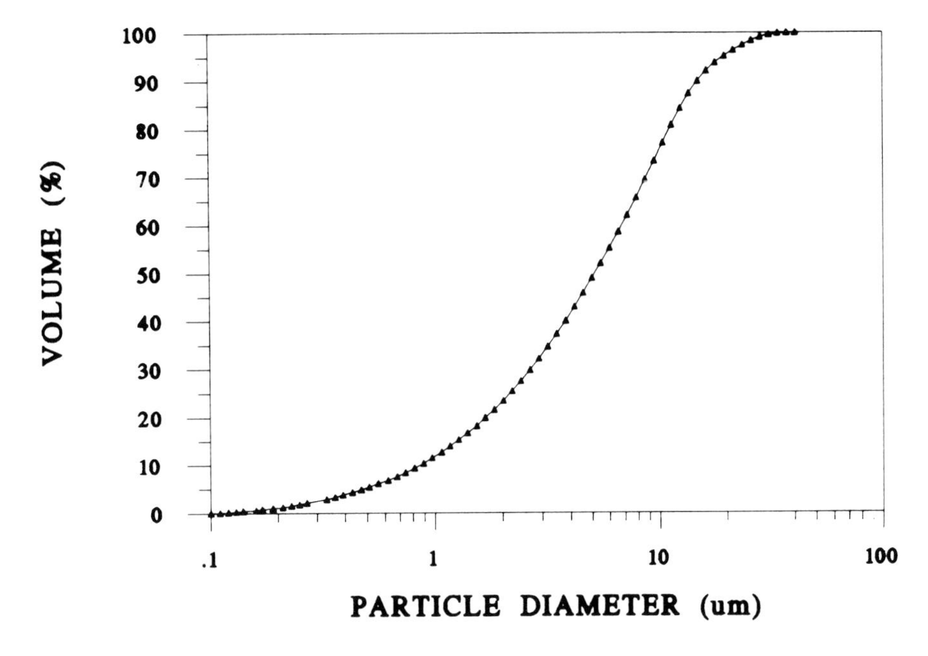

Figure 1 Typical particle size distribution of milled glasses.

During crushing and milling the product became contaminated with iron. This was removed magnetically. To do this, the milled slurries were passed through a mesh of magnetic stainless steel wire housed in a plastic tube placed between the poles of a powerful permanent magnet (flux density 1.5 T). The slurry was passed 25 times through the tube and after each pass the tube was removed from the magnet and washed with water to remove the captured particles of steel and iron oxide.

The slurries were placed in an oven at 105°C to evaporate the water. The powder cake was broken up in an agate mortar using an agate pestle.

The powder densities of the glasses were determined using a specific gravity bottle with water as the displacement liquid.

To prepare a press body for pellet formation, 2 wt% of Carbowax (20 M Union Carbide) was incorporated into the powder. This was achieved by making the powder into a paste with the required amount of distilled water into which the Carbowax had been dissolved. The paste was oven dried in the agate mortar and occasionally mixed with the agate pestle to retain a uniform distribution of the binder.

2.2 Pellet Formation

Pellets 10 mm diameter and approximately 3 mm in height were formed by single-ended pressing in a steel die using 40 MPa compaction pressure.

2.3 *Substrate Coating*

Stainless steel (AISI 430) sheet typically 25 mm × 25 mm × 1 mm were coated with a slurry of the glass powder in distilled water. The slurry had a solids content of about 60 wt%. It was applied to the steel using a syringe. The coated substrates were dried slowly at room temperature to avoid cracking the powder layer.

2.4 *Firing*

The binder was removed from the pellets before the sintering heat treatment by firing a large number of pellets at a time in a muffle kiln to 600°C for 2 h.

The sintering heat treatment for pellets and coated substrates was performed in a horizontal tube furnace (Isoheat, Worksop) controlled by a Eurotherm system. Sintering temperatures of 780, 800, 825 and 850°C were employed.

The pellets and substrates were placed in an alumina boat on alumina powder. This was done quickly for the pellets, after withdrawing the boat from the hot zone of the furnace which was at the required temperature for the sintering trial. The boat was then pushed back into the hot zone to be adjacent to the measurement thermocouple. This caused the temperature recorded by the thermocouple to drop by some 15°C. The sintering time was considered to commence from when the thermocouple recorded a temperature 5°C below the set temperature. The temperature reached this value in approximately 1 min. After the required sintering time the boat was withdrawn from the hot zone.

2.5 *Density of Pellets*

The densities of the pellets were determined using a mercury densitometer. Three pellets that had been given nominally the same heat treatment were measured to obtain the average value.

2.6 *Results of Sintering Trials*

The effect of zinc oxide content on the densification behaviour of pellets sintered at 850°C is shown in Fig. 2. The figure shows the relative density of the pellets as a function of sintering time. The relative density of a pellet is given by the bulk density divided by the powder density of the glass of which it is made. Pellets of all three compositions bloated, but the bloating decreased as zinc oxide content increased.

The effect of sintering temperature on the densification behaviour of pellets with the lowest zinc oxide content is shown in Fig. 3. For all sintering temperatures a maximum relative density greater than 95% was achieved. The relative density of the pellets made with the glass having the highest zinc oxide content did not decrease rapidly with time. However, this glass crystallised only slowly and required about 60 min to become highly crystallised. This is required to prevent circuit elements from sinking into the surface of coated substrates. Heating for

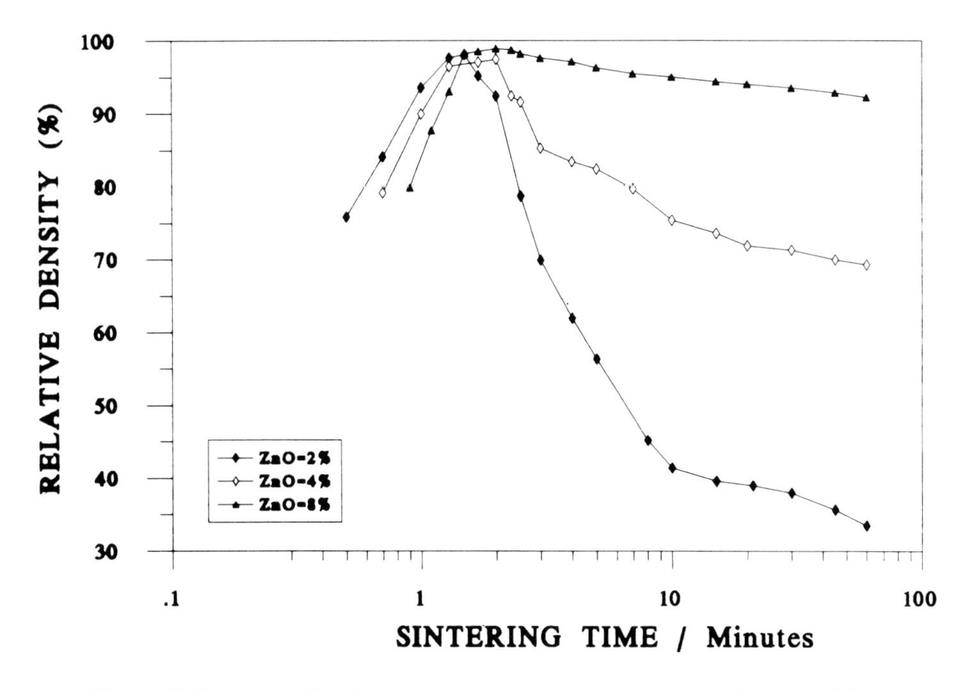

Figure 2 The effect of ZnO content on the sintering results of glasses at 850°C.

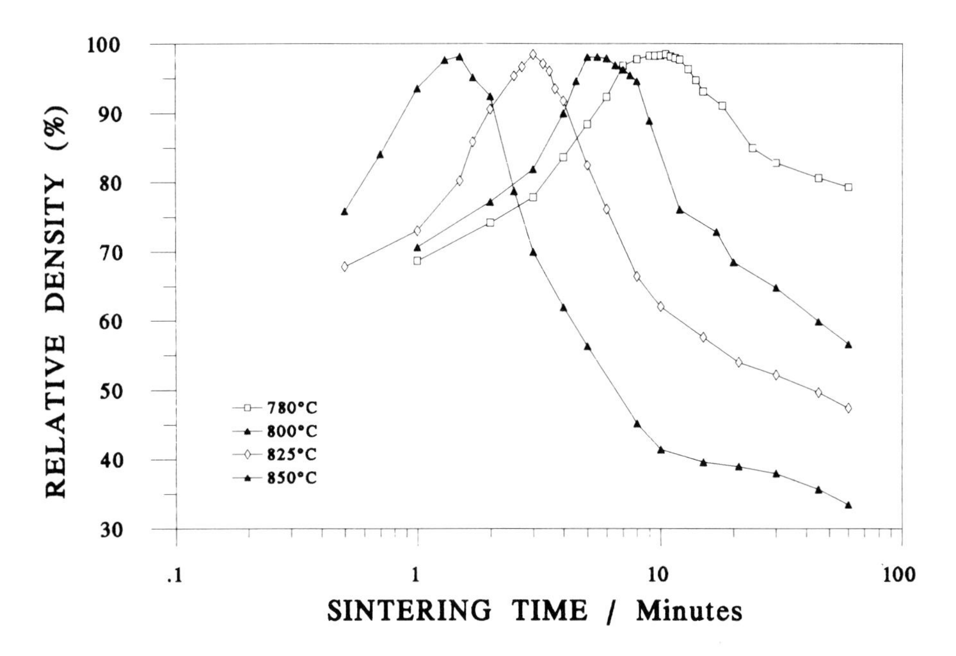

Figure 3 Effect of temperature on the sintering behaviour of glass containing 2 wt% ZnO.

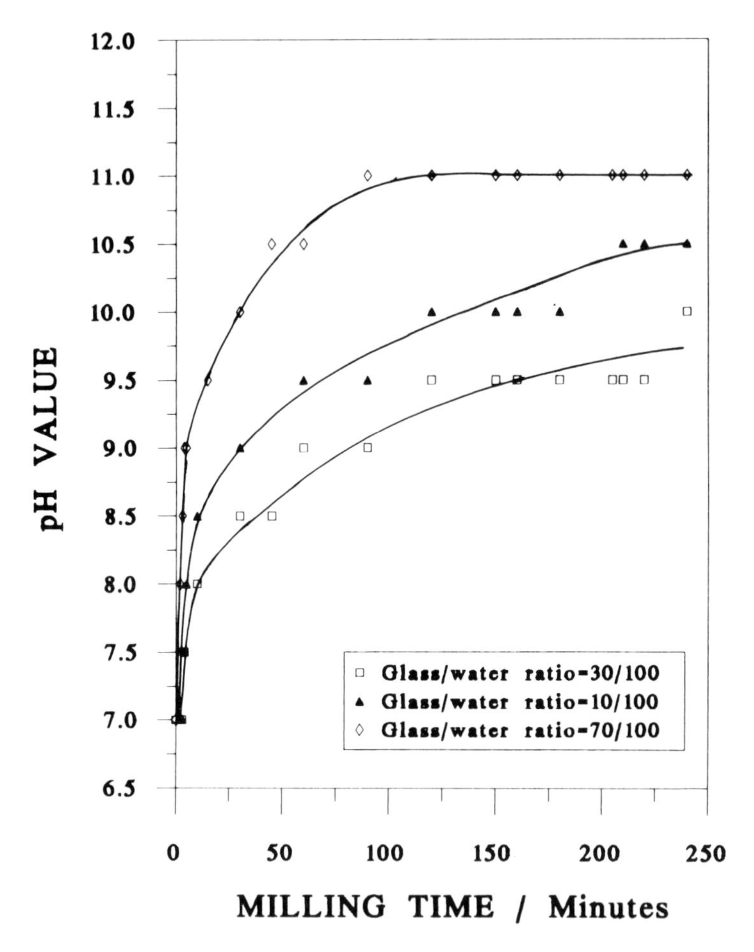

Figure 4 The pH values during milling glass containing 2 wt% ZnO with different glass/water ratios.

60 min is not compatible with the use of the belt furnaces that would be employed in commercial manufacture. A heat treatment of about 15 min at 850°C would be appropriate. The glass with the lowest zinc oxide content (CAM-6: 2 wt% ZnO) could be highly crystallised in 15 min and so would be suitable for use if the tendency to bloat could be reduced.

3. INVESTIGATION OF BLOATING

It was postulated that bloating might have arisen as a result of reaction between the glass powder and water during milling. This was investigated by determining the pH of the suspension during size reduction. The glass was ground in water in an electrically driven agate mortar with an agate pestle to allow easy access to

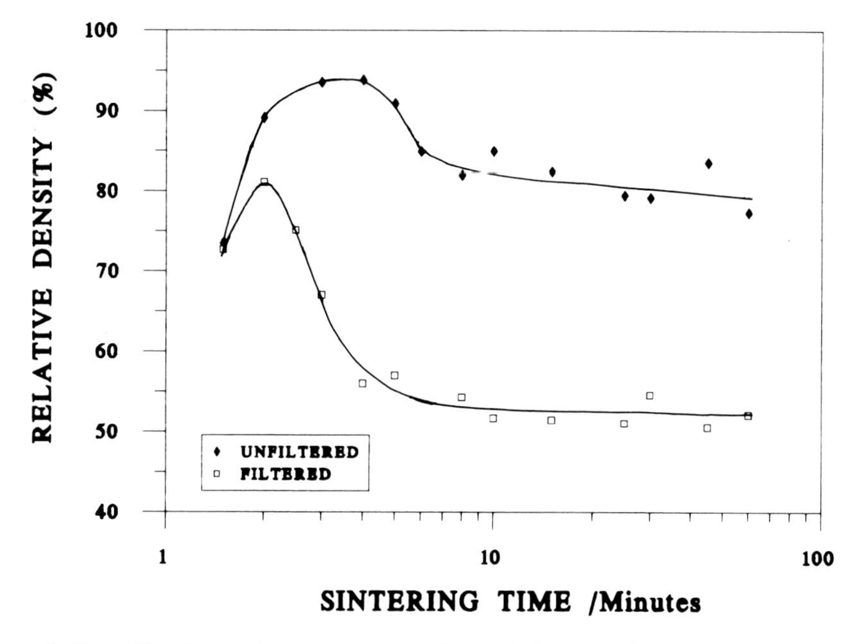

Figure 5 The pH values of aqueous suspensions of glass powder-containing 2 wt% and 8 wt% ZnO as a function of milling time.

determine the pH. The pH value was estimated using indicating paper. Results were obtained for several ratios of glass powder to water, and for glasses with different ZnO contents. Figure 4 shows how the pH value changed with the milling time for ratios of glass powder to water of 10, 30 and 70 g of powder to 100 g of water. The pH value appears to reach an equilibrium value within the time of the experiment for the highest glass powder to water ratio. This equilibrium value depended on the ZnO content, as shown in Fig. 5. It appears that the reaction with water was reduced as the ZnO content increased.

This was confirmed by the analysis of the solutions obtained by filtering the suspensions. These were analysed using the inductively coupled plasma emission spectroscopy (ICP) technique. The analyses for CAM-6 (2 wt% ZnO) and CAM-7 (8 wt% ZnO) are given in Table 3.

Table 3
Chemical analysis results of filtered solutions

Glass code	Elements			
	Ba μ g/ml	Si μ g/ml	B μ g/ml	Al μ g/ml
CAM-6	172	63	16	1.7
CAM-7	149	36	13	<1

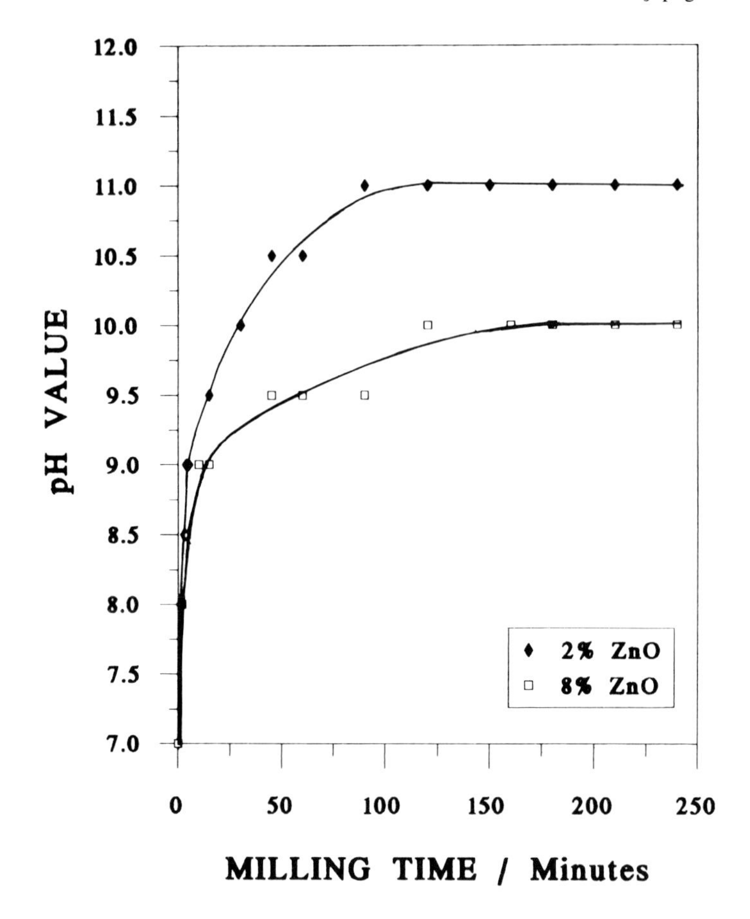

Figure 6 Effect of filtering on the sintering behaviour of glass containing 2 wt% ZnO.

The residue, resulting when the solution obtained after milling CAM-6 was evaporated, was analysed by X-ray diffraction. Three crystalline phases were identified. α-$Ba(OH)_2$, $(SiO_2).(0.2H_2O)$ and $3BaO.3B_2O_3.2SiO_2$.

It is considered that the hydrated compounds could break down during the sintering process to evolve water vapour that could cause the glass to bloat. In an attempt to reduce bloating, the powder suspensions were vacuum filtered prior to drying. This was very effective, as can be seen in Fig. 6. This shows the densification behaviour of pellets of CAM-6 (2 wt% ZnO) as a function of time, for filtered and unfiltered powders.

4. CONCLUSIONS

Dense, crystallised coatings on steel of a glass-ceramic based on barium disilicate have been produced using heat treatments that are suitable for commercial manufacture. This was made possible by reducing the tendency of the material to bloat during sintering. Bloating was considered to be caused by the breakdown of hydrated compounds resulting from the reaction of glass with water. These compounds could largely be removed by filtering the suspension prior to drying.

ACKNOWLEDGEMENT

This work is sponsored in the form of a scholarship by the Turkish Ministry of Education (A.Ç.)

REFERENCES

1. J. G. Collins: PhD Thesis, University of Sheffield, 1989.
2. W. Tie: Unpublished work, University of Sheffield.

The Use of Nonlinear Ferroelectric Ceramic Dielectrics in High-Voltage Pulsed Power

S. A. FAIRLIE and C. E. LITTLE

J. F. Allen Research Laboratories, Dept. of Physics and Astronomy, University of St Andrews, North Haugh, St Andrews, Fife, KY16 9SS, Scotland

ABSTRACT

Capacitors containing ferroelectric ceramic dielectrics are widely used in the generation of electrical pulses at voltages up to a few tens of kV with peak currents in the kA range. Generally the composition of each ceramic is chosen so that the operating temperature is close to the Curie temperature in order to maximise its relative dielectric constant. Operation close to the Curie temperature, however, makes the dielectric constant of these materials highly sensitive to both temperature and applied electric field. Much work has been done by commercial capacitor manufacturers to minimise these variations in capacitance, although in recent years it has been realised that the nonlinearity in dielectric constant with applied electric field can be exploited for the shaping of electrical pulses. If an electrical pulse is fed into a transmission line containing a nonlinear dielectric then different parts of the pulse will experience different dielectric constants and hence propagate with different phase velocities. In particular, if the dielectric undergoes a fall in dielectric constant with increasing electric field, then the sharpening of rising voltage edges is possible. Both experimental and theoretical studies are presented which lead to a greater understanding of this process including the identification of the material properties required for optimum performance in pulsed power systems.

1. INTRODUCTION

The work described here concerns the generation and manipulation of high-voltage electrical pulses. In this context, 'high voltage' means tens of kV with pulse durations of a few tens of ns. In general, source impedances lie between 5 and 50 Ω implying peak currents of up to several kiloamps.

It is well known that ferroelectric ceramics have relative dielectric constants that are dependent on both temperature and applied electric fields. While most commercial manufacturers of ferroelectric ceramic capacitors strive to produce materials with dielectric constants that are insensitive to applied electric fields it has recently been realised that this nonlinear behaviour can be usefully exploited. If a material with a relative dielectric constant that falls with applied electric field is used as the dielectric in a high-voltage transmission line, it is found that the leading edge of pulses propagating along the line become sharper. In effect the nonlinear behaviour of the dielectric allows the formation of electrical shock wave fronts.

To date, most high-voltage pulse-sharpening work has been done with discrete component transmission lines, i.e. using ladder networks with separate capacitor and inductor combinations.[1,2] This approach has limitations however as the component values chosen impose a 'quantisation' which limits the maximum degree of pulse sharpening that can be achieved. Discrete component ladder networks have a cutoff frequency which defines the upper limit beyond which frequency components cannot exist. To overcome this it has been necessary to use distributed transmission lines where the individual capacitors and inductors are replaced by a strip line configuration with a continuous piece of dielectric sandwiched between conducting tracks. Predicting the behaviour of these lines depends therefore on both geometry and an understanding of the material properties of ferroelectric ceramic dielectrics.

1.1 Pulse-Sharpening Theory

In order to produce pulse-sharpening action, the dielectric used in a transmission line must meet certain basic criteria. It is desirable that the material be a good insulator and have a low loss factor, but most importantly it must have a relative dielectric constant that falls with increasing applied electric field.

Pulse sharpening occurs because different parts of the pulse travel with different phase velocities. This arises since different parts of the pulse experience different propagation characteristics due to the voltage dependence of the dielectric constant of the material, ε. It is possible to define a catch-up time in terms of the amount by which the peak of the voltage pulse catches up with the more slowly propagating base of the pulse. In general:

$$\text{Phase velocity} = \frac{1}{\sqrt{LC(v)}} \tag{1}$$

$$\text{Catch-up time} = \sqrt{LC_0} - \sqrt{LC_s} \tag{2}$$

where L is the inductance per unit length of line and C_0 and C_s are the unstressed and stressed capacitances per unit length respectively.

Alternatively for a strip-line geometry the catch-up time per unit length of line may be expressed purely in terms of material properties:

$$\text{Catch-up per unit length} \quad \Delta T = \frac{1}{c}(\varepsilon_{r0} - \varepsilon_{rs})^{1/2} \tag{3}$$

Where the subscripts 0 and s refer to the unstressed and stressed states respectively and c is the velocity of light. It must be remembered however that ε_{rs} depends on the maximum applied electric field which is itself dependent on both geometry and the specified operating voltage.

The maximum applied field is very important as it determines the nonlinear properties of the material. The electric field must be large enough to operate the dielectric in its nonlinear regime. The maximum applied field is ultimately limited by the breakdown strength of the material so the thickness of the dielectric must be chosen in order to produce the maximum electric field across the dielectric without risk of breakdown at the chosen operating voltage.

1.2 Material Considerations

Much work remains to be done to find a practical material for high-voltage pulse sharpening but certain criteria have been deduced. For ferroelectric ceramic dielectrics in the frequency range of consideration here the dominant contribution to the overall dielectric constant comes from the presence of switchable electric dipoles. It is the failure of these dipoles to respond in a linear manner to changes in the applied field that results in a nonlinear dielectric constant. It has been found that choosing a material with a Curie temperature, T_c, just below the operating temperature of the system can result in a strong nonlinearity as the ceramic starts off in the paraelectric phase, and ferroelectric-type behaviour is then induced by the applied E field. The variation in dielectric constant for a typical barium titanate ceramic is shown in Fig. 1. It should be noted, however, that not all materials follow this same general form; some show a small initial rise in permittivity, especially at low applied fields.

For typical barium titanate ceramic compositions the small signal dielectric constant is also frequency dependent with a rapid fall-off at a frequency somewhere in the range of 10^9–10^{11} Hz depending upon material composition. This is well known and arises from the inability of the dipoles within the material to respond to rapid changes in the applied electric field. This has important consequences for pulse-sharpening action as it is believed to limit the ultimate degree to which a pulse can be sharpened.[3] Analysis by Katayev[4] of the analogous process of magnetic pulse sharpening in transmission lines filled with nonlinear magnetic materials shows that, given a long enough line with negligible losses, the leading

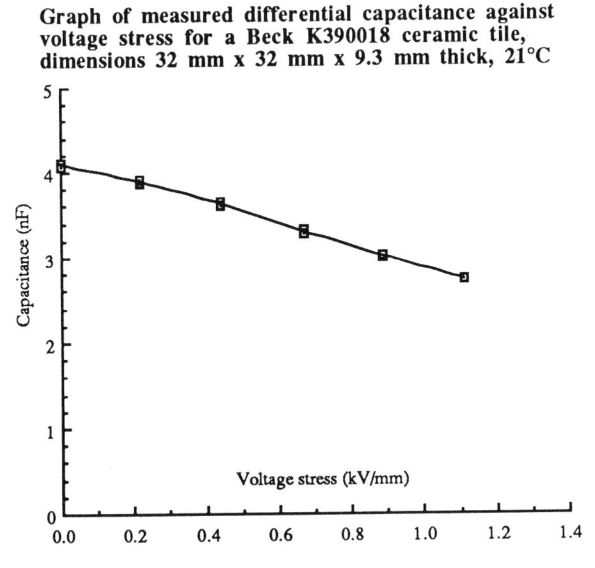

Figure 1 Graph of fall in relative dielectric constant with applied electric field for a barium titanate ceramic dielectric.

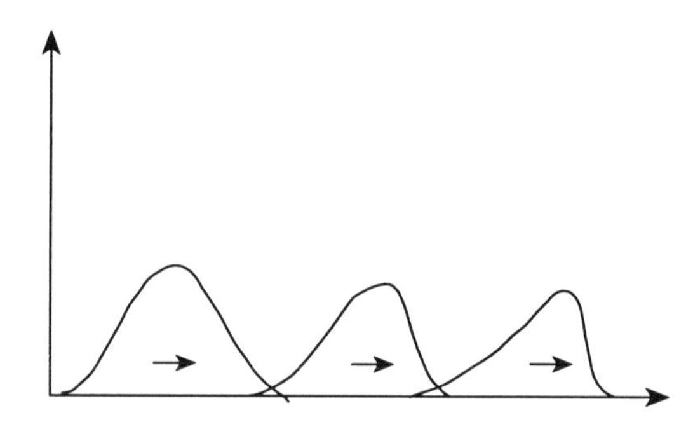

Figure 2 Progress of a voltage pulse along a nonlinear transmission line.

edge of the incident pulse becomes steeper, eventually forming a shock front. Gradually this shock front grows to envelop the whole of the initial leading edge with the gradient of the shock front determined by the response time of the switchable magnetic moments in the material. Recent analysis indicates that an analogous mechanism operates in nonlinear ferroelectric filled lines and that the ultimate degree of pulse sharpening achievable in these lines is similarly dependent on the frequency response of the material.[5] The development of a shock front is shown diagrammatically in Fig. 2.

2. EXPERIMENTAL DESCRIPTION

The properties of a number of electroceramic compositions have been measured to determine their suitability for this work. So far these have mainly been commercial grades of barium strontium titanate, but some PZT compositions have also been investigated. Having chosen a suitable material for investigation a quantity of blocks is then obtained with typical dimensions of 40 mm × 20 mm × several mm thick, the thickness being determined by the desired maximum applied electric field which is itself dictated by the dielectric strength of the material. These blocks are painted with silver conducting tracks then placed end to end, and clamped between clear Perspex plates to form a strip-line configuration. The characteristic impedance of this strip line is given approximately by the formula

$$Z_0 = \sqrt{\frac{L}{C}} = \frac{377d}{\sqrt{\varepsilon_r}(d+w)} \tag{4}$$

assuming $w \gg d$ where d is the dielectric thickness and w the strip-line track width. With real materials having dielectric constants in the range 10^3 to 10^4, lines typically have characteristic impedances of less than 50 Ω. The degree of pulse sharpening that can be expected is determined from equation (3) and the length of line is chosen accordingly. Typical line lengths are of the order of 1 m.

To prevent unwanted pulse reflections, the pulse-sharpening lines are terminated

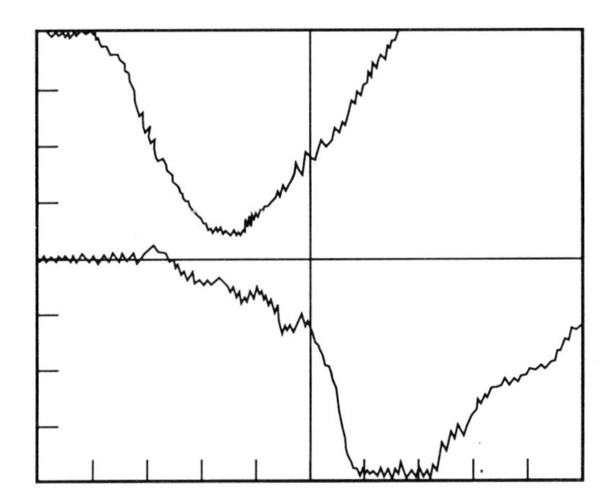

Figure 3 Oscillograms showing input and output voltage waveforms (upper and lower traces respectively) for a nonlinear transmission line with dielectric of nominal dielectric constant of 4000 terminated in a 9 Ω resistive load, line dimensions 3 mm wide × 9 mm thick × 1200 mm long. Vertical scale 5 kV/div, time base 50 ns/div.

by resistive loads equal to the characteristic impedance of the line. It can be seen, however, that the nonlinearity of the dielectric constant leads to a characteristic impedance of the line which is similarly voltage dependent. It has been found that termination with a load equal to the stressed impedance of the line provides the best available compromise. The lines are operated under insulating oil and pulsed from a conventional thyratron switched, pulse-forming circuit capable of generating pulses of up to 40 kV amplitude with a source impedance that can be chosen to be as low as 5 Ω.

3. RESULTS

Several commercial ceramic dielectrics based on barium titanate have been formed into strip lines and tested for pulse sharpening action. As a typical example the behaviour of the material whose nonlinear characteristics are given in Fig. 1 is described here. This particular composition, with a nominal (unstressed) relative dielectric constant of around 4000, was formed into a strip line with dimensions 3 mm thick × 9 mm wide and 1200 mm long. This was terminated in a 9 Ω resistive load and driven with a pulse of amplitude 20 kV. The input and output pulses as measured with Tektronix P6015A high voltage probes are given in Fig. 3.

It is apparent from Fig. 3 that a shock front has formed at the higher voltage part of the leading edge of the pulse. This is to be expected, as Fig. 1 shows the nonlinearity of the material increasing with voltage, i.e. the fall in dielectric constant with increasing voltage is not a straight line. The steepness of the shock front seems to be limited, however, by a dispersion mechanism and there is also a slowly rising 'pedestal' at the low voltage start of the pulse. These features are typical over the range of materials tested and are discussed further in the following section.

4. DISCUSSION

It is clear from the results obtained that there are frequency dispersion mechanisms operating in opposition to the pulse-sharpening effect. Two possible sources of dispersion have been identified, that arising from the strip-line geometry and that due to material characteristics.

Taking geometry effects first, the high permittivity of the nonlinear dielectrics used dictates that to produce lines with characteristic impedances in the range of interest 5 Ω–50 Ω, the track width, w, needs to be fairly narrow, of the order of a few mm. It is important for voltage transfer that the pulse-sharpening line be impedance matched to the pulse-generating circuit. The minimum source impedance of the pulse generator is 5 Ω. In many cases, especially with the higher dielectric constant materials, w has not therefore been significantly greater than the dielectric thickness, d, so that the assumptions of equation (4) do not hold. A consequence of having narrow track widths is that edge effects are significant, which result in a non-uniform electric field through the dielectric. This is important as the applied electric field determines the permittivity of the dielectric and hence the phase velocity of the propagating wave front. If the applied electric field falls off towards the edge of the strip line, the edges of the wave front will propagate more slowly than at the centre, leading to spatial dispersion, i.e. different parts of the incident wave front will arrive at the load at different times. Quite clearly this unwanted effect can be minimised by using strip-line geometries with wide tracks, however, the high permittivities of currently available materials dictate that these correspond to unsuitably low characteristic impedances.

Another source of frequency dispersion arises from the limited frequency response of the material itself. As stated earlier, most ferroelectric ceramic dielectrics have small signal permittivities that are fairly constant up to high frequencies of the order of 10^9–10^{11} Hz; above this threshold however the permittivity becomes frequency dependent as the material can no longer respond fast enough to the applied electric field. It is possible to measure the small signal frequency response of dielectrics and an indication of the dispersion that can be expected along a length of transmission line can be found by using standard techniques such as time domain reflectometry (TDR). The frequency response of a material subjected simultaneously to both high frequencies and high fields is more difficult to determine but it is certain that with large amplitude signals the usable bandwidth will be reduced. It can be inferred, however, that materials with a high, small-signal bandwidth will tend to have greater bandwidths when subjected to large signals.

It is clear from these experiments that improving the performance of high voltage pulse sharpening is dependent on the development of electroceramic materials with improved characteristics. These material development objectives can be summarised thus:

1. Lower nominal dielectric constant
   To permit the building of lines with higher impedances and wider tracks giving better field uniformity.
2. Better high-frequency response
   To reduce the fastest risetimes that can be achieved.
3. More nonlinear with applied E field
   To allow shorter lines with the same sharpening effect.
4. Lower losses
   To reduce pulse attenuation and unwanted heating at high repetition rates.

ACKNOWLEDGEMENT

The authors wish to thank DRA, St Andrews Road, Great Malvern, Worcestershire, WR14 3PS, for financially supporting this work.

REFERENCES

1. C. R. Wilson, M. M. Turner and P. W. Smith: *Appl. Phys. Lett.*, 1990, **56** (24), 2471–2473.
2. C. R. Wilson, M. M. Turner and P. W. Smith: Proc. 19th IEEE Power Modulator Symposium, San Diego, CA, 1990.
3. G. Branch and P. W. Smith: IEE Colloquium ‘Pulsed Power ’93’, Digest No: 1993/043, IEE, Savoy Place, London, 1993.
4. I. G. Katayev: *Electromagnetic Shock Waves*, Iliffe Books Ltd, London, 1966.
5. J. E. Dolan: UWCC, Cardiff, Private Communication.

© Crown Copyright 1994.

D.C. Pre-Breakdown Photon Emission from an Alumina Insulator in Vacuum

B. M. COAKER,* N. S. XU,* F. J. JONES† and R. V. LATHAM*
*Surface Science Group, Department of Electronic Engineering and Applied Physics, Aston University, Birmingham, B4 7ET, UK
†Cold Cathode Engineering, EEV Limited, Carholme Road, Lincoln, LN1 1SF, UK

ABSTRACT
A cylindrical alumina (94% Al_2O_3) insulator was operated under ultra-high vacuum ($<10^{-8}$ mBar) and subjected to radial D.C. electric fields, with current magnitudes limited to pre-breakdown values of less than 50 μA. Luminescence from the alumina was imaged using a high-resolution CCD camera, and the spectrum of the emitted light was quantified using a high-speed scanning monochromator and photo-multiplier tube detector. Video images of the observed luminescence are presented, together with wavelength spectra recorded at fixed values of D.C. bias voltage. An apparent time dependence of the photonic emission intensity is correlated with an observed time variation in the D.C. current/voltage characteristic of the bridged alumina insulator. The nature of the spectra is discussed in terms of solid-state photon-emission arising from radiative electron-recombination processes at impurity and structural defect centres within the surface layers of the solid insulator.

1. INTRODUCTION

At the present time, there is much discussion concerning the physical mechanism(s) involved in the surface flashover of solid insulators in vacuum:[1] a long-established 'desorbed gas' model proposes the injection of electrons from the cathode triple junction, with a cascade of electrons 'hopping' along the insulator surface desorbing gas as they impact the dielectric, and ionising the desorbed gas molecules.[2] However, two recent models suggest solid-state mechanisms of flashover, namely the explosive destabilisation of trapped charge in the dielectric,[3] and electron avalanche processes within the surface layers of the solid dielectric arising from the injection of electrons at the cathode triple junction.[4]

In previous work,[5] pre-breakdown electrical and optical phenomena observed on a bridged alumina dielectric in vacuum were presented. Significantly, a transparent anode technique[6] was used to determine the spatial location of electron emission from the surface of this alumina structure, and thus a spatial correlation was established between electron emission sites and the location of luminescent centres at the insulator surface. This finding indicates that a strong relation exists between the pre-breakdown electrical processes and the associated optical activity. More importantly, this localised pre-breakdown electron emission could be responsible for initiating a subsequent breakdown event or surface flashover.

This early study has, therefore, prompted a further detailed investigation of the optical emission process, aimed at improving the present understanding of the physical processes involved. Thus, a high-resolution video camera system was employed to record real-time sequences of optical-emission images. In order to obtain quantitative information relating to such optical emission, a high-speed optical spectrum analyser was used to measure the energy distribution of the emitted photons. Concurrent measurements of the pre-breakdown electrical characteristics of the alumina specimen augment the recorded optical phenomena.

The nature of the optical emission will be discussed in terms of charge transport processes associated with the alumina insulator. Particular emphasis is placed upon the physical mechanisms of charge injection into the solid dielectric regime, and how this may initiate optical luminosity of the dielectric.[7] The possible role of defect centres within the dielectric in promoting the observed pre-breakdown electrical and optical phenomena will also be discussed.

2. EXPERIMENTAL

2.1 System

The specimen module is identical to that described elsewhere,[5] and consists of an alumina (Al_2O_3, 94% purity, 10 μm mean grain size) tube, with two concentric rings of Mo–Mn metallised onto one end face of the alumina sample: a nominal spacing of 300 μm separated the two metallised rings. Figure 1 shows the general arrangement of the existing experimental system,[5] including the mounting of the metallised ceramic specimen in a stainless steel vacuum chamber. A system pressure of less than 1×10^{-8} mBar was used for this investigation, with the alumina specimen initially baked for 120 min at 400°C. Electrical connection was made between the inner metallised ring of the alumina sample and a high-voltage D.C. power supply, with the outer metallised ring brazed to a copper tag that was terminated to 0 V. A 30 MΩ resistor was connected in series with the high-voltage power supply, thereby limiting currents through the alumina specimen to pre-breakdown values of $<10^{-4}$ A.

A monochrome CCD camera (768 pixels × 576 lines) was used in conjunction with a 100 mm photographic macro lens to image the metallised alumina specimen through a glass viewport in the vacuum chamber wall. Figure 2 presents a plan view of the end face of the alumina tube, where background illumination was used to highlight the physical features of the specimen. The camera was interchangeable with a lens fixture, which coupled light from the specimen into a scanning monochromator via a fibre-optic light-guide; thus the CCD camera facilitated the spatial imaging of optical events associated with the metallised alumina specimen, whilst the scanning monochromator allowed a quantitative measurement of the spectral data associated with optical events on the alumina specimen.

2.2 *Procedures and Measurements*

2.2.1 *Electrical measurements and optical imaging*

A key objective of this study was the recording of the pre-breakdown electrical characteristics of the alumina specimen, with concurrent observation of the optical

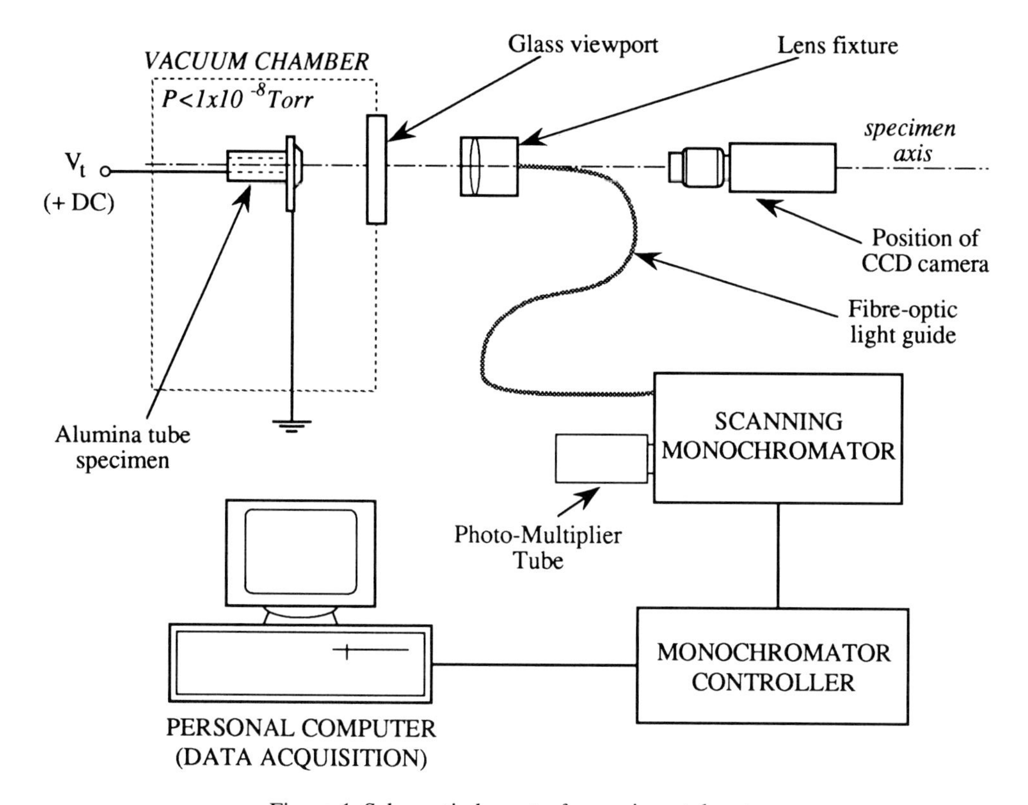

Figure 1 Schematic layout of experimental system.

processes associated with the insulator. Thus a positive D.C. voltage V_t was slowly applied to the inner electrode ring of the specimen up to a maximum amplitude of +5 kV, with corresponding values of pre-breakdown current I_t measured using an electrometer picoammeter. At an applied voltage of +2.5 kV, a ruby-red glow was seen to initiate in the alumina region between the ring electrodes; increasing V_t beyond 2.5 kV caused the intensity of this glow to increase. When V_t reached +4 kV, several bright-blue 'spots' were simultaneously observed in the alumina gap near to the inner ring electrode. The electrical I_t–V_t characteristics of the alumina specimen are given in Fig. 3(a), together with images of the optical events at the end face of the alumina sample corresponding to applied D.C. voltages of +2.5 kV and +4 kV.

Owing to its annular appearance, the ruby-red glow was termed 'doughnut' emission. Image-processing techniques were used to quantify the spatial intensity distribution of the doughnut and the bright-blue spots. Holding V_t at a steady value of +5 kV, the doughnut was observed for a further 30 min. During this time the intensity of the doughnut was seen to diminish, and at the end of the 30 min period the doughnut was no longer visible, although the bright-blue spots persisted.

2.2.2 Optical spectrometry

Following a 24 h rest period, +5 kV D.C. was again slowly applied to the inner electrode ring of the alumina specimen: the second I_t–V_t characteristic depicted in

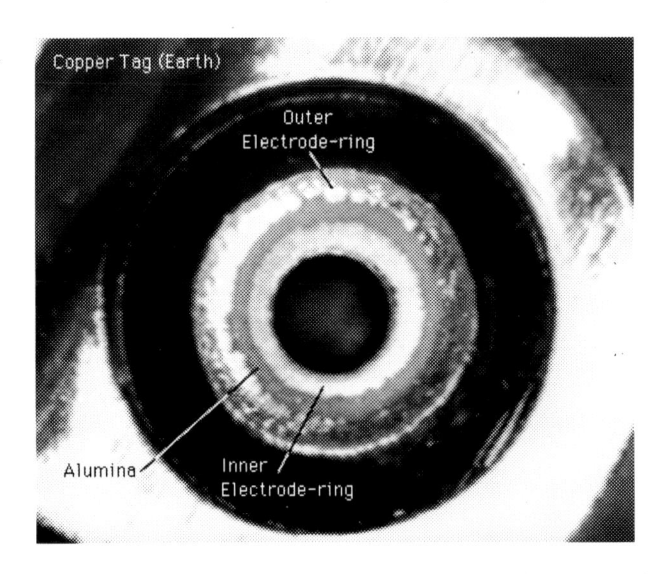

Figure 2 (a) Plan view of alumina specimen with metallised ring electrodes (illuminated image). (b) Schematic outline of alumina specimen.

(a)

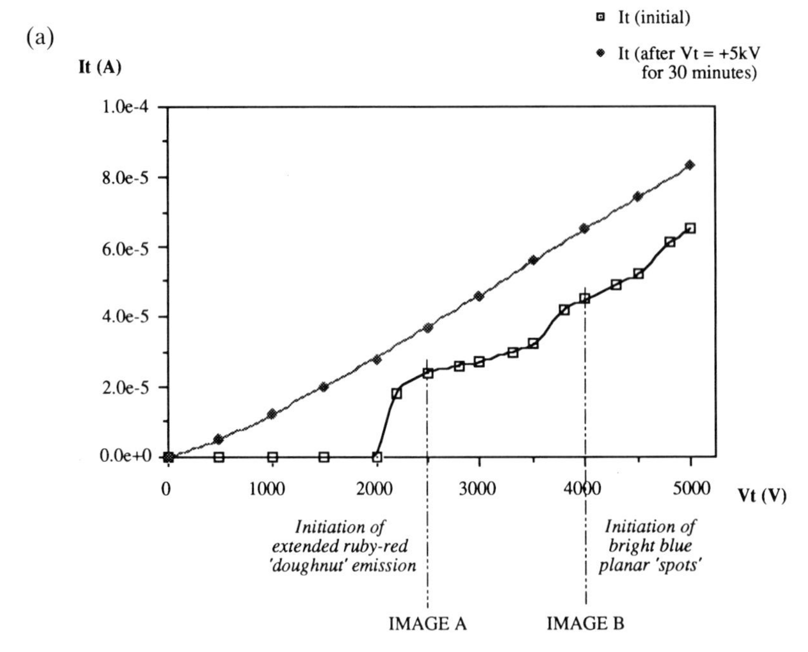

Figure 3 (a) Planar *I–V* characteristics of alumina specimen, recorded for (i) initial measurements, and (ii) following the exposure of specimen to +5 kV D.C. for 30 min. (b) Captured image of 'doughnut' emission from 300 μm-wide alumina region between the inner and outer ring electrodes (V_t = +2.5 kV: Image A). *Inset*: schematic section of end face of bridged alumina specimen, showing origin of the doughnut emission from 300 μm alumina gap. (c) Captured image of localised bright-blue 'spots' and 'doughnut' emission on end face of the alumina tube (V_t = +4 kV: Image B).

(b)

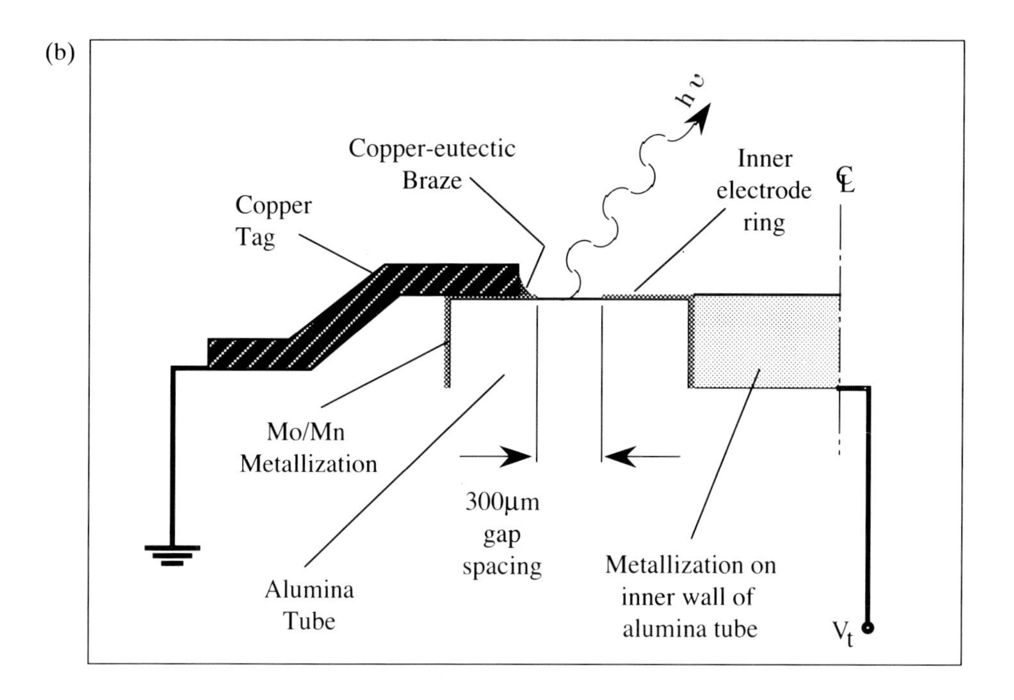

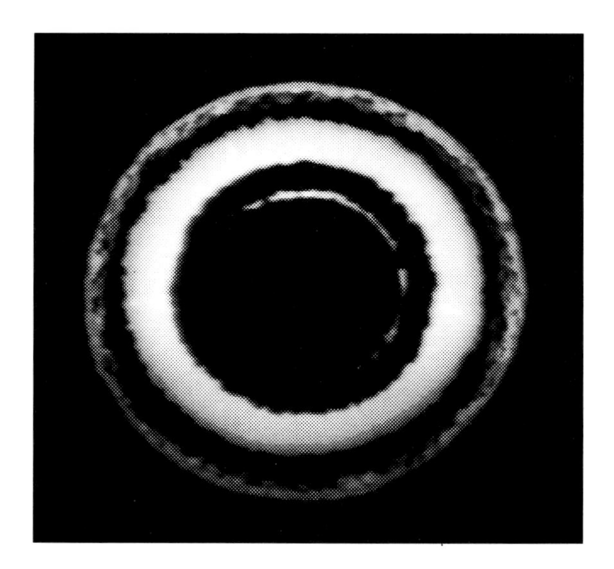

Figure 3—*contd.*

(c)

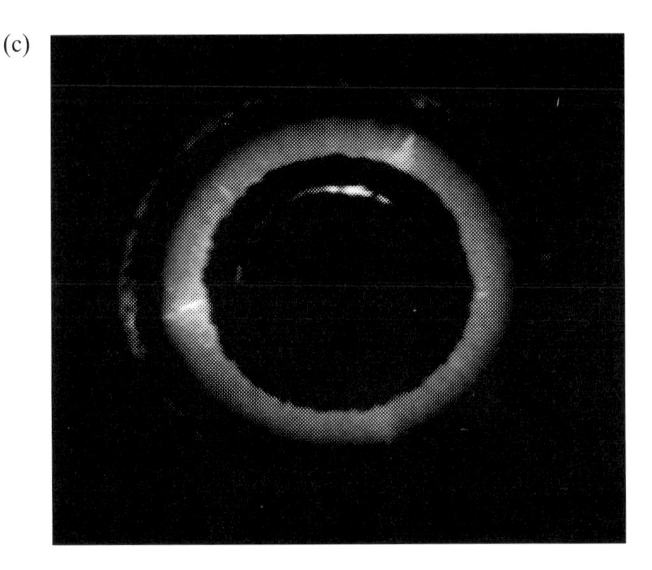

Figure 3—*contd.*

Fig. 3(a) was recorded, showing a *linear* form, and higher pre-breakdown current magnitudes compared with those of the initial test outlined in section 2.2.1. As V_t was increased, localised blue spots were again seen to 'switch on' in the 300 μm alumina gap near to the inner electrode ring, but the ruby-red 'doughnut' emission was not observed. V_t was subsequently increased slowly from +5 kV to +8 kV, and the doughnut emission was re-established.

The light emitted from the alumina specimen was then focussed through a lens fixture into a glass fibre-optic light-guide, which was connected to the input of a high-speed scanning monochromator; 180 μm slits at the monochromator input limited the spectral resolution of the instrument to 5 nm. Two hundred spectral scans were then made over the wavelength range 3000–8500 Å, each scan having 80 ms duration, and the average of these readings recorded. V_t was then increased to +10 kV, and a second optical spectrum recorded. The spectral data recorded for V_t values of +8 kV and +10 kV are presented in Fig. 4(a); similar forms of luminosity spectra to those in Fig. 4(a) have also been recorded for alumina by other workers.[7,8,9]

Increasing the value of V_t from +8 kV to +10 kV led to an increase in the intensity of the light emitted from the sample, although the wavelength characteristics of the spectrum remained unchanged. A broad band of emission was observed around 4100 Å, with a short wavelength cutoff at 3800 Å arising from the glass materials of the UHV viewport and the fibre-optic light-guide. A major emission peak could be seen at 6940 Å, with weak emission peaks at 6720 Å and 7150 Å: an expansion of the wavelength plot around the 6940 Å peak is shown in Fig. 4(b).

Finally, argon (Ar) gas was bled into the vacuum system, in order to investigate the effects of a low-pressure gas ambient upon the pre-breakdown behaviour of the

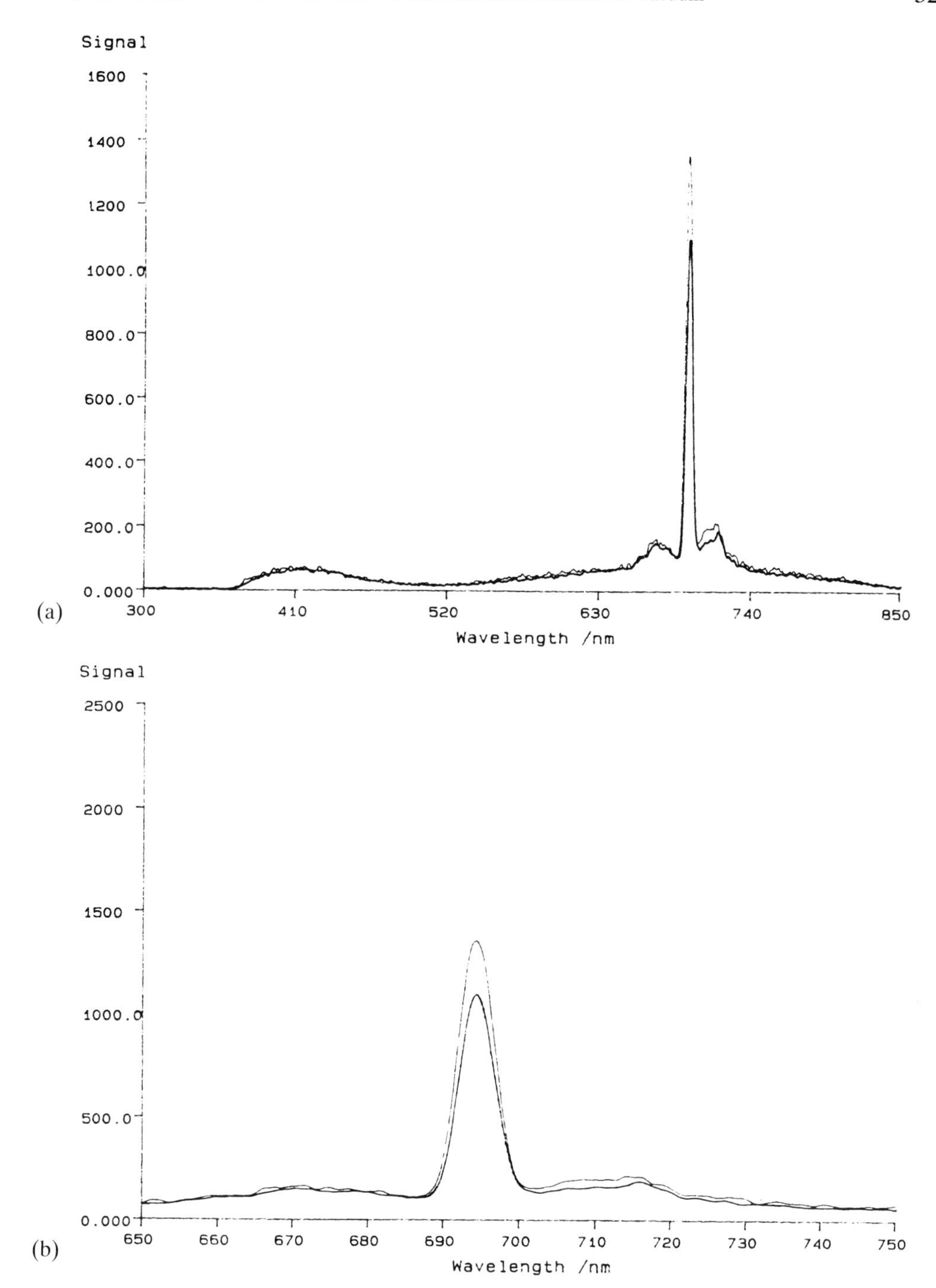

Figure 4 (a) Optical spectrum of doughnut and localised spot emission observed on end face of alumina tube specimen, for applied voltages (V_t) of +8 kV and +10 kV (200 spectral scans (averaged), 5 nm half-power bandwidth). (b) Expansion of optical spectrum wavelength axis around principal emission peak of 694 nm.

alumina insulator. V_t was slowly applied to the alumina specimen, over the voltage range $0 \leq V_t \leq 10$ kV, at system pressures of 10^{-6}, 10^{-5} and 10^{-4} mBar. No change was seen in the observed phenomena, either in the electrical characteristics or in the optical signature of the alumina specimen at each of these pressures, compared with earlier specimen operation under $<10^{-8}$ mBar (ultra-high) vacuum.

Continued application of a D.C. voltage (V_t) of +10 kV to the inner electrode ring of the sample, under vacuum, led to an eventual dimming of the doughnut emission after a period of four hours, although the bright-blue spots persisted for a further two hours before they were no longer visible. The measured I_t–V_t characteristics of the alumina specimen remained unchanged, in keeping with similar observations made in earlier work.[5]

3. DISCUSSION

Two types of D.C. pre-breakdown photon-emission processes have been recorded from an alumina specimen: a broad-area, deep-red doughnut emission (see Fig. 3(b)), and localised bright-blue spots (Fig. 3(c)). A concurrent measurement of the electrical (I_t–V_t) characteristic of the alumina insulator displayed features associated with the onset of these two optical phenomena. For clarity of presentation, the pre-breakdown optical and electrical characteristics associated with the alumina specimen will be discussed separately, with particular attention to the physical nature of the optical and electrical findings. Later discussion and conclusions will then draw upon both of these areas, and present a unified model to describe the optical and electrical phenomena associated with the D.C. pre-breakdown of an alumina insulator in vacuum.

3.1 Origin of Optical Emission

Considering the recorded images depicted in Figs 3(b) and 3(c), the optical emission spectrum recorded for the alumina specimen (Fig. 4(a)) shows predominant optical emission from the alumina in a narrow band of wavelengths around 694 nm, with an additional broad emission band centred around 410 nm extending into the UV region. The dominant spectral lines around 694 nm are characteristic of the observed deep-red doughnut emission, whilst the emission band centred around 410 nm is believed to be associated with the localised bright-blue spots observed on the alumina specimen (Fig. 3(c)): these phenomena represent the key findings of this work.

Two possible models may explain the physical nature of the optical emission from the alumina insulator, namely the desorption of gas from the insulator surface by 'hopping' electrons,[2] leading to a gas discharge above/over the surface of the insulator, or the injection of electrons from the cathode into the surface layers of the alumina,[4] giving rise to solid-state luminescence at defect sites within the solid dielectric:[9] these solid-state processes will be discussed later. With reference to Figs 4(a) and 4(b), the spectrum of the optical emission did not indicate any strong atomic spectral lines, with little evidence to suggest any atomic line spectra from ionised desorbed gas.[10] To consolidate this assumption, an experimental study of the behaviour of the photon emission in a residual atmosphere of argon gas showed no effect upon the optical signature of the photonic emission from the alumina sample.

Considering the material composition of the alumina ceramic, the specimen was of 94% purity (Al_2O_3) by weight, with an average grain size of 10 μm. A remaining fraction of oxide glass locked these alumina grains into a ceramic matrix; this glassy phase consisting of alumina, silica (SiO_2) and CaO, with trace (few ppm) concentrations of metal oxides including sodium, titanium and chromium.[11] Thus the presence of these material impurities, together with structural defects in the ceramic, will be considered alongside the findings of this work, and the possible sources of the observed optical emission phenomena will be discussed in detail.

3.1.1 Nature of 'doughnut' emission (λ = 694 nm)

Referring to the expanded wavelength-plot of the doughnut emission shown in Fig. 4(b), the form of this spectrum is in close agreement with the ruby emission spectra ($Al_2O_3{:}Cr^{3+}$), characterised by a narrow emission doublet (R_1,R_2) of two closely-spaced spectral lines, centred around 693.5 nm;[12] a half-power bandwidth of 5 nm at the monochromator input meant that a fine spectral structure such as these two closely-spaced lines could not be resolved in this experiment. This hypothesis, together with the ruby line (R line) nature of the recorded spectrum, indicates a significant luminescence *within* the bulk of the alumina dielectric, associated with optical emission from chromium (Cr) impurity,[8] according to the following radiation transition process:[12]

$$Cr^{4+} + e^- \rightarrow (Cr^{3+})^* \rightarrow Cr^{3+} + h\nu(R_1, R_2)$$

Figure 3(b) illustrates the doughnut emission as a *distributed* phenomenon around the entire 300 μm alumina region between the metallised ring electrodes: this finding is consistent with a distributed impurity population of chromium throughout the *bulk* of the alumina dielectric. Impurity concentrations of Cr in undoped Al_2O_3 are reported as typically 1 ppm[12] to 3 ppm.[13]

3.1.2 Nature of localised bright-blue 'spots' (λ = 410 nm)

Photon emission from alumina in this broad 410 nm band is characteristic of radiation transition associated with negative ion (oxygen (O^{2-})) vacancies:[14,15] these negative ion vacancies are termed *Farbzentrum*, or F centres.[16] This F centre emission process may be represented thus:

$$F^+ + e^- \rightarrow (F)^* \rightarrow F + h\nu(410\,\text{nm})$$

Similarly, emission from an F^+ centre (an oxygen vacancy that has trapped a single electron) has a characteristic emission band around 300 nm;[14,17] such emission would have been masked by the 380 nm cutoff presented by the glass optics of the experimental system. *F* centres are associated with defect sites within the Al_2O_3 crystal structure:[9,14] porosity in alumina is concentrated at grain corners and grain edges,[18] and defect centres are also associated with these grain boundaries.[4,16]

3.2 Electrical Characteristics

Both the initial and subsequent electrical (I_t–V_t) characteristics of the alumina specimen show measurable current flow through the insulator; the initial electrical characteristic of the alumina specimen (Fig. 3(a)) shows two current thresholds, associated with applied D.C. voltages of 2.5 kV and 4.0 kV (with applied electric field magnitudes of ~8 MVm^{-1} and ~13 MVm^{-1} respectively). The first threshold (corresponding to $V_t = +2.5$ kV) coincides with the initiation of the ruby-red doughnut glow within the 300 μm alumina region on the end face of the alumina specimen, whilst the second threshold marked the appearance of two bright-blue spots between the electrode rings in the alumina region.

A further feature of the electrical characteristics of the alumina specimen given in Fig. 3(a) is the non-reversible transformation of the material behaviour of the alumina following a sustained application of D.C. voltage across the specimen: the I_t–V_t characteristic of the alumina is modified following the application of D.C. voltage (V_t), from an initial *semi-insulating* state to a *semi-resistive* condition, which is characterised by a near-linear variation of current with applied D.C. voltage across the alumina specimen. This apparent decrease in the resistivity of the alumina may be explained by a migration of metal ions, such as sodium ions (Na^+) within the glassy phase of the ceramic, which may migrate in a D.C. field.[19] Vacant oxygen sites (*F* centres) may also migrate, within the crystalline phases of the ceramic.[19] These field-induced changes in the ceramic material, where the migration of metal ions or defects leads to a redistribution of these defects throughout the ceramic matrix, may in turn lead to local variations in the resistivity of the solid insulator, and consequently modify the electrical (I_t–V_t) behaviour of the ceramic.

3.3 Physical Processes of Photon Emission

Drawing upon the findings and observations described in sections 3.1 and 3.2, a qualitative account can be given of the physical processes believed to be associated with the observed optical phenomena:

3.3.1 Initiation and persistence of 'doughnut' emission

The initial state of the virgin alumina ceramic may be viewed as a distribution of empty charge-traps, associated with vacancy (*F* centre) and impurity (Cr^{3+}) defects in the alumina.[20] As voltage is applied to the specimen, measurable current flows through the ceramic, indicating significant charge-transport processes as a proportion of injected electrons reach the anode, with a further proportion of injected charge being trapped at defects in the insulator. This charge injection process may be accelerated by increasing the applied electric field, such that a threshold field is reached for electrons to tunnel from emission sites on the cathode into the conduction band of the ceramic insulator (charge injection may be further assisted by local enhancement (β) of the electric field). At this threshold point, the magnitude of the injected current is increased, as electrons tunnel into the ceramic and occupy vacant traps, allowing a higher proportion of injected electrons to reach the anode as more traps are filled.

This behaviour is consistent with both of the threshold currents shown in the initial *I–V* characteristic of Fig. 3(a), where the first current threshold coincided with the initiation of the doughnut emission (illustrated in Fig. 3(b)) as electrons filled the 1.79 eV traps at chromium impurity sites, followed by the second current threshold at which bright-blue emission was initiated at localised spots (Fig. 3(c)), with electrons filling the deeper 3.0 eV traps of the *F* centres at an increased value of applied electric field.

When the doughnut emission occurs, three electron-recombination processes may be involved: the ruby-red component of the observed emission can be described by a radiative transition through recombination between a conduction electron and a chromium impurity site (Cr^{4+} centre), as shown by process **1** in the band diagram of Fig. 5. A further contribution to this R_1, R_2 line emission can arise from *internal photoluminescence*, whereby the the 410 nm photons emitted by electron recombination at oxygen vacancies (*F* centres) in the alumina (process **2**) excite the absorption band of Cr^{3+} traps,[12] exciting trapped electrons into the

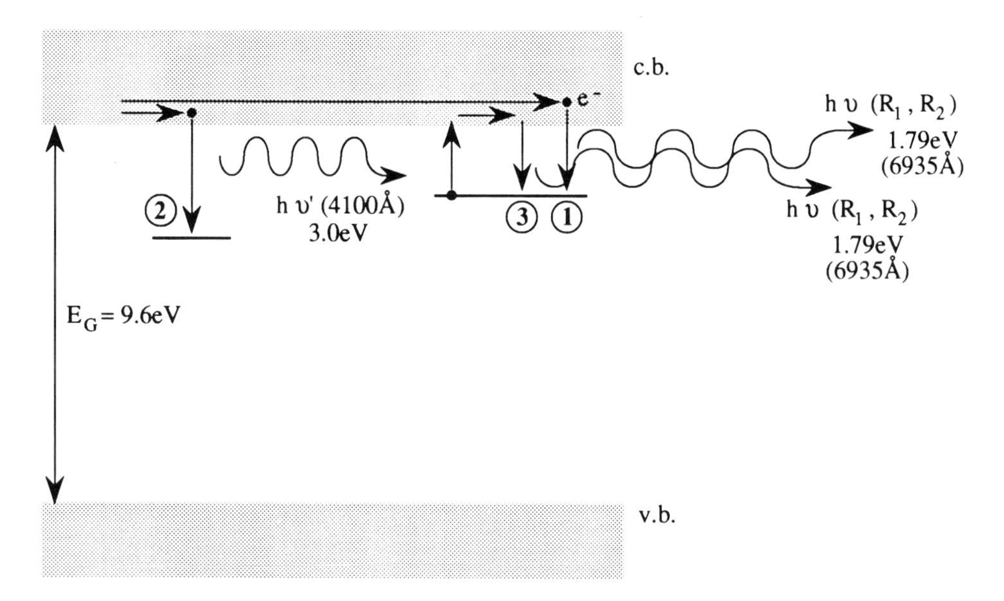

Figure 5 Energy-band diagram, illustrating photon emission by three solid-state mechanisms at defects within an alumina insulator.

conduction band of the alumina matrix; these electrons may subsequently recombine at a Cr^{3+} trap (process **3**), releasing a further R_1,R_2 photon, i.e. as shown in Fig. 5.

However, it is also important to note the observed finite *lifetime* of the doughnut emission, at a given applied voltage V_t (see Section 2.2.1). This finding may be explained by the eventual filling of the majority of Cr^{3+} traps, where the total F centre emission from the ceramic (and excitation of electrons from Cr^{3+} traps by these 410 nm photons) is not sufficient to balance the filling of the chrome traps. (Hence only a *quasi*-equilibrium exists between electron excitation into the conduction band, and electrons being trapped, at Cr^{3+} defects).

Filling of traps throughout the ceramic on a macroscopic scale increases the mean free path available to the injected electrons,[21] increasing the probability of electron flow from cathode to anode, yielding the increase in measured current over a 30 min time period as depicted in Fig. 3(a). In consequence, fewer electrons are trapped by Cr^{3+} defects since the majority of these traps are filled, and so the observed R line (694 nm) emission from these chrome centres is reduced, and the observed ruby-red doughnut emission is seen to diminish.

3.3.2 Onset and persistence of bright-blue 'spots' (λ = 410 nm)

Figure 3(c) shows a co-existence of the distributed doughnut emission with several localised bright-blue spots. These spots have also been observed in isolation, without the presence of the doughnut, both in Section 2.2.2 of this work, and in previous studies.[5] Filling of the deep (3.0 eV) F centre traps associated with these spots may be promoted by localised electric field enhancement, in the vicinity of filled Cr^{3+} traps, whereby the filling of the chrome-defect traps causes local distortions of the applied D.C. field, enhancing the local field at the F centre. A more noticeable observation of this field effect is the distinct 'turning-on' of the bright-blue spots when the applied voltage, V_t, is increased beyond a second threshold value of +4 kV (Fig. 3(a)).

The dimming of these spots after several hours of voltage application[5] may be explained by the eventual filling of traps at F centre defects, leaving fewer empty traps available for electron recombination, thereby reducing the photon emission from bright spots on the alumina specimen. Any increase in measured current flow, arising from the greater electron mean free path created by the macroscopic filling of the F centre traps,[21] is likely to be incremental upon the marked current increase associated with the filling of Cr^{3+} traps (Fig. 3(a)), given the dominance of these Cr^{3+} traps in the electro-optical behaviour of the alumina ceramic. This hypothesis is evidenced by the predominance of the 694 nm line associated with Cr^{3+} defects (Fig. 4(a)), and the marked increase in measured current (I_t) following the reduction in the intensity of the doughnut for V_t = 5 kV (Fig. 3(a)), whereas no noticeable change in I_t has been observed following the extinction of the bright-blue spots.[5] Thus the bright-blue spots appear to be of lesser significance than the doughnut emission in the D.C. pre-breakdown optical behaviour of the alumina ceramic.

4. CONCLUSIONS

For the metal–insulator–metal (M–I–M) regime studied in this work, consisting of the end face of a 94% alumina (Al_2O_3) tube bridged by two planar concentric rings of Mo/Mn metallisation:

- the photon emission associated with D.C. pre-breakdown of the alumina is consistent with radiative electron-recombination processes within the insulator, with a deep-red emission (λ = 694 nm) originating from impurity centres (in this case chromium) within the glassy phase of the ceramic[8,12] and a bright-blue (λ = 410 nm) emission from negative ion vacancy defects (*F* centres) within the crystalline alumina phase.[9,14]
- measurement of the electrical characteristics of the virgin insulator sample showed two current thresholds, coincident with the initiation of the deep-red and bright-blue emissions. Onset of these current-thresholds was attributed to the filling of charge-traps by injected electrons, giving rise to the observed radiation of photons, and causing the measured current to increase due to a lengthening of the mean free path available to injected electrons as more trap sites were filled.[20]
- subjecting the alumina specimen to D.C. voltage stress over a time period of several hours produced an irreversible increase in the measured resistivity of the insulator. This behaviour was consistent with the macroscopic filling of traps associated with defect centres in the alumina, thereby increasing the electron mean free path, increasing the proportion of electrons flowing as measured D.C. current to the anode;[20] such a hypothesis was evidenced by the observed decrease in the luminescence from the alumina, indicating a reduction in the number of injected electrons being trapped and thereby radiating photons. Another possible modification mechanism of the alumina resistivity is the migration of defects under the influence of the applied electric field, where the redistribution of the defect centres results in a change in the measured resistivity.[19]

ACKNOWLEDGEMENTS

The authors wish to thank Anne Burns of Rees Instruments Limited, for her provision of a Monolight high-speed scanning monochromator, and her invaluable assistance in the spectroscopic investigation. Discussions between one of the authors (BMC) and Andrew Duncan and Dr Brian Franklin of Morgan Matroc Limited (Rugby Division) were very helpful. Some of the authors (BMC, NSX and RVL) also wish to thank EEV Limited for their continued help and support in this work, with particular thanks to Graham Jex and Richard Keyse (EEV, Lincoln) for the supply of metallised alumina specimens. This work was carried out under the Science and Engineering Research Council (SERC) Total Technology studentship scheme.

REFERENCES

1. H. C. Miller: in *XVth International Symposium on Discharges and Electrical Insulation in Vacuum*, 165–174, D. König ed. (Darmstadt, Germany, 1992.
2. R. A. Anderson and J. P. Brainard: *J. Appl. Phys.*, 1980, **51**, 1414–1421.
3. G. Blaise and C. LeGressus: *C.R. Acad. Sci. Paris*, 1992, **314**, 1017–1024.
4. N. C. Jaitly and T. S. Sudarshan: *J. Appl. Phys.*, 1988, **64**, 3411–3418.
5. B. M. Coaker, N. S. Xu, F. J. Jones and R. V. Latham: in *XVth International Symposium on Discharges and Electrical Insulation in Vacuum*, D. König ed., 46–50. Darmstadt, Germany, 1992.
6. R. V. Latham, K. H. Bayliss and B. M. Cox: *J. Phys. D: Appl. Phys.*, 1986, **19**, 219–231.
7. T. Asokan and T. S. Sudarshan: *IEEE Trans. Elec. Insul.*, 1993, **28**, 192–199.
8. Y. Saito, N. Matuda, S. Anami, A. Kinbara, G. Horikoshi and J. Tanaka: *IEEE Trans. Elec. Insul.*, 1989, **24**, 1029–1032.
9. N. C. Jaitly and T. S. Sudarshan: *J. Appl. Phys.*, 1986, **60**, 3711–3719.
10. G. R. Harrison: *Massachusetts Institute of Technology Wavelength Tables*, 28–175, Wiley, New York, 1939.
11. A. Duncan: private communication, 1993.
12. D. Lapraz, P. Iacconi, D. Daviller and B. Guilhot: *Phys. Stat. Sol. (a)*, 1991, **126**, 521–531.
13. F. G. Will, H. G. deLorenzi and K. H. Janora: *J. Am. Ceram. Soc.*, 1992, **75**, 295–304.
14. K. H. Lee and J. H. Crawford: *Phys. Rev. B*, 1979, **19**, 3217–3221.
15. P. Kulis, Z. Rachko, M. Springis, I. Tale and J. Jansons: *Radiation Effects and Defects in Solids*, 1991, **119–121**, 963–968.
16. N. W. Ashcroft and N. D. Mermin: Solid State Physics, 622–635. Saunders, Philadelphia, 1976.
17. K. J. Caulfield, R. Cooper and J. F. Boas: *J. Nucl. Mater.*, 1991, **184**, 150–151.
18. Y. Liu and B. R. Patterson: *J. Am. Ceram. Soc.*, 1992, **75**, 2599–2600.
19. A. J. Moulson and J. M. Herbert: *Electroceramics — Materials, Properties, Applications*, 182–192, Chapman & Hall, London, 1990.
20. G. Blaise: *Le Vide, les Couches Minces — Supplément au n° 260*, 1992, 417–426.
21. C. LeGressus and G. Blaise: *IEEE Trans. Elec. Insul.*, 1992, **27**, 472–481.

Monitoring the Integrity of MOS Gate Oxides

M. J. TUNNICLIFFE, L. DONG and V. M. DWYER
International Electronics Reliability Institute, Department of Electronic and Electrical Engineering, Loughborough University of Technology, Loughborough, Leicestershire, LE11 3TU, UK

ABSTRACT
Inter-sample variability of time-dependent dielectric breakdown in MOS gate oxides was experimentally examined. MOS-C and MOSFET structures were subjected to constant-voltage stress and the resulting times-to-breakdown were recorded. Failure distributions were found to be bi-modal, with distinct 'early' and 'late' failure categories. The ratio of early failures to late failures increased with increasing stress voltage. The late failure distributions were approximately Weibull in shape, indicating a weak-link failure mechanism. Although pre-breakdown tunnelling was a function of electric field, the average time-to-breakdown appeared to depend solely upon voltage. Little correlation existed between the times-to-breakdown of MOSFET and MOS-C structures fabricated upon the same die. Analysis suggests that the oxide breakdown is in fact principally uni-modal, and that the apparent bi-modal behaviour is caused by interference from the transient voltage drop across the series substrate resistance.

1. INTRODUCTION

This paper describes part of an ongoing study of the nature and mechanisms of time-dependent dielectric breakdown (TDDB) in metal oxide semiconductor (MOS) gate oxide layers. The general aims are:

1. The advancement of dielectric breakdown physics,
2. The development of an accurate dielectric breakdown model, and
3. The development of a quick and efficient oxide characterisation test for use during integrated-circuit fabrication.

So far, the investigations have focused on the times-to-breakdown (t_{bd}) of MOS devices subjected to constant voltage stress. Preliminary results, reported earlier this year,[1] showed that:

1. Very little correlation exists between the times-to-breakdown of MOS field effect transistors (MOSFETs) and MOS capacitors (MOS-Cs) fabricated on the same die.
2. Variations in the time-to-breakdown and the pre-breakdown oxide tunnelling current are statistically unrelated.

This latter observation is important since it conflicts with the generally-accepted model of TDDB: namely that some form of 'damage' is continuously inflicted upon the oxide by the tunnelling current. Breakdown occurs when this 'damage' reaches some critical level.[2,3] This picture has become widely accepted, and has formed the basis for several elaborate failure models (e.g. Ref. 4). Any doubt about its validity must therefore be thoroughly investigated.

The present paper seeks to extend these earlier studies by the acquisition of further experimental data and the development of a theoretical model to explain the results.

2. TEST SAMPLES AND APPARATUS

The device samples used throughout these experiments were small-dimension enhancement/depletion model MOSFETs and wide-area ($215 \times 268\,\mu m$) MOS-C structures, fabricated on a 4″ ⟨100⟩ p-type silicon wafer. All devices had gate oxides of nominal thickness 40 nm and n^+-doped polysilicon gate electrodes.

The MOSFETs on each die were interconnected with common source and gate contacts and individual drain terminals, such that gate stress was applied to the entire network at once. The total gate area A of the network was approximately $3.10^{-4}\,cm^2$. The MOS-C gate inputs were isolated from all other devices on the die.

The variation in local oxide thickness (T_{ox}) for each die was gauged by measuring the capacitance (C_{ox}) of an additional wide-area MOS-C structure ($416 \times 553\,\mu m$) which was reserved for this purpose. Measurement was performed at 10 kHz using a Wayne–Kerr 420 LCR meter, with a -10 V bias potential to ensure carrier accumulation at both oxide surfaces. The oxide thicknesses, estimated from the formula $T_{ox} = \varepsilon_0 \varepsilon_{ox}\,.\,\text{Area}/C_{ox}$ (where $\varepsilon_0 \varepsilon_{ox}$ = oxide permittivity), had a mean value of 41.42 nm and a standard deviation of 0.161 nm.

The series substrate resistances (R_b) for the MOSFET and MOS-C structures were obtained by measuring the device impedances at 10 MHz and extracting the in-phase components. R_b was found to equal $250 \pm 10\,\Omega$ for the MOS-Cs and $79 \pm 10\,\Omega$ for the MOSFET networks.

The wafers were mounted on a 4″ brass chuck and held in place by a vacuum suction system. A $20\,\mu m$ tip manually adjustable microprober was used to probe the device gate bond pads under microscopic observation. Constant-voltage stress was supplied by a Hewlett Packard 4145B parametric analyser and applied to the device under test (DUT) between gate and substrate. The stress voltage was accurate to within ± 10 mV and had a rise time of approximately $800\,\mu s$. Throughout the experiments, the gates were stressed negatively with respect to the substrate.

The HP4145B was configured to monitor the oxide current as a function of time, and the device voltage was independently monitored by a Hewlett Packard 54111D digital storage oscilloscope. For long time scales ($t_{bd} > 1$ sec), breakdown was easily detected by the HP4145B as a rapid rise in the injection current. For shorter time-scales ($t_{bd} < 1$ sec), the most accurate indication of breakdown was a sudden drop in the device voltage as measured by the oscilloscope.

3. EXPERIMENTAL PROCEDURES AND RESULTS

A 15×13 matrix of die was selected in the centre of the wafer, and results are reported for those devices only. Approximately equal numbers of MOS-Cs and

MOSFET networks were subjected to constant-voltage stresses of −43.5 V, −44.0 V and −44.5 V. Devices stressed at each particular voltage were spread as uniformly as possible across the wafer surface in order to eliminate any positional dependencies. The time-to-breakdown (t_{bd}) was recorded for each device, together with the maximum tunnelling current density (J_{max}) and the tunnelling current density at breakdown (J_{bd}). (Note: current measurement was only possible for $t_{bd} > 0.1$ s due to the finite integration time of the parametric analyser.) Room temperature (28°C) was maintained throughout all the experiments, and the devices were constantly illuminated by the microscope illumination system.

The time-to-breakdown results are shown in Fig. 1, plotted in the Weibull format, i.e. $\ln(-\ln(1-F(t)))$ against $\log(t)$ (where $F(t)$ = failure probability at time t). The distributions are clearly bi-modal, consisting of 'early' failures ($t_{bd} < 100$ ms) and 'late' failures ($t_{bd} > 100$ ms). The ratio of the numbers of early and late failures generally increases with the stress voltage, V.

The late failure distributions are generally linear, indicating a Weibull distribution of the form: $F(t) = 1 - \exp[-(t/\tau)^{\beta}]$ (where β is the distribution slope and τ is the characteristic lifetime). The Weibull distribution is related to extreme-value statistics, and may be used to model failures at weak spots in a device.[2,5] The nature of the 'early' failure distribution is difficult to determine since its lower tail is dominated by the stress voltage risetime.

Figure 2 compares the values of $\log(t_{bd})$ obtained from MOSFET and MOS-C structures on the same die. (Devices on any given die were always stressed at the same voltage.) The broken diagonal lines indicate the results which would be expected if the failure times for a given die were equal. The high degree of scatter indicates that this is not generally true. Of those die which exhibited early failure, only 36% suffered it in both MOS-C and MOSFET networks. This (together with the results of earlier studies)[1] demonstrates the limitations of the MOS-C as a test vehicle for monitoring in-wafer variations of oxide integrity.

Figure 3 shows the values of log (J_{max}) measured for the MOSFET and MOS-C structures plotted against the corresponding values of $\log(t_{bd})$. Although both depend upon voltage, no obvious relationship appears to exist between them within a given data set.

Figure 4 shows $\log(J_{max})$ and $\log(t_{bd})$ plotted against the corresponding electric fields $(V - AJ_{max}R_b)/T_{ox}$ and $(V - AJ_{bd}R_b)/T_{ox}$. The data can be analysed either as a single data set or as a series separate data sets, each associated with a particular stress voltage. In the former case, current clearly rises and t_{bd} falls with increasing field. In the latter case, although a distinct current rise is observed with increasing field, no corresponding t_{bd} decrease is obvious. If anything, the time-to-breakdown appears to *increase* with field (this is most obvious in the −44.0 V data). However, insufficient data are available to confirm such a trend.

4. DISCUSSION AND MODELLING OF RESULTS

Bi-modal failure distributions have been observed by many earlier workers (e.g. Refs 2, 5) and have normally been attributed to 'intrinsic' and 'extrinsic' failures.

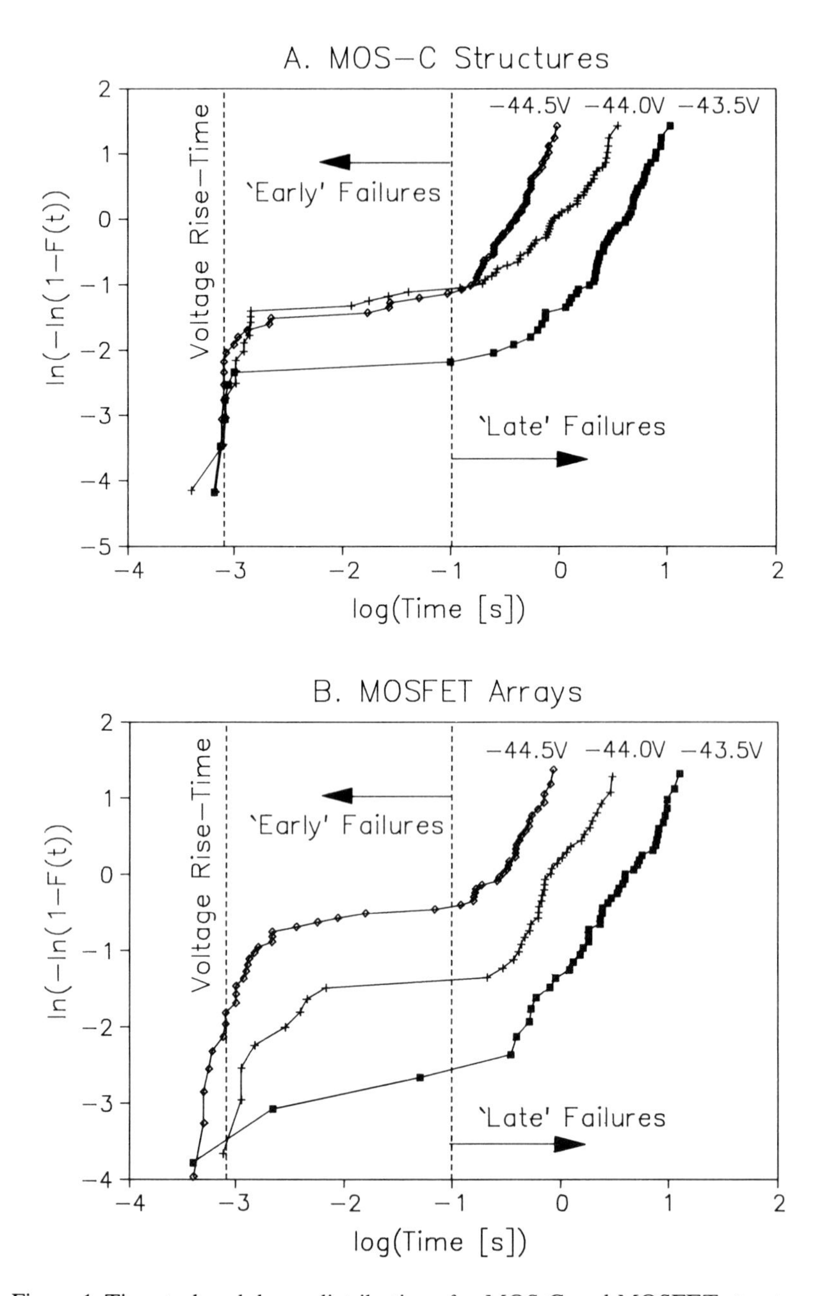

Figure 1 Time-to-breakdown distributions for MOS-C and MOSFET structures.

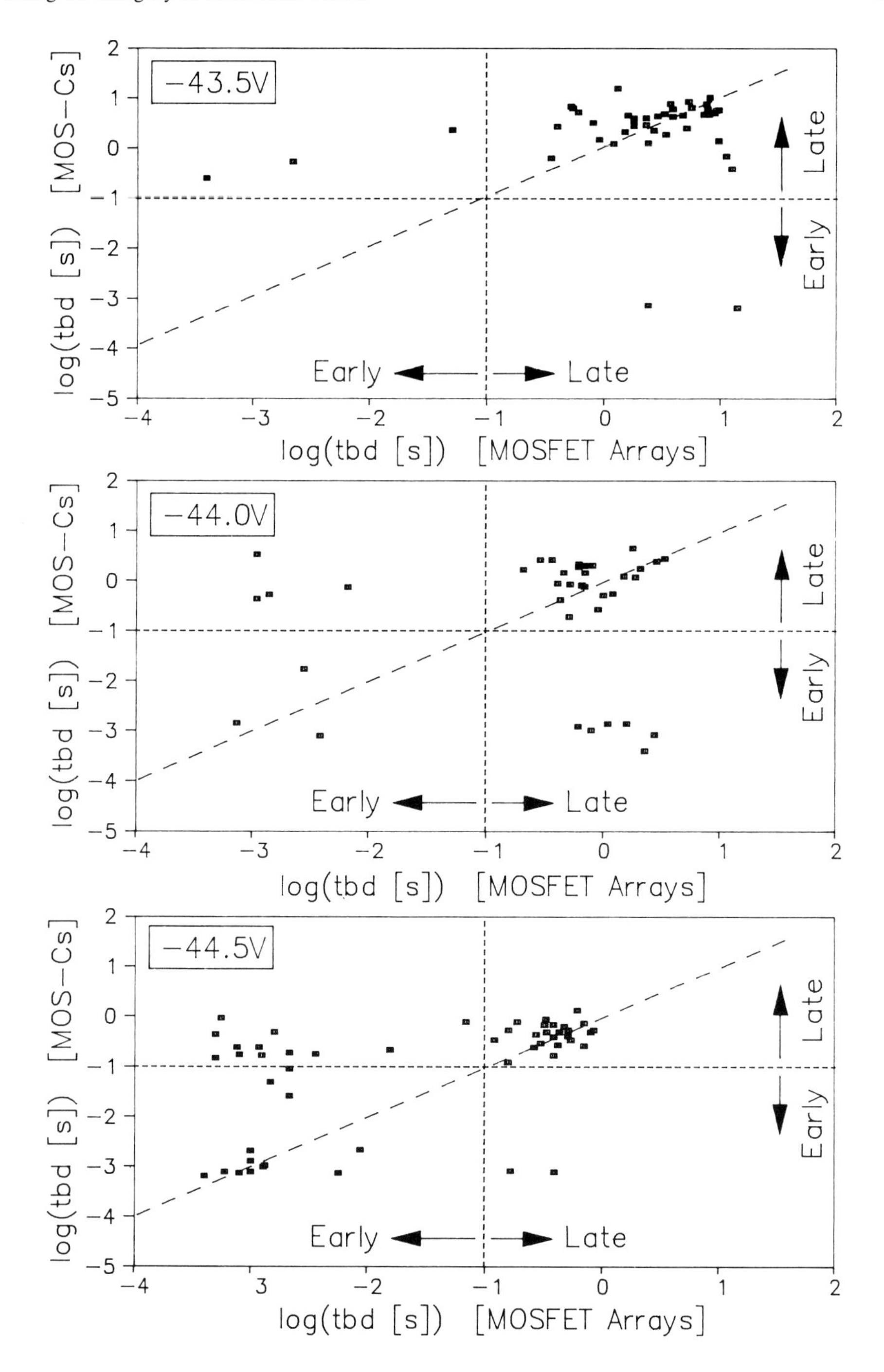

Figure 2 Comparison between breakdown times of MOSFET and MOS-C structures fabricated on same die.

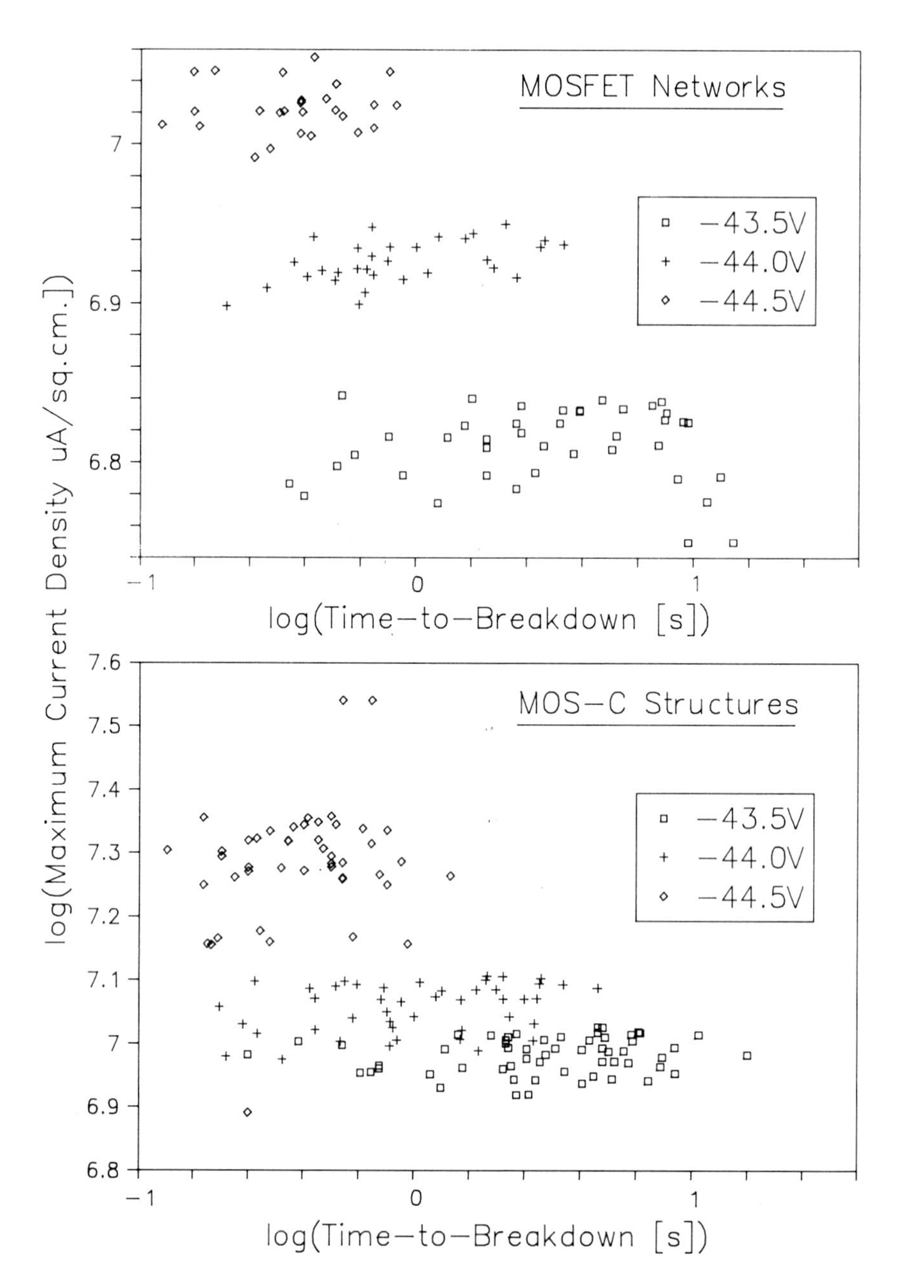

Figure 3 Maximum injection current *v* time-to-breakdown for MOSFET and MOS-C structures.

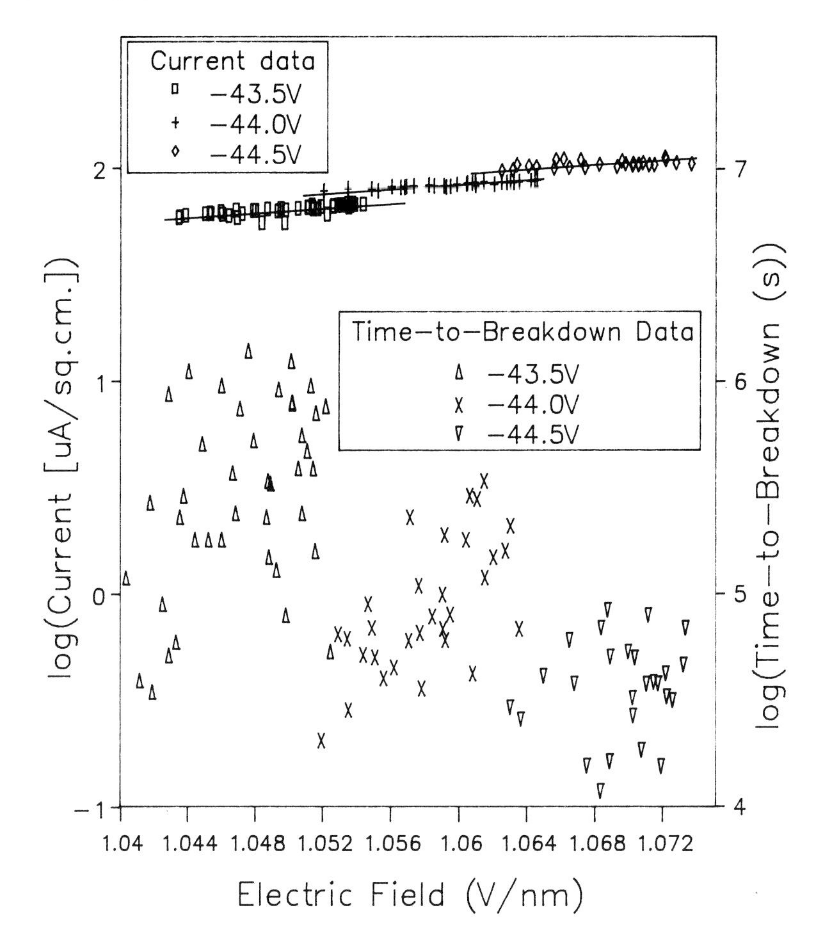

Figure 4 Field dependencies of maximum injection current and time-to-breakdown in MOSFET structures.

Intrinsic failures are associated with the inherent oxide properties, while extrinsic failures are caused by structural defects and/or chemical impurities. The latter occur much more rapidly than the former, and form a separate statistical distribution.

The application of this model[2] to Fig. 1 requires a voltage-dependent extrinsic defect density. It is possible that some extrinsic defects are only activated above a particular field, such that their apparent population increases with voltage. An alternative scheme, based upon the time dependence of the pre-breakdown tunnelling current, is presented below.

Consider first the pre-breakdown tunnelling current transient. Figure 5 shows typical waveforms obtained from four different MOS-C structures at four different voltages. (Currents were monitored by measuring the voltage across a 771.4 Ω resistor, connected in series with the oxide.) In all four cases, the oxide current

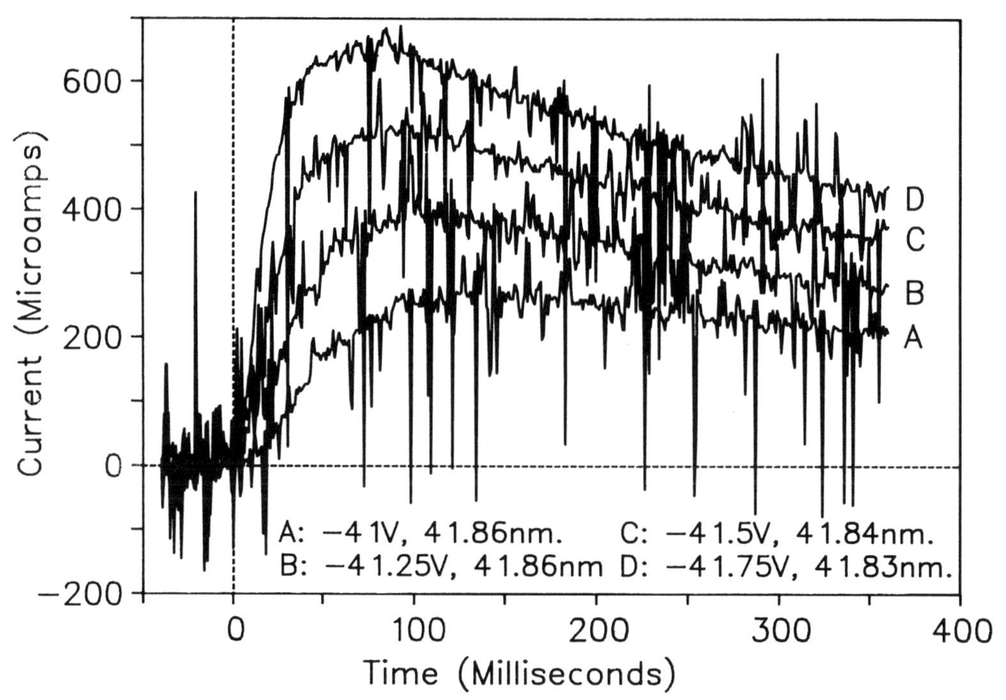

Figure 5 Typical injection current waveforms for MOS-C structures.

magnitude ($I_{ox}(t)$) increases from zero to a maximum value at about 100 ms, beyond which it decays with time. Since this current generates a voltage drop across the bulk resistance ($R_b = 250\,\Omega$), the *oxide* voltage magnitude ($V_{ox} = V - I_{ox}R_b$) experiences a 'dip' at approximately 100 ms.

In order to continue, we make the following reasonable assumptions:

1. Oxide breakdown is caused by a continuous, non-reversible accumulation of damage inflicted during the pre-breakdown period. Breakdown occurs when the damage reaches a critical level (hereafter called the 'damage-to-breakdown'). This assumption is consistent with previous models.[2,3]
2. The rate at which oxide damage accumulates increases rapidly with oxide voltage and is independent of the oxide tunnelling current. Although this differs from existing models (which generally require *damage rate* $\propto$ exp(-*constant/field*),[3] it is justified by the results in Figs 3 and 4 which show t_{bd} dependent on V only.
3. The damage-to-breakdown varies randomly between devices according to a

uni-modal distribution. This eliminates the need to invoke any voltage-dependent 'extrinsic' defects.

When time ≪100 ms, the oxide voltage magnitude is high, together with the rate of oxide wearout. Many of the devices at the lower end of the damage-to-breakdown distribution fail in this region, creating the 'early' failure mode. As time increases to ~100 ms, the voltage magnitude decreases together with the wearout rate. Very few devices fail in this region. When the current decays and the voltage rises once more, the remainder of the devices break down to form the 'late' failure distribution.

The above model would clearly account for the voltage dependence of the early failure distribution: As the stress voltage increases, a greater number of oxides reach their critical damage level in the $t < 100$ ms region and fewer survive to fail in the late distribution.

Finally, it is necessary to consider the physical implications of this work. The fact that t_{bd} depends upon voltage rather than field suggests that wearout is governed by the *absolute* energy of the charge carriers, rather than their rate of acceleration as was previously believed.[2] In addition to this, the apparent independence of the maximum current as a function of t_{bd} suggests that the injected electrons are not responsible for the oxide wearout. Damage may therefore be caused by some other mechanism such as mobile ion drift or substrate hole injection.

5. CONCLUSIONS

This paper presents an experimental study on the inter-sample variability of oxide wearout, together with a qualitative theoretical model to explain the observed failure distributions. The fundamental conclusions are as follows:

1. Time-dependent dielectric breakdown under constant voltage stress exhibits a bi-modal distribution. The 'late' (i.e. long time-scale) mode is governed approximately by the Weibull model. The ratio of 'early' (short time-scale) failures to 'late' (long time-scale) failures increases with the stress voltage V.
2. While the pre-breakdown oxide tunnelling current is a function of electric field, the time-to-breakdown appears to depend solely upon the absolute voltage. For a given voltage, the tunnelling current and the time-to-breakdown are statistically unrelated.
3. There is little or no correlation between the time-to-failure of MOSFET and MOS-C structures fabricated on the same die. This supports earlier claims that the MOS-C has limited use as an oxide integrity test vehicle.[1]

A framework has been developed to explain the apparently bi-modal time-to-breakdown distributions in terms of a single uni-modal damage-to-breakdown distribution. Although the model is qualitative, a quantitative version is currently under development.

REFERENCES

1. M. J. Tunnicliffe and V. M. Dwyer: presented at EPMS'93 Conference, 1993, Sheffield Hallam University, Sheffield, UK and accepted for the International Journal of Electronics, 1993.
2. D. R. Wolters and J. J. Van der Schoot: *Phillips J. Res.*, 1985, **40**, 115–192.
3. I. C. Chen, S. Holland and C. Hu: *IEEE Trans. Electron. Dev.*, 1985, **ED32**, 413–422.
4. C. K. Chan and M. B. Carey: *IEEE Trans. on Reliability*, 1992, **41**, 414–420.
5. R.-P. Vollertsen: *Microelectron. Reliab.*, 1993, **33**, 1665–1677.

Index

Author Index